T0140216

Springer Theses

Recognizing Outstanding Ph.D. Research

Aims and Scope

The series "Springer Theses" brings together a selection of the very best Ph.D. theses from around the world and across the physical sciences. Nominated and endorsed by two recognized specialists, each published volume has been selected for its scientific excellence and the high impact of its contents for the pertinent field of research. For greater accessibility to non-specialists, the published versions include an extended introduction, as well as a foreword by the student's supervisor explaining the special relevance of the work for the field. As a whole, the series will provide a valuable resource both for newcomers to the research fields described, and for other scientists seeking detailed background information on special questions. Finally, it provides an accredited documentation of the valuable contributions made by today's younger generation of scientists.

Theses are accepted into the series by invited nomination only and must fulfill all of the following criteria

- They must be written in good English.
- The topic should fall within the confines of Chemistry, Physics, Earth Sciences, Engineering and related interdisciplinary fields such as Materials, Nanoscience, Chemical Engineering, Complex Systems and Biophysics.
- The work reported in the thesis must represent a significant scientific advance.
- If the thesis includes previously published material, permission to reproduce this must be gained from the respective copyright holder.
- They must have been examined and passed during the 12 months prior to nomination.
- Each thesis should include a foreword by the supervisor outlining the significance of its content.
- The theses should have a clearly defined structure including an introduction accessible to scientists not expert in that particular field.

More information about this series at http://www.springer.com/series/8790

Ahmed Tarek Abouelfadl Mohamed

Measurement of Higgs Boson Production Cross Sections in the Diphoton Channel

with the full ATLAS Run-2 Data
and Constraints on Anomalous Higgs Boson
Interactions

Doctoral Thesis accepted by
CNRS (LPNHE-Paris), Paris, France

 Springer

Author
Dr. Ahmed Tarek Abouelfadl Mohamed
Department of Physics and Astronomy
Michigan State University
East Lansing, MI, USA

Supervisor
Giovanni Marchiori
CNRS (LPNHE-Paris)
Paris, France

ISSN 2190-5053   ISSN 2190-5061   (electronic)
Springer Theses
ISBN 978-3-030-59518-0   ISBN 978-3-030-59516-6   (eBook)
https://doi.org/10.1007/978-3-030-59516-6

This Springer imprint is published by the registered company Springer Nature Switzerland AG
The registered company address is: Gewerbestrasse 11, 6330 Cham, Switzerland

"The gnostic lives in this world of change,
Aware constantly of the Return,
The Return which the origin of true life is.
He sees in Autumn's majesty, that Return for
which he yearns,
Which for him is the Spring of heavenly life.
If Spring be the origin of life below,
Autumn is the Spring of eternal life,
That life for the Return to which the gnostic
lives here on earth.
And so for him Autumn is the Spring of life
divine,
Heralding the Return to that life that never
ends."

Autumn—Sayyed Hossin Nasr

To my family.

Supervisor's Foreward

It is an honour and a pleasure to write this foreword to the Ph.D. thesis manuscript of my former graduate student, Ahmed Tarek, on the measurement of the Higgs boson production cross sections in the diphoton channel and their interpretation in terms of anomalous Higgs boson interactions. The Higgs boson, discovered only eight years ago by the ATLAS and CMS Collaborations at the Large Hadron Collider (LHC), is at the same time the cornerstone of the Standard Model (SM) of particle physics, our current theory of the elementary particles and their fundamental interactions, and one of the least experimentally studied pieces of that puzzle. Its observed properties, so far in agreement with the predictions, may deviate—with more data and more accurate measurements, in particular regions of the phase space—from them and provide insights in what a more fundamental theory Beyond the SM (BSM) would be. The study of the Higgs boson could then be the portal to the holy grail of BSM physics that particle physicists around the world have been looking for since decades, to overcome the known limitations of the SM itself. It is with this spirit of exploring uncharted territories and the aim of extending further the knowledge on the properties of the Higgs boson and the constraints that they could set on any BSM model, that Ahmed Tarek worked during his thesis. The work presented in this dissertation focuses on measurements of the Higgs boson properties using its decays to two photons, one of the most sensitive "channels", with which the Higgs boson was discovered in 2012. The measurements are based on the proton—proton collisions collected by the ATLAS detector at the LHC during the full second period (Run 2) of data taking (2015–2018). In view of precise measurements of the Higgs boson properties with these decays, Ahmed has performed a meticulous work of (re-)calibration of the ATLAS electromagnetic calorimeter after the LHC Long Shutdown 1 (LS1) during 2013–2014, focusing in particular on the precise characterization of its pre-sampler, the intercalibration of its various longitudinal layers, and the full characterization of the material upstream of the calorimeter. The results obtained by Ahmed were a crucial part of the complex analysis chain used in ATLAS to calibrate the photon energy scale and resolution, and allowed a first measurement of the Higgs boson mass with a partial ATLAS Run 2 dataset, reaching—in combination with an analogous measurement in the Higgs to four lepton channel—the same precision of the combined ATLAS+CMS Run 1 result. The logical evolution of Ahmed's path was to move from detector work to physics analysis and interpretation. After the work on the calibration of the photon candidates reconstructed by the ATLAS detector, Ahmed developed the core elements for the measurements of the fiducial inclusive and differential cross sections of Higgs to diphoton events. In particular, he studied the isolation requirements to be applied both at particle-level (theory) and detector-level (experiment), evaluated all the uncertainties in the measurement and developed the criteria to select the unfolding algorithm used to obtain measurements at the particle-level from the detector-level signal yields. These measurements were then compared with multiple theory predictions and combined with results from the analysis of the decay of Higgs to four leptons. Finally, Ahmed performed a state-of-the-art model-independent interpretation of the results, using them to set for the first time—within an LHC experimental collaboration—constraints on the anomalous Higgs boson interactions with gauge bosons in the context of the Standard Model Effective Field Theory (EFT). This manuscript describes clearly and in a pedagogical way these measurements and interpretations. The chapters on the photon energy calibration (Chap. 4), the Higgs boson cross measurements (Chap. 5) and EFT interpretation (Chap. 6) are properly complemented by three introductory chapters on the Standard Model (Chap. 1), BSM theories and the EFT approach (Chap. 2), and the experimental apparatus (the LHC and the ATLAS detector, Chap. 3), providing a complete, self-contained description of Ahmed's work, as well as of the theoretical and experimental tool on which it is based, that can constitute a useful reading for a student starting to work on these topics and be of interest even to a non-expert reader. The more technically minded ones will find more details on the techniques for the statistical interpretation of the results and for the unfolding in the two Interludes that complete the manuscript. To conclude on a personal note, I would like to stress the excellent scientific and human qualities of Ahmed, who worked along the three years of his Ph.D. with an exemplary dedication, ingenuity and openness towards his colleagues. All these ingredients in a single person, with the support of all the persons working close to him, led to the excellent results presented in this manuscript. I would also like to warmly thank Sandrine Laplace, the initial supervisor of Ahmed, who trained him excellently during his first year, and gave me the opportunity to supervise Ahmed for the remainder of the thesis when she embarked on a new adventure in another field: it was a pleasure and a rewarding experience to supervise, work and discuss with Ahmed in the following two years.

Paris, France   Giovanni Marchiori
September 2020

Abstract

This thesis presents the measurement of the fiducial integrated and differential cross sections for the production of the Higgs boson decaying to two photons. The measurement is performed using 139 fb^{-1} of proton–proton collision data at $\sqrt{s} = 13$ TeV collected by the ATLAS experiment at the Large Hadron Collider. The cross section measurement in the two-photon final state is performed in a fiducial region closely matching the experimental selection and compared to Standard Model (SM) predictions. The measurement is performed for the inclusive fiducial region and differentially for the Higgs boson transverse momentum, the Higgs boson absolute rapidity, and various jet variables including the jet multiplicity, the transverse momentum of the leading jet, the invariant mass of the two leading jets and Azimuthal-angle separation of the two leading jets. These variables probe precisely the Higgs boson kinematic and CP properties. This resulted in a measurement of the inclusive $H \rightarrow \gamma\gamma$ fiducial cross section of 65.2 ± 7.1 fb which is in excellent agreement with the SM predictions of 63.6 ± 3.3 fb. The measured differential cross sections are used to probe the strength and tensor structure of the anomalous Higgs boson interactions using an effective Lagrangian, which introduces additional CP-even and CP-odd interactions.

Preface

Throughout history, attempts were carried out to understand the fundamental building blocks of matter and the forces that govern their interactions. This evolved from a naive 4-element description of nature, through theories and discoveries on atoms, arriving at the sub-atomic world with one of the most accurate scientific theories known, the Standard Model (SM) of particle physics [1–5]. The Standard Model of particle physics is a theory that describes three of the fundamental interactions between the elementary particles that compose matter: the electro-magnetic, strong and weak forces. The Standard Model was developed in the early 1970s, and it has successfully explained all experimental results from particle colliders. Furthermore, the Standard Model predicted the existence of an additional fundamental particle, the Higgs boson, to explain how fundamental particles acquire their masses. The Higgs boson particle was discovered in July 2012, by the ATLAS [6] and CMS [7], experiments at the Large Hadron Collider (LHC) [8].

The Higgs boson is a massive scalar particle, that was predicted in the context of spontaneous electroweak symmetry breaking (EWSB) theory, proposed by P. Higgs [9], R. Brout and F. Englert [10], and G. Guralnik, C. Hagen and T. Kibble [11] to generate masses for the massive force carriers. Since the discovery of the Higgs boson, several studies were performed to measure the different properties (mass, spin and parity couplings) of the Higgs boson. These measurements rely on proton-proton collision data from the LHC, collected at a center of mass energy of 7 and 8 TeV during the LHC Run-1 (2011–2012), corresponding to an integrated luminosity of approximately 5 and 20 fb^{-1} respectively, and at a center of mass energy of 13 TeV during the LHC Run-2 (2015–2018), corresponding to an inte-grated luminosity of approximately 139 fb^{-1}. In Run-2, the significant increase of the collected data and the increased center-of-mass energy allowed performing different measurements of the Higgs boson properties at unprecedented precision [12–14]. These measurements probed the different Higgs boson production and decay modes. Among those decay modes, is the decay in the two-photon channel $H \rightarrow \gamma\gamma$. Despite its small branching ratio (around 0.2%), this was one of the main discovery channels in 2012, thanks to the excellent photon reconstruction and identification efficiency and energy resolution at ATLAS and CMS.

This thesis presents the precise measurement of the Higgs boson cross section measured inclusively in a fiducial region defined by the detector acceptance, and differentially as a function of several kinematic variables. These measurements are performed in a model-independent manner and corrected for detector effects, which allows for a direct comparison with different theory predictions and with results of other experiments or decay channels. The results presented in this thesis are based on the full Run-2 data collected by the ATLAS experiment. These measurements are sensitive to (i) the cross section ratios between the different Higgs boson production modes, (ii) the amount of additional radiation produced in association with the Higgs boson and (iii) the charge Conjugation and Parity (CP) properties of the Higgs boson. The measurements show excellent agreement with the Standard Model predictions.

Despite the successes of the Standard Model, there are a few experimental signatures, mostly from non-collider physics experiments, that do not fit within the framework of the Standard Model, detailed in Sect. 1.4. These signatures hint at physics Beyond the Standard Model (BSM). Numerous approaches are pursued to search for BSM physics with the LHC data, namely a direct and an indirect approach. The direct approach typically searches for signatures associated with the presence of new particles, such as resonant excesses in the invariant mass distribution of the decay products of these new particles. The indirect approach, on the other hand, looks for hints of BSM physics by searching for deviations between the measurements of quantities that are precisely predicted by the SM, and the theoretical values of such quantities. The measured deviations are then interpreted in terms of concrete BSM models or, as is done in this thesis, using an Effective Field Theory (EFT) approach. This approach is based on the decoupling theorem [15]. It extends the Standard Model Lagrangian with higher-order operators. These operators modify the SM Higgs boson couplings and kinematics, and therefore, one can use precision Higgs boson measurement to constrain these operators experimentally. The EFT approach is a model-independent approach, assuming there are now light resonances, as these additional generic terms can be translated to model-dependent parameters in a separate step. In this thesis, the EFT approach was used to constrain anomalous Higgs boson couplings using the precise measurement of the Higgs boson cross section in the two-photon channel. The precise measurements of the differential Higgs boson production cross sections obtained in this channel allowed us to set tight limits on the strengths of several anomalous interactions between the Higgs boson and the gauge bosons, either *CP*-conserving or *CP*-violating ones.

Overview of the Manuscript and Personal Contributions

The manuscript is divided into three parts: the first part includes the description of the theoretical framework underlying this work; the second part describes the experimental apparatus, with particular emphasis on the calibration of electron and photon energies; the last part illustrates, in detail, the measurement of the Higgs boson cross sections and their interpretation in an EFT framework. The manuscript includes, as well, two interludes describing the statistical treatment used throughout the manuscript and the unfolding procedure for the Higgs boson cross section measurement. The three parts are formed by the following chapters:

- Chapter 1 begins with an overview of the Standard Model and the spontaneous electroweak-symmetry breaking (the Higgs mechanism). The chapter then focuses on the Higgs boson phenomenology at the LHC, summarizing the different production and decay modes and the state-of-the-art SM calculations of the relevant processes, including their main sources of theoretical uncertainties. This chapter includes, as well, a review of the latest Higgs boson properties measurements.
- Chapter 2 is dedicated to the theoretical foundations of the Effective Field Theory (EFT) framework. It includes motivations and details for the different EFT bases used in this thesis. The phenomenology of these bases is also described.
- Chapter 4 describes the experimental setup. It begins with a description of the LHC machine and performance. The second part of the chapter focuses on the ATLAS experiment, detailing its different sub-detectors, with more emphasis on the sub-detector most relevant to this thesis, namely the liquid-Argon electromagnetic calorimeter. The chapter concludes with an overview of the reconstruction and identification techniques used for the different particles produced in the final state of the processes under study in this manuscript.
- Chapter 5 describes the calibration of electron and photon energies. It begins with a description of the full calibration chain of the response of the ATLAS electromagnetic calorimeter. The chapter emphasizes the calibration of the energy response of the different longitudinal layers of the calorimeter. In particular, the calibration of the pre-sampler layer, that I have worked on during the first year of my doctoral studies. The calibration of the energy response of the pre-sampler is a complex procedure, exploiting the sensitivity of different particles to non-active material in the detector. I have extracted the pre-sampler energy scale corrections and computed the relative systematic uncertainties using 36 fb^{-1} of 13 TeV pp collisions collected in 2015–2016. This work was documented in a paper, published in 2018 [16], detailing the different calibration ingredients. Also, I was one of the authors of the Internal ATLAS support note for the layer calibration [17].
- Chapter 7 details the measurement of the Higgs boson production cross section in the two-photon channel. The inclusive cross section and the differential cross sections as a function of six kinematic variables ($p_T^{\gamma\gamma}$, $|y_{\gamma\gamma}|$, $p_T^{j_1}$, N_{jets}, m_{jj} and $\Delta\phi_{jj,\text{signed}}$) are measured. All measurements are performed in fiducial regions defined at the particle-level by criteria that closely match the experimental requirements, to minimize theory uncertainties from extrapolation to the full phase space. In particular, photons are required to be isolated at particle-level from nearby hadronic activity. All the results shown in this chapter are based on my work. In particular:

- I have performed the selection of the fiducial region and studies of the particle-level isolation requirement, including the evaluation of an alternative matching of the detector-level isolation to particle-level isolation based on a veto of jets in a cone of radius $\Delta R_{\gamma j}$ around the photon direction.
- I have optimized the choice of the binning of the differential cross section measurements.
- I have implemented and executed the maximum likelihood fit to the invariant mass distribution of the two photons in the selected events to determine the number of signal events, based on some analytical models for the signal and background distributions.
- I have estimated the expected impact of the experimental and theoretical systematic uncertainties on the Higgs boson signal yield and diphoton invariant mass distribution and implemented them as nuisance parameters in the fit. In particular, I developed a new technique for the estimation of uncertainties due to jets from pileup events using Monte Carlo-based subtraction.
- I have contributed to the studies leading to the choice of the analytical background model and the estimation of the systematic uncertainty related to this choice.
- I have performed extensive comparisons of different unfolding methods to find the method leading to the best compromise between statistical uncertainty and systematic bias

I was a co-author of the internal ATLAS note documenting this analysis. These results were published in a conference note released for 2019 EPS-HEP conference [12]. Previously, in 2018, a cross section measurement as a function of fewer kinematic variables was performed using a subset of the full Run-2 data (80 fb^{-1} collected between 2015–2017) and published in a conference note released for the 2018 ICHEP conference [18]. For that measurement, I was a co-author of the ATLAS supporting analysis note and gave major contributions to all the different stages of the analysis. I have also contributed during the first year to the measurement of the cross sections using 36 fb^{-1} of data collected in 2015–2016, and published in Phys. Rev. D [19]. I was granted exceptional authorship for my contributions.

- Chapter 8 concludes the manuscript with an interpretation of the measured cross sections of Chap. 7 in an effective field theory framework. The work presented in the chapter was all done by myself. The interpretation was performed using two EFT bases: the basis of the Strongly Interacting Light Higgs (SILH) lagrangian [20], also used previously to interpret the results of Ref. [19], and—for the first time in ATLAS—the Warsaw basis of the SMEFT lagrangian [21]. These results were also included in the 2019 EPS-HEP conference note [12].

East Lansing, USA   Dr. Ahmed Tarek Abouelfadl Mohamed

References

1. Gell-Mann M (1964) A schematic model of baryons and mesons. In: Phys. Lett. 8:214–215 10.1016/S0031-9163(64)92001-3 (p 17, 24)
2. Zweig G (1964–1978) An SU(3) model for strong interaction symmetry and its breaking. Version 2. In: Lichtenberg D B, Rosen S P (eds) Developments in the quark theory of hadrons, vol 1. pp 22–101 (p 17, 24)
3. Sheldon L. Glashow (1964) Partial-symmetries of weak interactions. Nucl. Phys 22(4): 579–588. ISSN: 0029-5582. https://doi.org/10.1016/0029-5582(61)90469-2. URL: http://www.sciencedirect.com/science/article/pii/0029558261904692 (p 17, 24)
4. Salam A (1969) The standard model. In: Svartholm N, Stockholm (ed) Elementary particle theory, p 367 (p 17, 24)
5. Weinberg S (1967) A Model of Leptons. Phys. Rev. Lett. 19 (21):1264–1266. 10.1103/PhysRevLett.19.1264. URL: https://link.aps.org/doi/10.1103/PhysRevLett.19.1264 (p 17, 24)
6. Georges A, et al (2012) Observation of a new particle in the search for the standard model higgs boson with the ATLAS detector at the LHC. Phys. Lett. B716, pp 1–29. 10.1016/j.physletb.2012.08.020. arXiv: 1207.7214 [hep-ex] (p 17, 41, 49, 73, 151)
7. Serguei C, et al (2012) Observation of a new boson at a mass of 125 GeV with the CMS experiment at the LHC. Phys. Lett. B716:30–61. 10.1016/j.physletb.2012.08.021. arXiv: 1207.7235 [hep-ex] (p 17, 41, 49, 151)
8. Lyndon E, Philip B (2008) LHC Machine. J Instrum. 3(08):S08001–S08001. 10.1088/1748-0221/3/08/s08001. URL: https://doi.org/10.1088
9. Higgs PW (1964) Broken symmetries, massless particles and gauge fields. Phys. Lett. 12:132–133. 10.1016/0031-9163(64)91136-9 (p 17, 24)
10. Englert F,Brout R (1964) Broken symmetry and the mass of gauge vector mesons. Phys. Rev. Lett. 13 :321–323. 10.1103/PhysRevLett.13.321 (p 17, 24)
11. Guralnik GS, Hagen CR, Kibble TWB (1964) Global conservation laws and massless particles. Phys. Rev. Lett. 13 (20):585–587. 10.1103/PhysRevLett.13.585. URL: https://link.aps.org/doi/10.1103/PhysRevLett.13.585 (p 17, 24)
12. Measurements and interpretations of Higgs-boson fiducial cross sections in the diphoton decay channel using 139 fb^{-1} of pp collision data at $\sqrt{s} = 13$ TeV with the ATLAS detector. Tech. rep. ATLAS-CONF-2019-029. Geneva: CERN, 2019. URL: https://cds.cern.ch/record/2682800 (p 17, 19)
13. ATLAS Collaboration. Measurements of the Higgs boson inclusive, differential and production cross sections in the 4ℓdecay channel at $\sqrt{s} = 13$ TeV with the ATLAS detector. Tech. rep. ATLAS-COMCONF-2019-045. Geneva: CERN, 2017. URL: https://cds.cern.ch/record/2680220 (p 17, 200, 201)
14. ATLAS Collaboration. Combined measurement of the total and differential cross sections in the H $\rightarrow \gamma\gamma$ and the H $\rightarrow$ ZZ* $\rightarrow 4\ell$ decay channels at $\sqrt{s} = 13$ TeV with the ATLAS detector. Tech. rep. ATLASCOM-CONF-2019-049. Geneva: CERN, 2019. URL: https://cds.cern.ch/record/2681143 (p 17, 201, 202)
15. Appelquist T, Carazzone J (1975) Infrared singularities and massive fields. Phys. Rev. D11: 2856. 10.1103/PhysRevD.11.2856 (p 18, 50)
16. Aaboud M, et al (2019) Electron and photon energy calibration with the ATLAS detector using 2015-2016 LHC proton-proton collision data. JINST 14(03):P03017. 10.1088/1748-0221/14/03/P03017. arXiv: 1812.03848 [hep-ex] (p 19, 93, 105–111, 129-133)
17. Laudrain A, et al (2017) Electromagnetic calorimeter layers energy scale determination. Tech. rep. ATLCOM-PHYS-2017-760. Geneva: CERN. URL: https://cds.cern.ch/record/2268812 (p 19)

18. Measurements of Higgs boson properties in the diphoton decay channel using 80 fb^{-1} of pp collision data at $\sqrt{s} = 13$ TeV with the ATLAS detector. Tech. rep. ATLAS-CONF-2018-028. Geneva: CERN, 2018. URL: https://cds.cern.ch/record/2628771 (p 19, 145, 196, 249)

19. Aaboud M, et al (2018) Measurements of higgs boson properties in the diphoton decay channel with 36 fb^{-1} of pp collision data at $\sqrt{s} = 13$ TeV with the ATLAS detector. Phys. Rev. D98:052005. 10.1103/PhysRevD.98.052005. arXiv: 1802.04146 [hep-ex] (p 19, 152, 196, 215, 232, 233, 249)

20. Alloul A, Fuks B, Sanz V (2014) Phenomenology of the higgs effective lagrangian via feynrules. JHEP 04:110. 10.1007/JHEP04(2014)110. arXiv: 1310.5150 [hep-ph] (p 19, 56, 249)

21. Brivio I, Trott M (2017) The standard model as an effective field theory. 10.1016/j.physrep.2018.11.002. arXiv: 1706.08945 [hep-ph] (p 19, 53–55, 249)

Acknowledgements

Needless to say, this work could not have been possible without the guidance and support of a lot of people, to whom I am very grateful.

Foremost, I would like to express my deepest gratitude to my supervisor, Giovanni Marchiori, for his continuous support and constant motivation during the Ph.D. Giovanni is a role model and an example of how a particle physicist should be. I would like to extend my gratitude to Sandrine Laplace for her supervision and guidance during the first phase of the Ph.D. and during the master internship. I consider myself to be very lucky to have had the chance to work closely with physicists of exceptional merit.

I also wish to thank the members of the jury. I especially thank the referees André David and Louis Fayard for their thorough review and insightful comments.

This work would not have been possible without the support of the LPNHE ATLAS group. I sincerely thank Ioannis Nomidis for his relentless support. It was a real pleasure to work closely with him. A special thanks to Lydia Roos, José Ocariz, Bogdan Malaescu and Bertrand Laforge for all the discussions and guidance.

I am grateful also for the CERN e/γ and *HGamma* groups for their contribution and coordination of various parts of this work. I especially thank Guillaume Unal and Maarten Boonekamp for their advice during the qualification task.

I would like to thank a lot my fellow Ph.D. students at LPNHE for being amazing colleagues during the Ph.D. (and the masters) journey. I am thankful for all the stimulating discussions, the enormous support and all the fun we had. A special thank you to Yee and Stefano for the great support and encouragement.

I am eternally indebted to my family for all the love and encouragement that is beyond words.

Contents

Part II Experiment

Part I
Theory

Chapter 1
The Standard Model of Particle Physics

Since the second half of the 20th century we have been witnessing the triumphs of the *Standard Model* of particle physics (SM). The Standard Model of particle physics refers mainly to a set of theories developed in the mid-1970s. The Standard Model describes the known fundamental constituents of the universe and their interactions. In this chapter, we will review the different concepts underlying the SM and will introduce the SM Lagrangian, with a focus on the Higgs sector. We will briefly review the different production and decay modes of the Higgs boson with particular emphasis on the decay in the diphoton channel as it is the decay channel used in this thesis to study the Higgs boson properties. The chapter also includes a review of the latest measurements of the Higgs boson properties and motivations for the searches of physics beyond the Standard Model (BSM).

1.1 Introduction

Fundamental concepts

One of the fundamental concepts in the formulation of the Standard Model is that of *symmetry*. Symmetry refers to the set of the transformations under which the physical system remains unchanged, or invariant. The importance of this concept comes from the Noether theorem [1], relating continuous symmetries of the action of a system to conserved quantities. For example, the invariance or symmetry under space (time) translations results in the conservation of momentum (energy).

In the context of quantum field theory,[1] one of the cornerstones in the construction of the Standard Model is that of *gauge* symmetry. To understand gauge symmetry,

---

[1] See Refs. [2, 3] for a comprehensive overview.

© The Editor(s) (if applicable) and The Author(s), under exclusive license
to Springer Nature Switzerland AG 2020
A. Tarek Abouelfadl Mohamed, *Measurement of Higgs Boson Production
Cross Sections in the Diphoton Channel*, Springer Theses,
https://doi.org/10.1007/978-3-030-59516-6_1

let us consider a massive complex scalar field ψ with mass m described by the Lagrangian density (simply called "Lagrangian" in the following):

$$\mathcal{L} = \partial^\mu \psi^\dagger \partial_\mu \psi - m^2 \psi^\dagger \psi \tag{1.1}$$

This Lagrangian is invariant with respect to a rotation with angle α in the complex plane:

$$\psi(x) \rightarrow \psi(x)e^{i\alpha} \quad , \quad \psi^\dagger(x) \rightarrow \psi^\dagger(x)e^{-i\alpha} \tag{1.2}$$

Therefore, this Lagrangian is said to be symmetric under *global* $U(1)$ transformations, with global denoting that α is the same for any space-time point x, and $U(1)$ is the one-dimensional group of unitary transformations. This $U(1)$ symmetry, however, is broken when α depends on x. Therefore, the complex scalar field Lagrangian does not respect *local* $U(1)$ symmetry. The invariance under local $U(1)$ transformation can be restored by introducing a new vector field $A_\mu(x)$ that transforms as $A_\mu \rightarrow A_\mu - \frac{1}{q}\partial_\mu \alpha(x)$ and by replacing the derivative in the Lagrangian of Eq. (1.1) with a *covariant* derivative D_μ defined as

$$D_\mu = \partial_\mu + iqA_\mu(x), \tag{1.3}$$

where q is the coupling strength between the scalar field ψ and A_μ. Using these definitions, the Lagrangian is now symmetric with respect to local $U(1)$ transformation. The field A_μ that is introduced to restore the local symmetry is then known as a *gauge field*, and the Lagrangian with these modifications is known as a *gauge theory*. The terminology *gauge* implies that the field theory admits different configurations of the fields which yield identical observables. The field A_μ, in this case, is thought of as an interaction mediator that couples to the scalar field ψ. The number of gauge fields required to restore a given local symmetry is related to the number of generators in the gauge symmetry group. In our example, it was the $U(1)$ symmetry group with one generator, and hence this resulted in a single gauge field to restore the local gauge symmetry.

1.1.1   Overview of the Standard Model

The Standard Model is a gauge theory that describes the elementary particles and their fundamental interactions (aside from gravity). The SM is built from three families of elementary spin-$\frac{1}{2}$ particles (fermions) classified in quarks and leptons with a total of 12 particles (with their anti-particle counterparts). The different properties (mass, charge, and spin) of these particles are summarized in Fig. 1.1. The SM describes the electromagnetic, weak, and strong nuclear interactions between these particles as follows:

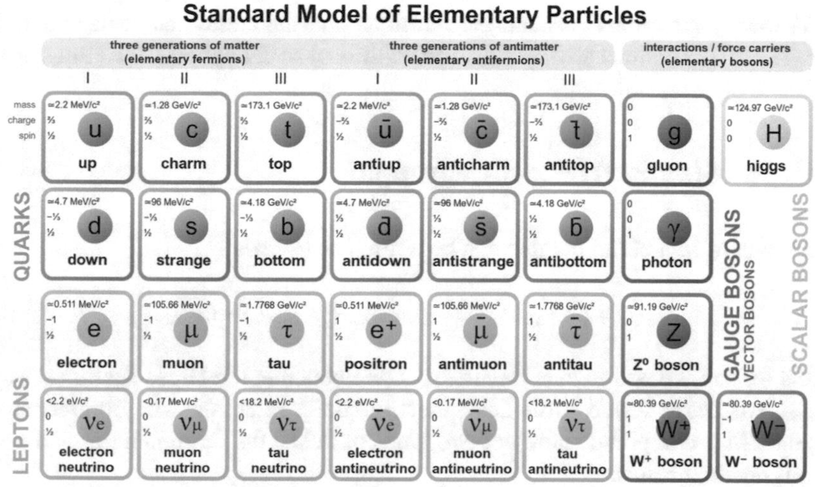

Standard Model of Elementary Particles

	three generations of matter (elementary fermions)			three generations of antimatter (elementary antifermions)			interactions / force carriers (elementary bosons)	
	I	II	III	I	II	III		
mass charge spin	≈2.2 MeV/c² ⅔ ½ **u** up	≈1.28 GeV/c² ⅔ ½ **c** charm	≈173.1 GeV/c² ⅔ ½ **t** top	≈2.2 MeV/c² -⅔ ½ **ū** antiup	≈1.28 GeV/c² -⅔ ½ **c̄** anticharm	≈173.1 GeV/c² -⅔ ½ **t̄** antitop	0 0 1 **g** gluon	≈124.97 GeV/c² 0 0 **H** higgs
	≈4.7 MeV/c² -⅓ ½ **d** down	≈96 MeV/c² -⅓ ½ **s** strange	≈4.18 GeV/c² -⅓ ½ **b** bottom	≈4.7 MeV/c² ⅓ ½ **d̄** antidown	≈96 MeV/c² ⅓ ½ **s̄** antistrange	≈4.18 GeV/c² ⅓ ½ **b̄** antibottom	0 0 1 **γ** photon	
	≈0.511 MeV/c² -1 ½ **e** electron	≈105.66 MeV/c² -1 ½ **μ** muon	≈1.7768 GeV/c² -1 ½ **τ** tau	≈0.511 MeV/c² 1 ½ **e⁺** positron	≈105.66 MeV/c² 1 ½ **μ̄** antimuon	≈1.7768 GeV/c² 1 ½ **τ̄** antitau	≈91.19 GeV/c² 0 1 **Z** Z⁰ boson	
	<2.2 eV/c² 0 ½ **νₑ** electron neutrino	<0.17 MeV/c² 0 ½ **ν_μ** muon neutrino	<18.2 MeV/c² 0 ½ **ν_τ** tau neutrino	<2.2 eV/c² 0 ½ **ν̄ₑ** electron antineutrino	<0.17 MeV/c² 0 ½ **ν̄_μ** muon antineutrino	<18.2 MeV/c² 0 ½ **ν̄_τ** tau antineutrino	≈80.39 GeV/c² 1 1 **W⁺** W⁺ boson	≈80.39 GeV/c² -1 1 **W⁻** W⁻ boson

QUARKS | LEPTONS | GAUGE BOSONS / VECTOR BOSONS | SCALAR BOSONS

Fig. 1.1 A sketch of the standard model ingredients [12]

- The description of the strong interaction is based on the theory of **Quantum Chromo–Dynamics (QCD)** [4, 5]. Quantum Chromo-Dynamics is a gauge theory with $SU(3)_C$ symmetry. It is based on the conservation of the strong charge, referred to as the *color* charge (hence the subscript C). The color charge comes in three forms: red, green, and blue. The strong force is mediated by the exchange of 8 massless vector bosons, known as gluons.

- The description of the electromagnetic and weak interactions is based on the **GWS electroweak** theory, unifying electromagnetism and weak interactions. The electroweak model, proposed by Glashow [6], Salam [7] and Weinberg [8], is based on the gauge symmetry group $SU(2)_L \otimes U(1)_Y$, where the subscripts L and Y refer to left-handed isospin and hypercharge. The electromagnetic and weak interactions are propagated by four-vector bosons: a massless photon and three massive bosons respectively, the charged $W^\pm$ bosons and the neutral Z boson.

The full gauge symmetry group of the SM is thus $SU(3)_C \otimes SU(2)_L \otimes U(1)_Y$.

Without additional fields, the gauge invariance of the SM is only possible for massless particles since explicit mass terms for fermions or gauge bosons would not be gauge invariant. However, massless gauge bosons will lead to long-range forces which contradicts various experimental evidence that the weak interactions have a very short range and are mediated by massive gauge bosons. This contradiction was the motivation for the spontaneous electroweak symmetry breaking (EWSB) theory, proposed by P. Higgs [9], R. Brout and F. Englert [10] and G. Guralnik, C. Hagen and T. Kibble [11] to generate vector boson and fermion masses. This is done by introducing an $SU(2)$ doublet of a complex scalar field with the neutral component undergoing a phase transition known as *spontaneous symmetry breaking* and developing a non-zero vacuum expectation value (VEV) that gives rise to the $W^\pm$ and Z bosons masses. The fermion masses, on the other hand, are generated by Yukawa interactions between the fermions and the additional scalar field. The complex $SU(2)$ doublet also results in an additional scalar particle: the Higgs boson.

1.2   The Standard Model Lagrangian

The Standard Model Lagrangian can be written as follows:

$$\mathcal{L}_{SM} = \mathcal{L}_{fermions} + \mathcal{L}_{gauge\ bosons} + \mathcal{L}_{Higgs} + \mathcal{L}_{Yukawa} \tag{1.4}$$

In this section we will review the first two components; the Higgs sector ($\mathcal{L}_{Higgs} + \mathcal{L}_{Yukawa}$) will be reviewed in a dedicated section. Before electroweak symmetry breaking, the SM is composed mainly of two kinds of fields: the fermion fields and gauge bosons detailed below.

1.2.1   The Fermion Fields

The first component of the SM Lagrangian, $\mathcal{L}_{fermions}$, describes the propagation and the interactions of fermions with gauge bosons. Fermions are spin $1/2$ particles that can be categorized into three types: Weyl (massless), Dirac (massive) and Majorana (each particle is its own anti-particle) fermions. All Standard Model fermions after the symmetry breaking are Dirac fermions (aside from neutrinos as their nature is not yet determined to be Dirac or Majorana). The kinetic term of a Dirac fermion field ψ is $i\bar{\psi}\gamma^\mu\partial_\mu\psi$, where γ^μ are the Dirac matrices [13]. In the SM, a distinction is made between left-handed and right-handed quarks and leptons. This distinction is a result of the field transformations under the $SU(2)_L$ symmetry group. The left-handed fermions are grouped in weak isospin doublets $f_L = \frac{1}{2}(1 - \gamma_5)f$. The right-handed fermions are weak isosinglets $f_R = \frac{1}{2}(1 + \gamma_5)f$. This is summarized below for the three generations [14]:

	Leptons		Quarks	

$$L_1 = \begin{pmatrix} \nu_e \\ e^- \end{pmatrix}_L \ , \ e_{R1} = e_R^- \ , \ Q_1 = \begin{pmatrix} u \\ d \end{pmatrix}_L \ , \ u_{R1} = u_R \ , \ d_{R1} = d_R$$

$$L_2 = \begin{pmatrix} \nu_\mu \\ \mu^- \end{pmatrix}_L \ , \ e_{R2} = \mu_R^- \ , \ Q_2 = \begin{pmatrix} c \\ s \end{pmatrix}_L \ , \ u_{R2} = c_R \ , \ d_{R2} = s_R \tag{1.5}$$

$$L_3 = \begin{pmatrix} \nu_\tau \\ \tau^- \end{pmatrix}_L \ , \ e_{R3} = \tau_R^- \ , \ Q_3 = \begin{pmatrix} t \\ b \end{pmatrix}_L \ , \ u_{R3} = t_R \ , \ d_{R3} = b_R$$

In the language of the electroweak unification, one can define the weak hypercharge Y_f in terms of the electric charge (Q_f) and the third component of the weak isospin I_f^3 by the equation:

$$Y_f = 2Q_f - 2I_f^3 \tag{1.6}$$

This leads to the hypercharges $Y_L = -1, Y_{Q_L} = \frac{1}{3}$ for the doublets and $Y_R = -2, Y_u = \frac{4}{3}, Y_r = -\frac{2}{3}$ for the singlets. One of the consequences of the left-handed components being $SU(2)_L$ doublets whereas the right-handed components are singlets is that it is impossible to include an explicit mass term $(-m_f \bar{\psi}\psi)$ for any of the fermions detailed above. The explicit mass terms mix the left- and right-handed components, which violate the isospin symmetry.

In addition to the electroweak interactions, the quark fields $Q_{(1,2,3)}, u_{R(1,2,3)}$ and $d_{R(1,2,3)}$ are charged under the $SU(3)$ color group. They are represented as triplets transforming with their corresponding fundamental representation of $SU(3)$, whereas the leptonic fields are color singlets, i.e. not charged under the strong interaction.

1.2.2 The Gauge Fields

The second component of the SM Lagrangian, $\mathcal{L}_{\text{gauge bosons}}$, describes the propagation of the gauge bosons. The gauge bosons are the spin-one bosons mediating the interaction and restoring the local gauge symmetry as detailed in the introduction. The gauge-fields Lagrangian follows:

$$\mathcal{L}_{\text{gauge boson}} = -\frac{1}{4}G_{\mu\nu}^a G_a^{\mu\nu} - \frac{1}{4}W_{\mu\nu}^a W_a^{\mu\nu} - \frac{1}{4}B_{\mu\nu}B^{\mu\nu}. \tag{1.7}$$

These fields correspond to the generators of their respective symmetry groups. The field B_μ correspond to the generator of the $U(1)_Y$ group, the fields $W_\mu^{a=1,2,3}$ correspond to the three generators of the $SU(2)_L$ group and the fields $G_\mu^{a=1..8}$ correspond to the eight generators of $SU(3)_C$. The field strength tensors of these fields are given by:

$$\begin{aligned}
G_{\mu\nu}^a &= \partial_\mu G_\nu^a - \partial_\nu G_\mu^a + g_s f^{abc} G_\mu^b G_\nu^c, \\
W_{\mu\nu}^a &= \partial_\mu W_\nu^a - \partial_\nu W_\mu^a + g_2 \epsilon^{abc} W_\mu^b W_\nu^c, \\
B_{\mu\nu} &= \partial_\mu B_\nu - \partial_\nu B_\mu,
\end{aligned} \tag{1.8}$$

where g_s and g_2 are the strong and weak coupling constants

The main differences between these fields result from the different nature of their corresponding symmetry groups. For example, the $SU(2)_L$ and $SU(3)_C$ groups are non-abelian groups. For the $SU(2)_L$ group, the fields $W_\mu^{a=1,2,3}$ correspond to the $SU(2)$ group generators T^a that follow the commutation relations $[T^a, T^b] = i\epsilon^{abc}T^c$ where ϵ^{abc} is the antisymmetric tensor. For $SU(3)_C$ the fields $G_\mu^{a=1..8}$ correspond to the $SU(3)$ generators λ^a that result in the commutation $[\lambda^a, \lambda^b] = if^{abc}\lambda^c$ where the tensor f^{abc} contains the structure constants of the $SU(3)$ group. The matrices λ^a are known as the Gell-Mann matrices [15]. The non-abelian group structure of $SU(2)$ and $SU(3)$ results in self-interactions between the W^μ and G^μ gauge fields. The behavior of the $SU(2)$ and $SU(3)$ groups is different from that of the $U(1)_Y$ group, where the fields commute, as the generators Y of $U(1)$ follow $[Y^a, Y^b] = 0$.

The coupling between the fermion fields and the gauge fields is established by defining the covariant derivative that preserves gauge invariance. The covariant derivatives D_μ for the case of quarks (that couple to all gauge bosons) is:

$$D_\mu \psi = \left(\partial_\mu - ig_s \frac{\lambda_a}{2} G_\mu^a - ig_2 T_a W_\mu^a - i\frac{g_1}{2} Y B_m u \right) \psi \qquad (1.9)$$

with g_s, g_2, g_1 the coupling constants of the $SU(3)_C$, $SU(2)_L$, and $U(1)_Y$ interactions respectively. From this definition, and by replacing the derivative in the Dirac Lagrangian with the covariant derivative, the different fermion and gauge boson interactions can be determined for the electroweak and QCD components.

Quantum Chromo-Dynamics (QCD) The couplings between quarks and gluons are defined from the $SU(3)_C$ covariant derivative. For example, for first generation quarks:

$$\mathcal{L}_{\text{QCD}}^{\text{interactions}} = \frac{g_s}{2} \bar{Q}_1 \gamma^\mu \lambda_a G_\mu^a Q_1 \qquad (1.10)$$

where g_s is the coupling constant of the strong interactions and λ_a are the $SU(3)$ generators. The constant g_s, or equivalently $\alpha_s = \frac{g_s^2}{4\pi}$, is the fundamental parameter of QCD, along with the quark masses. Using these parameters one can evaluate the QCD scattering amplitudes in powers of α_s. However, loop graphs are divergent and need to be fixed using a renormalization procedure with a renormalization scale μ_R that is typically close to the scale of the momentum transfer of the process. In this case, the dependence of the coupling constant on the renormalization scale μ_R is derived using the renormalization group equations (RGE). For QCD, one finds the following relation [16, §9.1.1]:

$$\mu_R^2 \frac{d\alpha_s}{d\mu_R^2} = \beta(\alpha_s) = -(b_0 \alpha_s + \ldots), \qquad (1.11)$$

where $b_0 = (33 - 2n_f)/(12\pi)$, and n_f denotes the number of quark flavors. b_0 is referred to as the 1-loop β-function coefficient. The dependence of α_s on μ_R, plotted in Fig. 1.2, defines the characteristic properties of QCD interactions:

- *Asymptotic freedom*, which refers to the fact that the strong interaction coupling becomes weak for processes with large momentum transfers (large Q^2) and small distances. This is a result of the minus sign in Eq. (1.11). Solving Eq. (1.11) for $\alpha_s(\mu_R^2)$ and using the experimental data in Fig. 1.2 one finds:

Fig. 1.2 Summary of
measurements of α_s as a
function of the energy scale
Q. The respective degree of
QCD perturbation theory
used in the extraction of α_s
is indicated in parenthesis
[16, §9.4.8]

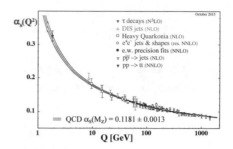

$$\alpha_s(\mu_R^2) = \frac{1}{b_0 \log(\frac{\mu_R^2}{\Lambda_{QCD}^2})}, \tag{1.12}$$

where Λ_{QCD} is the fundamental scale of QCD at which the coupling constant
diverges and is estimated to be $\Lambda_{QCD} \simeq 220$ MeV. This scale defines the infrared
cutoff of QCD, meaning that the for $\mu_R \gg \Lambda_{QCD}$ the coupling constant becomes
small ($\alpha_s \ll 1$) and perturbation theory is valid.

- *Quark confinement*, which refers to the fact that at small Q^2 (large distances), the
  coupling between the quarks becomes strong and prevents quarks from existing
  as isolated particles. The increasing potential energy due to the increasing separation between the quarks is large enough that it produces quark and anti-quark
  pairs in a process known as *hadronization*. This results in the quarks forming
  colorless hadrons: baryons (qqq) or mesons ($\bar{q}q$). These final state hadrons will
  appear predominantly in collimated bunches, which are generically called *jets*. In
  this case, the momentum transfer Q^2 is similar to the QCD scale, $Q^2 \approx \Lambda_{QCD}^2$,
  hence perturbation theory is not valid. This problem is tackled using the property
  of *factorization* of the non-perturbative hadronization process from the perturbative hard process as they occur at different time scales. The predictions of the
  non-perturbative component are based on Monte-Carlo simulations using *parton-
  shower* event generators with different hadronization models such as HERWIG [17]
  and PYTHIA [18].

Electroweak theory The couplings between fermions and electroweak bosons can be
determined by expanding the covariant derivative of Eq. (1.9) in the Dirac Lagrangian.
For example, for the first generation leptons:

$$\mathcal{L}_{EW}^{\text{interactions}} = \overbrace{-i\bar{L}_1\gamma^\mu \left(\frac{iY_L g_1}{2}B_\mu\right)L_1 - i\bar{e}_{R1}\gamma^\mu \left(\frac{iY_R g_1}{2}B_\mu\right)e_{R1}}^{U(1)\text{ terms}} \overbrace{- i\bar{L}_1\gamma^\mu \left(\frac{iT_a g_2}{2}W_\mu^a\right)L_1}^{SU(2)\text{ terms}},$$
$$\tag{1.13}$$

using $\bar{L}\gamma^\mu L = \bar{\nu}_L\gamma^\mu\nu_L + \bar{e}_L\gamma^\mu e_L$ and the full expression of the $T_a W_\mu^a$ in $SU(2)$
terms:

$$T_a W_\mu^a = \begin{pmatrix} W_\mu^3 & W_\mu^1 - i W_\mu^2 \\ W_\mu^1 + i W_\mu^2 & -W_\mu^3 \end{pmatrix} = \begin{pmatrix} W_\mu^3 & -\sqrt{2} W_\mu^+ \\ -\sqrt{2} W_\mu^- & -W_\mu^3 \end{pmatrix}, \qquad (1.14)$$

where $W_\mu^\pm = \frac{1}{\sqrt{2}}(W_\mu^1 \mp i W_\mu^2)$ are the physical fields of the charged $W_\mu^\pm$ bosons. The total Lagrangian of the electroweak interactions for first generation leptons follows:

$$\mathcal{L}_{EW}^{interactions} = -\frac{g_1}{2} \left[Y_L \left(\bar{\nu}_L \gamma^\mu \nu_L - \bar{e}_L \gamma^\mu e_L \right) - Y_R \bar{e}_R \gamma^\mu e_R \right] B_\mu$$
$$+ \frac{g_2}{2} \left[\bar{\nu}_L \gamma^\mu \nu_L W_\mu^3 - \sqrt{2} \bar{\nu}_L \gamma^\mu e_L W_\mu^+ - \sqrt{2} \bar{e}_L \gamma^\mu \nu_L W_\mu^- - \bar{e}_L \gamma^\mu e_L W_\mu^3 \right] \quad (1.15)$$

From this equation, one can deduce the charged current weak interactions, mediated by the $W^\pm$ bosons, coupling electrons and neutrinos. Equation (1.15) includes as well terms coupling electrons (and neutrinos) with themselves. One can then interpret the combinations $g_2 B_\mu + g_1 W_\mu^0$ and $-g_1 B_\mu + g_2 W_\mu^0$ as the physical photon A_μ and neutral Z_μ boson fields, defined as:

$$Z_\mu = \frac{-g_1 B_\mu + g_2 W_\mu^0}{\sqrt{g_1^2 + g_2^2}} \qquad (1.16)$$

$$A_\mu = \frac{g_2 B_\mu + g_1 W_\mu^0}{\sqrt{g_1^2 + g_2^2}}, \qquad (1.17)$$

One can then define the mixing angle:

$$\theta_W = \sin^{-1} \frac{g_1}{\sqrt{g_1^2 + g_2^2}}, \qquad (1.18)$$

also known as the Weinberg weak mixing angle, by which the rotation of the electroweak fields W^0 and B^0 occurs and produces the Z^0 boson and the photon after spontaneous symmetry breaking.

These relations define as well the relation between the electric charge e (or the electromagnetic couplings constant $\alpha = \frac{e^2}{4\pi\epsilon_0 \hbar c} \simeq \frac{1}{137}$) and the electroweak SU(2) and U(1) couplings g_1 and g_2:

$$e = g_1 \cos \theta_W. \qquad (1.19)$$

The experiments have confirmed the existence of the $W_\mu^\pm$ and Z_μ bosons, with masses of $m_W = 80.363 \pm 0.020$ GeV and $m_Z = 91.1876 \pm 0.0021$ GeV respectively [16]. This explains the short range of the weak interaction ($\sim 10^{-18}$ m): accounting for the masses m_V in the propagator of these gauge bosons results in a potential $U(r) = e^{-r m_V}$ that falls quickly with the distance r. However, including a mass term $\frac{1}{2} m_V^2 W_\mu W^\mu$ in the SM Lagrangian will violate the local $SU(2) \times U(1)$ gauge invariance. This problem, in addition to the mass problem of fermions, were the motivations for the development of the theory of spontaneous electroweak symmetry breaking, commonly known as the Higgs mechanism, which is detailed in the next section.

1.2.3   Electroweak Spontaneous Symmetry Breaking

As detailed in the previous section, explicit mass terms can not be used for fermions and massive gauge bosons. Therefore, attempts to solve this problem were undertaken using the concept of *spontaneous symmetry breaking*. This concept originated in the field of condensed matter physics. A famous example of this concept is from Landau's theory of phase transitions [19]. For a ferromagnetic material in the absence of external fields and at high temperatures, the spatial average of the magnetic moment (or magnetization) is zero. This results in a global $O(3)$ rotational symmetry for this system, as shown in the left plot of Fig. 1.3. However, it was found that when the system cools down below a certain *critical* temperature $T < T_C$, the system undergoes a phase transition and the magnetization becomes non-zero as all of the spins line up along a single direction, as shown in the right plot of Fig. 1.3. This phase transition breaks the global $O(3)$ rotational symmetry. The symmetry, in this case, is said to be "spontaneously" broken as there is nothing in the Lagrangian describing the system that can explain such a symmetry breaking.

A similar idea was brought to the field of particle physics to solve the problem of masses of gauge bosons. The simplest of such solutions was the introduction of a complex $SU(2)$ doublet scalar field Φ in the SM lagrangian:

$$\Phi = \begin{pmatrix} \phi^+ \\ \phi^0 \end{pmatrix} = \frac{1}{\sqrt{2}} \begin{pmatrix} \phi_3 + i\phi_4 \\ \phi_1 + i\phi_2 \end{pmatrix} \tag{1.20}$$

This additional scalar field is defined to have weak hypercharge $Y_L = +1$. The gauge-invariant Lagrangian term of this scalar is:

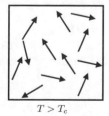
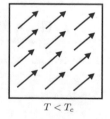

$$T > T_c \qquad\qquad\qquad T < T_c$$

Fig. 1.3 An example of the Landau theory using a piece of ferromagnetic material. Left: for temperatures $T > T_c$, the average magnetization (magnetic moment) is zero. The zero magnetization is the same for any angle of rotation. Right: for temperatures $T < T_c$, the spins align at a particular direction breaking the rotational symmetry of the system

Fig. 1.4 The potential $V(\Phi)$ for a scalar field ϕ with $\mu^2 > 0$ (left plot) and with $\mu^2 < 0$ (right), giving rise to the typical "Mexican hat" shape

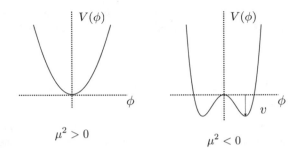

$$\mathcal{L}_{\text{Higgs}} = (D^\mu \Phi)^\dagger (D_\mu \Phi) - V(\Phi) , \quad V(\Phi) = \mu^2 \Phi^\dagger \Phi + \lambda (\Phi^\dagger \Phi)^2, \tag{1.21}$$

where the covariant derivative is defined as

$$D_\mu \Phi = \left(\partial_\mu \Phi - i g_2 T_a W_\mu^a - \frac{i g_1}{2} B_\mu \right) \Phi. \tag{1.22}$$

For $\mu^2 > 0$ the potential $V(\Phi)$ has a minimum at $\phi = 0$, as shown in the left plot of Fig. 1.4. The minimum corresponds to the ground state, also known as the vacuum, $\langle 0|\phi|0 \rangle = 0$. The Lagrangian $\mathcal{L}_{\text{Higgs}}$ in this case is that of a scalar particle of mass μ. However, for $\mu^2 < 0$, the potential has now a "Mexican-hat" shape, as shown in the right plot of Fig. 1.4, with a continuum set of minima satisfying the criterium:

$$|\langle \Phi \rangle_0|^2 = |\langle 0|\Phi|0 \rangle|^2 = v^2/2 \text{ with } v = \left(\frac{-\mu^2}{\lambda} \right)^{1/2} \tag{1.23}$$

In the Higgs potential, the self-coupling parameter λ is chosen to be positive to ensure that the potential is bound from below. The minimum of the potential after the spontaneous symmetry breaking occurs for the neutral component of the doublet to preserve the $U(1)$ symmetry. The non-zero vacuum expectation value results in the breaking of the symmetry:

$$SU(2)_L \times U(1)_Y \rightarrow U(1)_{EM} \tag{1.24}$$

Using Eq. (1.23), one can rewrite the field Φ in terms of physical fields using the unitary gauge:

$$\Phi = \frac{1}{\sqrt{2}} \begin{pmatrix} 0 \\ v + H \end{pmatrix}, \tag{1.25}$$

where H is the physical Higgs field. Using this definition in Eq. (1.22), one finds immediately the terms of the vector bosons masses in terms of the non-zero vev (v) as:

$$(D^\mu \Phi)^\dagger (D^\mu \Phi) = \frac{1}{2} \partial_\mu H \partial^\mu H + \frac{(v+H)^2}{8} \left(2g_1^2 W_\mu W^\mu + (g_1^2 + g_2^2) Z_\mu Z^\mu \right)$$

(1.26)

This leads to the vector boson masses:

$$m_W^2 = \frac{1}{4} g_1^2 v^2, \quad m_Z^2 = \frac{1}{4} (g_1^2 + g_2^2) v^2, \quad m_A = 0.$$

(1.27)

Therefore, by the introduction of the Higgs field, three of the Goldstone bosons [20] that result from the breaking of the continuous symmetry were absorbed by the $W^\pm$ and Z bosons to form their longitudinal components and give them masses. The $U(1)_{EM}$ symmetry is still unbroken and hence the photon remains massless. From Eq. (1.27), the vacuum expectation value is related to the W boson mass, which in turn is related to the Fermi coupling constant by the relation [21]:

$$m_W = \frac{g_1}{2\sqrt{\sqrt{2} G_F}},$$

(1.28)

where $G_F = 1.166 \times 10^{-5} \, \text{GeV}^{-2}$ is the Fermi coupling constant. From this relation, the value of the Higgs vacuum expectation value is:

$$v = \sqrt{\frac{1}{\sqrt{2} G_F}} = 246.22 \, \text{GeV}.$$

(1.29)

Using the scalar doublet, we can generate the fermion mass terms by introducing the $SU(2)_L \times U(1)_Y$ invariant Yukawa Lagrangian:

$$\mathcal{L}_{\text{Yukawa}} = -\lambda_e \bar{L} \Phi e_R - \lambda_d \bar{Q} \Phi d_R \lambda_u \bar{Q} \Phi u_R + h.c.$$

(1.30)

Similar to Eq. (1.26), by replacing the scalar field as an expansion around the vacuum expectation value in the unitary gauge one can obtain fermion mass terms as follows (for example, for electrons):

$$\mathcal{L}_{\text{Yukawa}} = \frac{-1}{\sqrt{2}} \lambda_e (v + H) \bar{e}_L e_R + \dots,$$

(1.31)

where the constant term in front of $\bar{e}_L e_R$ is identified with the fermion masses:

$$m_e = \frac{\lambda_e v}{\sqrt{2}}, \quad m_u = \frac{\lambda_u v}{\sqrt{2}}, \quad m_d = \frac{\lambda_d v}{\sqrt{2}}.$$

(1.32)

Therefore, using the same scalar doublet one can obtain the fermion masses by breaking the $SU(2)_L \times U(1)_Y$ electroweak gauge symmetry while preserving the $U(1)_{EM}$ and the $SU(3)_C$ symmetries. The neutrinos remain massless, but one could account for their masses by including right-handed partners in the model, and the corresponding Higgs boson Yukawa interactions. The upper limits on the neutrino masses would imply that the Yukawa couplings for the neutrinos would be very small, of the order of 10^{-12}. To summarize, the vector boson and fermion masses are included in the SM by introducing the Higgs field that undergoes a phase transition giving non-zero vacuum expectation value, resulting in masses for fermions and gauge bosons. From a cosmological point of view, the universe was cooling down after the Big Bang till approximately 10^{16} K (10^{-12} s after the Big Bang) when the electroweak phase transition took place, breaking the $SU(2)_L \times (U)_Y$ symmetry. Before the spontaneous symmetry breaking, particles were massless, and the universe respected the $SU(2)_L \times (U)_Y$ symmetry.

1.2.4 The SM Higgs Boson

As detailed in the previous section, the remaining degree of freedom of the added scalar doublet is what we identify as the Higgs boson particle. The kinematic propagation of the Higgs boson is derived directly from the covariant derivative in Eq. (1.26), whereas the mass m_H and the self-coupling λ are from the scalar potential $V(\Phi)$ giving, the following Higgs Lagrangian:

$$\mathcal{L}_{\text{Higgs}} = \frac{1}{2}(\partial^\mu H)^2 - \lambda v^2 H^2 - \lambda v H^3 - \frac{\lambda}{4} H^4 \tag{1.33}$$

The mass of the Higgs boson is read directly from this Lagrangian: $m_H^2 = 2\lambda v^2$. The value of the Higgs mass is a free parameter of the SM since it depends on the unknown parameter λ. Therefore, the mass of the Higgs boson is determined experimentally. Nevertheless, there are theoretical constraints on the values of the Higgs bosons mass related to unitarity requirement from W boson scattering, setting an upper limit $m_H < 700$ GeV; triviality and vacuum stability bounds also limit the Higgs boson mass to the range 70 GeV $\leq m_H \leq 180$ GeV [14]. The SM Higgs boson is a CP-even particle, as the CP-even component of the scalar doublet is the component that acquires the non-zero vev.

In addition, the Lagrangian shows also Higgs boson triple and quartic self-interaction vertices:

$$g_{H3} = 3i\frac{m_H^2}{v}, \qquad g_{H4} = 3i\frac{m_H^2}{v^2}. \tag{1.34}$$

The Higgs boson couplings to fermions arise from the same Yukawa interactions giving rise to the fermion mass terms and are proportional to their masses, as can be derived from in Eq. (1.32); the couplings to the gauge bosons arise from the terms corresponding to the gauge interactions introduced by the covariant derivatives, after spontaneous symmetry breaking, and are proportional to the square of the masses of the weak bosons, as one can deduce from Eq. (1.27):

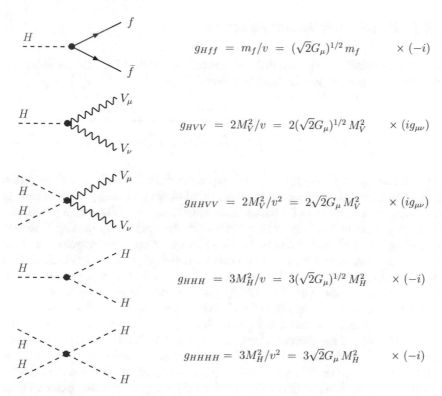

Fig. 1.5 The Higgs interactions with fermions (first diagram from the top) and gauge bosons (second and third diagrams), in addition to the Higgs self-interactions (fourth and fifth diagrams), as well as the corresponding couplings [14]

$$g_{Hff} = i\frac{m_f}{v}, \quad g_{HVV} = -2i\frac{m_V^2}{v}, \quad g_{HHVV} = -2i\frac{m_V^2}{v^2} \quad (1.35)$$

The different vertices of the Higgs boson couplings are summarized in Fig. 1.5.

1.3 Phenomenology of the SM Higgs Boson at the LHC

In this section we will review the main production and decay modes of the SM Higgs boson at the Large Hadron Collider. The review is preceded by a brief introduction to initial state and distribution functions of the partons in the colliding protons.

1.3.1 Parton Distribution Functions

In a hadron-hadron collision (such as the pp collisions at the LHC), the cross-section (σ) for a given process is calculated as follows [22]:

$$\sigma = \overbrace{\int dx_1 f_{q_i/p_1}(x_1, \mu^2)}^{parton-1} \overbrace{\int dx_2 f_{q_j/p_2}(x_2, \mu^2)}^{parton-2} \overbrace{\hat{\sigma}(x_1 p_1, x_2 p_2, \mu^2)}^{hard\ process}, \quad (1.36)$$

where the integrals $\int dx_{1,2} f_{q/p}(x_{1,2}, \mu^2)$ represent the probability of a given parton q_i to carry a fraction x of the initial momentum of the hadron $p_{1,2}$. The functions $f_{q/p}(x_{1,2}, \mu^2)$ are known as the parton distribution functions (PDF) with μ denoting the energy scale of the hard process. PDFs are the probability density functions for finding a particle with momentum fraction x of the colliding partons at a given momentum scale μ. The hard process cross section $\hat{\sigma}(x_1 p_1, x_2 p_2, \mu^2)$ is factorized from the computed parton-level cross section. The factorization procedure is a result of the complex structure of the colliding protons, that are made by three valence quarks as well as the strong gluon field between them, leading to the creation of a *sea* of virtual quark and anti-quark pairs. Therefore, the precise knowledge of the PDFs is an essential ingredient for the computation of the hard process cross section. The shape of the PDFs, shown in Fig. 1.6 at different energy scales, is estimated from a fit to experimental data, mostly from deep inelastic scattering data at the Hadron-Electron Ring Accelerator (HERA) at DESY [23] but also including other collider data, and evolved to different energy scales using the DGLAP equations [24–26].

The determination of the PDFs and their uncertainties is performed by different independent collaborations, such as the CT14 [27], MMHT2014 [28], and NNPDF3.0 [29] ones. For the calculation of the expected cross sections at the LHC, a combination of the results from the different collaborations is performed and a unified set of uncertainties is provided, following the PDF4LHC15 recommendations [23]. An example of the results from the different collaborations is shown in Fig. 1.7. The combination of the PDF uncertainties resulted in two sets of uncertainties with a total

Fig. 1.6 Examples of parton distribution functions obtained in NNLO NNPDF3.0 global analysis at different scales **a** $\mu^2 = 10$ GeV^2 and **b** $\mu^2 = 10^4$ GeV^2 [16]

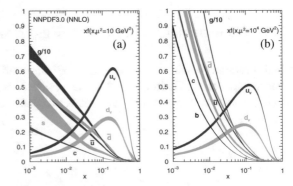

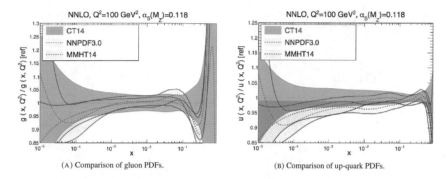

(A) Comparison of gluon PDFs.

(B) Comparison of up-quark PDFs.

Fig. 1.7 Examples of **A** the gluon and **B** the up-quark PDFs determined by three different global PDF fits. The PDFs are evaluated for a scale $Q^2 = 100$ GeV^2 and are normalized to the central values of CT14 [23]

of 100 and 30 eigenvariations that affect the PDFs and are propagated to the cross section computations. In this thesis, the 30 eigenvariation set is used to estimate the uncertainties on the cross sections. An additional uncertainty that affects these computations is that in the value of the strong coupling constant α_s. The PDF4LHC15 recommendations suggests to use a value of $\alpha_s(m_Z^2) = 0.1180 \pm 0.0015$. The α_s uncertainty is propagated to the procedure of determining the PDFs from the experimental data, and a combined PDF+α_s set of uncertainties is provided.

1.3.2 Main Higgs Boson Production Modes

In the LHC, the main production modes for the Higgs boson are: (i) gluon-gluon fusion (ggF), (ii) vector boson fusion (VBF), (iii) associated production with W/Z boson (VH) and (iv) associated production with heavy quarks, $t\bar{t}H$ and $b\bar{b}H$. The Feynman diagrams for these production modes are shown in Fig. 1.10. In this section, we will briefly review the main production mechanisms and their state-of-the-art cross section calculations. This review is preceded by a summary of the QCD uncertainties and EW corrections affecting such calculations (Fig. 1.8).

QCD uncertainties For various Higgs boson production modes, the production involves strongly interacting particles. In these cases, the calculation of the production cross section is affected by large corrections when going from the QCD leading-order calculations (LO) to higher-orders (HO). These corrections are also called K-factors:

$$K = \frac{\sigma_{HO}(pp \to H + X)}{\sigma_{LO}(pp \to H + X)}. \tag{1.37}$$

The value of the K-factor depends on the process and is larger when there is color annhilation [30], i.e. when colored objects (such as gluons) result in color singlets

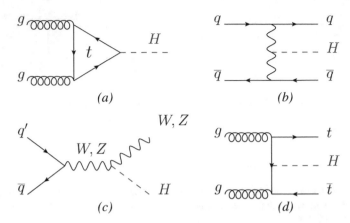

Fig. 1.8 Feynman diagrams of the main Higgs boson production mechanisms in the LHC: **a** gluon-gluon fusion, **b** vector boson fusion, **c** associated production with vector boson and **d** associated production with a $t\bar{t}$ pair [16]

(such as the Higgs boson). When going to higher orders in the perturbative expansion of the cross section calculation, the K-factor tends to stabilize, i.e. the corrections get smaller and smaller from one order to the next one [14]. In most cases, QCD processes are already rather well-calculated with the inclusion of next-to-leading order (NLO) QCD corrections.

Any calculation is affected by uncertainties related to the missing HO corrections that are not included in the truncation of the infinite expansion of the perturbative series. From the previous discussion, it is clear that these uncertainties tend to become smaller when higher orders are included in the calculation. The accuracy of a particular fixed-order calculation is estimated by varying the following scales [14] used in the calculation:

- The *renormalization* scale μ_R at which one defines the strong coupling constant.
- The *factorization* scale μ_F at which the matching between the perturbative calculation of the matrix element and non-perturbative part in the PDFs.

The uncertainty on a given prediction is estimated by varying these scales simultaneously around a central value Q assumed to be representative of the physical scale of the process: $Q/2 < \mu_R, \mu_F < 2Q$ resulting in 9-point variations. These uncertainties only give an estimate of the effect of the missing HO corrections, but not necessarily the full uncertainty from the missing HO, as sometimes it is observed that the relative K-factor between calculations at different order of accuracy are larger than the uncertainties estimated with this procedure [16, §9.2.4] as new channels might be available at HO resulting in convergence problems.

Additional sources of uncertainties affect calculations that involve resummation associated to the choice of the argument of the logarithms being resummed. For non-perturbative QCD corrections, the uncertainties are estimated by comparing the different Monte-Carlo simulations or different tunes of a given generator [16, §9.2.4].

Electroweak corrections In addition to the higher-order QCD calculations, higher-order electroweak corrections (EWC) are needed to account for additional EW radiative processes and loop corrections. These corrections are needed, for example, to precisely extract the SM parameters from measured observables such as the mass of the W boson, m_W, or when predicting its mass from Eq. (1.28). EW radiative corrections, however, are more complicated to calculate than QCD ones because of renormalization and the treatment of massive unstable particles [31]. Therefore, EWC are currently computed only up to next-to-leading order in the electromagnetic coupling constant α. The impact of these corrections is typically expressed in terms of relative corrections δ_{EW} defined as:

$$\delta_{\text{EW}} = \frac{\sigma_{NLO}^{EW}}{\sigma_{LO}} \qquad (1.38)$$

EWC are most relevant for the Higgs boson production with vector bosons (detailed below). EWC can be numerically at least as important as NNLO QCD corrections, and for certain processes and in certain kinematic regions (such as large transverse momentum) they may be the dominant corrections. Large EW corrections can account for the cases when the uncertainties of LO QCD calculations do not cover the NLO predictions, as there is an important interplay between the two corrections (for example when a photon is emitted after QCD radiation) [32].

Gluon-Gluon Fusion

Gluon-gluon fusion, $gg \rightarrow H$, is the main Higgs boson production mode at the LHC. This production mode is mediated by triangular loops of virtual heavy quarks, mainly the top quark and to a lesser extent the bottom quark. This is due to the large Higgs Yukawa couplings to the heavy quarks. Despite this production mode being generated via loops, it is the most dominant one at the LHC since gluons make up most of the proton momentum. The inclusive cross section of this production mode has been estimated at N^{3}LO accuracy in QCD and NLO in EW accuracy, and at $\sqrt{s} = 13$ TeV has a predicted value [31] of:

$$\sigma_{ggF}^{N3LO} = 48.6 \text{ pb}_{-3.3 \text{ pb}(-6.7\%)}^{+2.2 \text{ pb}(+4.6\%)}(\text{QCD}) \pm 1.6 \text{ pb } (3.2\%)(\text{PDF}+\alpha_s). \qquad (1.39)$$

The predictions are improved by *resumming* the soft collinear gluon contribution to the cross section at N^{3}LL logarithmic accuracy and matched to state-of-the-art QCD calculations [33, 34]. An additional source of improved accuracy for the calculation is from the setting the top-quark mass to its pole mass value, rather than using the $m_t \to \infty$ approximation.

For gluon-fusion production at leading order, the produced Higgs boson has no transverse momentum. The transverse momentum p_T of the Higgs boson is generated with higher orders as the additional radiation balances the Higgs boson p_T. Similarly, the Higgs boson rapidity distribution is affected by the inclusion of additional partons. The distribution peaks at $y_H = 0$ and decreases at larger rapidity, reaching almost zero for $|y_H| \geq 4$ due to the restriction of the available phase space [14].

Vector Boson Fusion

Vector boson fusion (VBF) is the Higgs boson production process with the second-largest cross section at the LHC. VBF proceeds by the scattering of two quarks mediated by $t-$ or $u-$ channel exchange of W or Z bosons, that interact producing a Higgs boson. Therefore, this process is generally known as the $qq \to qqH$ process as the scattered quarks give rise to two hard jets in the forward and backward regions of the detector. Also, the color singlet nature of the W and Z bosons suppresses additional gluon radiation from the central rapidity regions. This particular topology can be used to suppress ggF $H + 2j$ background and obtain a sample that is relatively pure in VBF-produced events. This production mode can be used to probe the Higgs boson couplings to the W and Z bosons. Besides, the angular distributions of the additional final-state quarks can be used to probe the CP-properties of the Higgs boson [35]. State-of-the-art VBF inclusive cross section calculations are performed at an approximate NNLO QCD and NLO EW accuracy [36],[2] giving the following value at $\sqrt{s} = 13$ TeV:

$$\sigma_{VBF}^{NNLO} = 3.92 \ \text{pb}_{-0.008\ \text{pb}(-0.2\%)}^{+0.02\ \text{pb}(+0.5\%)}(\text{QCD}) \pm 0.074 \ \text{pb} \ (1.9\%)(\text{PDF}+\alpha_s) \quad (1.40)$$

Associated Production with a Vector Boson

Associated production of the Higgs boson with a W or Z boson is the Higgs boson production with the third-largest cross section at the LHC. This production mode is known as the VH production mode ($V = W, Z$) or Higgsstrahlung. In this production mode a quark and anti-quark pair interact giving an off-shell vector boson that then goes on shell radiating a Higgs boson, $q\bar{q} \to V^* \to VH$. In addition, for the ZH associated production, and not WH due to charge conservation, an additional contribution comes from $gg \to ZH$ through top-quark loops, as shown in Fig. 1.9.

These production modes can be identified at the LHC using the final state decay products of the vector bosons. For instance, in the case of leptonic decays of the vector bosons, the process to reconstruct is $pp \to WH \to \nu_\ell \ell H$ and $pp \to ZH \to \ell^+\ell^-/\nu_\ell \bar{\nu}_\ell H$. The computation of these cross sections is performed up to NNLO in QCD accuracy and NLO in EW resulting in the following cross sections at $\sqrt{s} = 13$ TeV [36]:

---

[2]An N^{3}LO QCD accuracy calculation was performed for VBF [37]. The calculation found per-mille corrections with respect to the NNLO, but resulted in a substantial decrease in QCD scale uncertainty. The approximate NNLO calculation is used, however, in this work.

Fig. 1.9 Feynman diagrams of the $gg \rightarrow ZH$ production mode. This process contributes at $\mathcal{O}(\alpha_s^2)$ order [14]

$$\sigma_{WH}^{NNLO} = 1.48 \ \text{pb}_{-0.01 \ \text{pb}(-0.7\%)}^{+0.007 \ \text{pb}(+0.5\%)} (\text{QCD}) \pm 0.03 \ \text{pb} \ (1.9\%)(\text{PDF}+\alpha_s) \quad (1.41)$$

$$\sigma_{qq \rightarrow ZH, gg \rightarrow ZH}^{NNLO} = 0.91 \ \text{pb}_{-0.025 \ \text{pb}(-2.7\%)}^{+0.028 \ \text{pb}(+3.2\%)} (\text{QCD}) \pm 0.014 \ \text{pb} \ (1.6\%)(\text{PDF}+\alpha_s) \quad (1.42)$$

Associated Production with Heavy Quarks
The Higgs boson production mode with the fourth largest cross section at the LHC is the associated Higgs boson production with heavy quarks: $t\bar{t}H$ and $b\bar{b}H$. These production modes are initiated by two initial gluons that produce two top or bottom quarks and the Higgs boson is emitted from the quark lines. These production modes allow probing the Higgs Yukawa coupling to the third generation quarks. However, $b\bar{b}H$ production has a kinematic configuration very similar to that of gluon fusion, which has a much larger cross section, and is therefore difficult to identify. The cross sections for these production modes were computed at NLO QCD and EW accuracy for $t\bar{t}H$ [38–40] and NNLO QCD accuracy for $b\bar{b}H$ [41, 42] at $\sqrt{s} = 13$ TeV:

$$\sigma_{ttH}^{NLO} = 0.51 \ \text{pb}_{-0.047 \ \text{pb}(-9.2\%)}^{+0.0285 \ \text{pb}(+5.8\%)} (\text{QCD}) \pm 0.018 \ \text{pb} \ (3.6\%)(\text{PDF}+\alpha_s) \quad (1.43)$$

$$\sigma_{bbH}^{NNLO} = 0.48 \ \text{pb}_{-0.115 \ \text{pb}(-23.9\%)}^{+0.096 \ \text{pb}(+20.1\%)} (\text{QCD}+\text{PDF}+\alpha_s) \quad (1.44)$$

Summary of the different production modes A summary of the different production modes and their theoretical uncertainties as a function of the pp center of mass energy is shown in Fig. 1.10. The figure shows a significant increase in production cross sections going from the LHC Run-1 energies ($\sqrt{s} = 7, 8$ TeV) to the Run-2 energy ($\sqrt{s} = 13$ TeV), by approximately a factor 2.3 for ggF up to a factor 4 for $t\bar{t}H$ (background cross sections also increased, though not with the same factors).

1.3.3 Main Higgs Boson Decay Modes

Using the knowledge of the different Higgs boson couplings detailed in Sect. 1.3.2, the branching ratios for the different decay modes can be determined. The Higgs boson couplings to the different particles (gauge bosons and fermions) are proportional to their mass and hence the Higgs boson will decay to the heaviest particle allowed by the phase space. The SM Higgs boson total width can be determined once its mass is known, as shown in Fig. 1.11a. The SM Higgs boson with $m_H = 125$ GeV

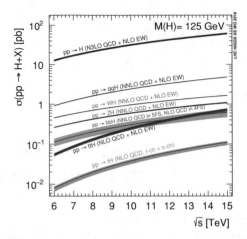

Fig. 1.10 Evolution of the cross section of the different Higgs boson production modes as a function of the *pp* center-of-mass [31]

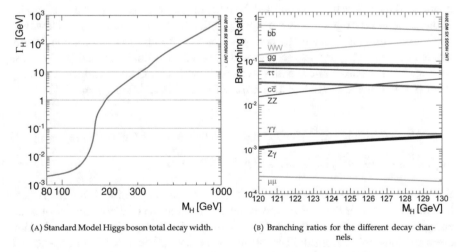

(A) Standard Model Higgs boson total decay width.   (B) Branching ratios for the different decay channels.

Fig. 1.11 **A** Total Higgs boson decay width as a function of its mass [36]. **B** SM Higgs boson branching ratios as a function of the Higgs boson mass [31]

has a width $\Gamma_H = 4.07 \pm 1.6$ MeV [31]. This decay width is significantly smaller than the experimental resolution, and hence the precise measurement of the Higgs boson properties will depend on the decay channel with the best experimental resolution. The partial widths of the different decay modes are computed with the HDECAY program [43], and PROPHECY4F for $H \rightarrow 4\ell$ decays [44].

The branching ratios of the different decay modes are summarized in Table 1.1 for a Higgs boson with $m_H = 125$ GeV. The variation of the branching ratios for different mass values are shown in Fig. 1.11b. The branching ratios vary as a function of the Higgs boson mass as other decay channels become kinematically available and hence the increasing trend with Higgs mass for some of the decay modes.

Table 1.1 The branching ratios and their relative uncertainties for a SM Higgs boson with $m_H = 125$ GeV [36]

Decay channels	Branching ratio	Rel. uncertainty
$H \to b\bar{b}$	5.84×10^{-1}	$+3.2\%$ -3.3%
$H \to W^+W^-$	2.14×10^{-1}	$+4.3\%$ -4.2%
$H \to \tau\tau$	6.27×10^{-2}	$+5.7\%$ -5.7%
$H \to \bar{c}c$	2.88×10^{-2}	$+5.45\%$ -1.97%
$H \to ZZ$	2.62×10^{-2}	$+4.3\%$ -4.1%
$H \to \gamma\gamma$	2.27×10^{-3}	$+5.0\%$ -4.9%
$H \to Z\gamma$	1.53×10^{-3}	$+9.0\%$ -8.9%
$H \to \mu^+\mu^-$	2.18×10^{-4}	$+6.0\%$ -5.9%

The dominant decay mode is $H \to b\bar{b}$, as the bottom quark is the heaviest particle accessible to the available phase space for $m_H = 125$ GeV. From the experimental viewpoint, this decay mode suffers from large QCD background from top quarks decaying to bottom quarks. The experimental mass resolution in this channel typically ranges from 10% to 15%.

The final state with the second-largest branching ratio is the W^*W one, where one of the W bosons is off-shell. This decay channel can be identified by tagging W boson decays to leptons, leading to a relatively clean final state. Nevertheless, since the W boson leptonic decays involve neutrinos which can not be detected, the resolution in this channel is quite poor (approximately 20%).

The decay channel with the third-largest branching ratio is the $H \to \tau\tau$, with a resolution similar to $H \to b\bar{b}$. This decay channel can be used to directly probe the Yukawa couplings to leptons as it is the only channel with direct lepton coupling and sizable branching ratio given the high mass of the tau lepton.

Other decay channels with smaller branching ratios include: $H \to c\bar{c}$, which suffers from large background mainly from $Z + jets$ and $t\bar{t}$ decays and small branching ratio (approximately 3%) making the direct search for this coupling very challenging experimentally with only limits set on such decay mode [45]; the $H \to \mu\mu$ decay channel which can be used to directly measure the Yukawa coupling to a second-generation fermion, this channel, however, has very high background from Drell-Yan processes and given the very small branching ratio only limits are set on this decay mode [46].

In addition to these decay channels, there are two additional channels with small branching ratios, but excellent experimental resolution (typically 1-2%) and efficiency leading to reasonable signal-over-background, which made them the main Higgs boson discovery channels or so-called "golden channels". They are the $H \to \gamma\gamma$ and $H \to ZZ^* \to 4\ell$ decays. These channels also provide an excellent measurement of the Higgs boson mass and provide means of measuring the Higgs boson production cross section as a function of several kinematic variables, as will be detailed in Chap. 7.

Fig. 1.12 Feynman
diagrams of the $H \to \gamma\gamma$
decay channel. The leading
contributions are from W
boson and top quarks
loops [47]

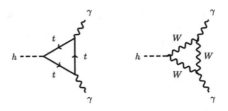

*$H \to \gamma\gamma$ **decay channel*** The $H \to \gamma\gamma$ decay is one of the main Higgs boson discovery channels. This decay proceeds via loop-induced diagrams resulting in two
photons, as shown in Fig. 1.12. This is because the Higgs boson does not couple
directly to the massless photons, and hence the branching ratio of this channel is
very small (Table 1.1). The dominant contribution comes from W-mediated loops.
The heavy fermions are the second contributing mediators in the loop diagrams.
Among them, the leading contribution arises from top-quark loops, resulting from
the Higgs boson coupling being proportional to the quark masses, and hence the
effect of the light fermions in the loop is very small. This channel has a high sensitivity to effects from particles beyond the Standard Model that are too heavy to be
directly produced but can affect the loop contributions.

The Higgs boson decay to diphotons has a very clean signature. The Higgs boson
signal appears as an excess of events in the diphoton invariant mass spectrum, $m_{\gamma\gamma}$,
on top of a monotonically decreasing background of SM diphoton, photon+jet, and
dijet events, where the (hadronic) jets are misidentified as photons. The non-resonant
SM diphoton production arises mainly from the following processes (see also the
Feynman diagrams in Fig. 1.13):

- The Born process $qq \to \gamma\gamma$.
- The box process $gg \to \gamma\gamma$, through a loop of quarks. Despite being loop induced,
  this process has a non-negligible cross section due to the large gluon luminosity
  at the LHC.
- The bremsstrahlung process $gq \to q\gamma\gamma$.
- Jet fragmentation, where photons are produced from the non-perturbative collinear
  fragmentation of a hard parton (typically a π^0).

1.3.4 Status of the Higgs Boson Properties Measurement

In July 2012, both the ATLAS [48] and CMS [49] collaborations announced the
discovery of a new scalar resonance with a mass of approximately 125 GeV and
properties consistent with the SM Higgs boson [50, 51]. Following the discovery,
extensive studies were performed in order to determine precisely the different parameters of the discovered Higgs boson. This included a precise mass measurement, and
measurement of its couplings to fermions and bosons, of its quantum numbers and

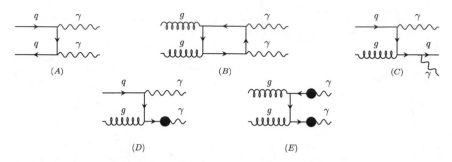

Fig. 1.13 Examples of SM non-resonant diphoton Feynman diagrams: **A** Born process $qq \to \gamma\gamma$, **B** Box process $gg \to \gamma\gamma$, **C** Bremsstrahlung process $qg \to q\gamma\gamma$ and jet fragmentation, resulting in **D** one and **E** two energetic photons

different kinematic properties. In this section, we will briefly review the status of the Higgs boson properties measurements.

Higgs boson mass As detailed in the previous section, the high precision Higgs boson mass measurement is driven by the excellent experimental resolution in the $H \to \gamma\gamma$ and $H \to 4\ell$ channels. The mass measurement is performed via a signal extraction from a signal+background data sample. The background in the $H \to \gamma\gamma$ was detailed in the previous section, whereas that of $H \to 4\ell$ is mainly from non-resonant ZZ^* SM production and to a lesser extent $Z + jets$ and $t\bar{t}$. The mass measurement relies on isolating parts of the phase space with the best resolution and signal-to-background ratios to achieve the best percision [52]. Examples of such measurements are shown in Fig. 1.14. In addition, a combination of the mass measurement is performed combining different channels of the same experiment and combining both ATLAS and CMS, as shown in Fig. 1.15. The latest ATLAS+CMS combination was performed using Run-1 data at $\sqrt{s} = 7, 8$ TeV, resulting in a mass measurement [53]:

$$m_H = 125.09 \pm 0.25 \text{ (stat.)} \pm 0.11 \text{ (sys.) GeV} \qquad (1.45)$$

In addition, a mass measurement was performed during the LHC Run-2 using the ATLAS data at $\sqrt{s} = 13$ TeV, as detailed in Sect. 5.7.

Higgs boson couplings The measurement of the Higgs boson coupling to W and Z boson is mainly driven by the measurement of Higgs boson decays to ZZ and WW as well as from the measurement of the VBF and VH production cross sections. The measurement of the Higgs boson coupling to fermions can be performed either directly through Higgs boson decays to second and third-generation fermions (b, τ, μ), or indirectly through gluon-fusion or $H \to \gamma\gamma$ loops, or via the Higgs boson production with top-quarks ($t\bar{t}H$). The experiments typically measure the product of a production cross section and the branching ratio for the decay under investigation, usually expressed in terms of the *signal strength* μ, defined as:

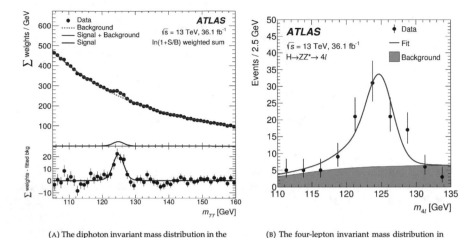

(A) The diphoton invariant mass distribution in the $H \to \gamma\gamma$ mass measurement

(B) The four-lepton invariant mass distribution in the $H \to 4\ell$ mass measurement

Fig. 1.14 **A** Diphoton invariant mass distribution of all selected data events, overlaid with the result of the fit (solid red line) using 36 fb^{-1} of pp collisions at $\sqrt{s} = 13$ TeV. Both for data and the fit, each category is weighted by a factor $\log(1 + S/B)$, where S and B are the fitted signal and background yields in a $m_{\gamma\gamma}$ interval containing 90% of the expected signal. The dotted line describes the background component of the model. The bottom inset shows the difference between the sum of weights and the background component of the fitted model (dots), compared with the signal model (black line). **B** Invariant mass distribution for the data (points with error bars) shown together with the simultaneous fit result to $H \to ZZ^* \to 4\ell$ candidates (continuous line) selected in 36 fb^{-1} of pp collisions at $\sqrt{s} = 13$ TeV. The background component of the fit is also shown (filled area). The signal probability density function is evaluated per-event and averaged over the observed data [52]

Fig. 1.15 Summary of the ATLAS and CMS mass measurement in the $\gamma\gamma$ and 4ℓ channels in Run-1 and Run-2 [16]

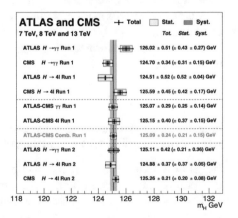

$$\mu = \frac{\sigma \times BR}{(\sigma \times BR)_{SM}}. \tag{1.46}$$

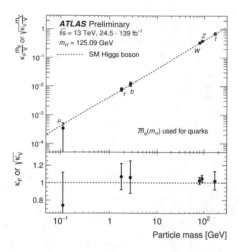

Fig. 1.16 Reduced coupling strength modifiers $\kappa_F = m_F/v$ for fermions ($F = t, b, \tau, \mu$) and $\sqrt{\kappa_V} m_V/v$ for weak gauge bosons ($V = W, Z$) as a function of their masses m_F and m_V, respectively, and for a vacuum expectation value of the Higgs field $v = 246$ GeV. The SM prediction for both cases is also shown (dotted line). The couplings strength modifiers κ_F and κ_V are measured assuming no BSM contributions to the Higgs boson decays, and the SM structure of loop processes such as ggF, $H \to \gamma\gamma$ and $H \to gg$. The lower inset shows the ratios of the values to their SM predictions [58]

The signal strenghts in different decay channels and production modes can then be interpreted in terms of the Higgs boson couplings to the various particles. The observation and measurement of the Higgs boson couplings to third-generation fermions (top and bottom quarks, tau leptons) was only possible using Run-2 data. This was the result of the complex algorithms that were developed to distinguish the signal from the very large QCD backgrounds [54, 55]. Higgs boson couplings to top quarks were measured by means of the associated production with a $t\bar{t}$ pair in various different Higgs boson decay channels ($\gamma\gamma$, $b\bar{b}$, 4ℓ) [56]. A summary of the measured Higgs boson couplings to the different fermions and bosons is shown in Fig. 1.16 as a function of the particle masses. The Higgs boson couplings are expressed in terms of couplings strength modifiers, κ [57]. The κ-framework was introduced to parameterize SM deviations in production cross section and decay width replacing SM couplings $g \to \kappa_g g$, for example for the gluon-fusion process $gg \to H \to \gamma\gamma$ the parameterizations in the κ-framework is:

$$(\sigma \cdot BR) = \sigma_{SM}(gg \to H) \cdot BR_{SM}(H \to \gamma\gamma) \cdot \frac{\kappa_g^2 \kappa_\gamma^2}{\kappa_H^2}, \qquad (1.47)$$

with $\kappa_i = 1$ denoting SM predictions, and κ_H corresponding to the modification of the Higgs boson total width.

The cross section of the different Higgs boson production modes is measured using the different decay channels summarized in Fig. 1.17. The production modes are identified – in addition to the particles produced by the Higgs boson decay – by

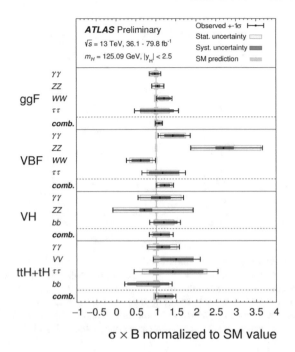

Fig. 1.17 Measured cross sections times branching fractions for ggF, VBF, VH and $\bar{t}tH + tH$ production in each relevant decay mode, normalized to their SM predictions. The values are obtained from a simultaneous fit to all decay channels. The cross section for the VH, $H \rightarrow \tau\tau$ process is constrained to its SM prediction. Combined results for each production mode are also shown, assuming SM values for the branching ratios into each decay mode. The black error bars, blue boxes and yellow boxes show the total, systematic, and statistical uncertainties in the measurements, respectively. The grey bands show the theory uncertainties in the predictions [58]

additional particles that are typical of one of the various production modes, such as high-p_T jets in VBF, leptons from vector bosons in VH, or b-jets in ttH events. From these measurements, a combination is performed to measure the global Higgs boson signal strength:

$$\mu = 1.13^{+0.09}_{-0.08} = 1.13 \pm 0.05 \text{ (stat.)} \pm 0.05 \text{ (exp.)}^{+0.05}_{-0.04} \text{ (sig. the.)} \pm 0.03 \text{ (bkg. th.)}, \tag{1.48}$$

where the total uncertainty is decomposed into components for statistical uncertainties, experimental systematic uncertainties, and theory uncertainties on signal and background modeling as detailed in [58]. The measured signal strength is consistent with the SM prediction with a p-value of 13%. More details in the measurement of the Higgs boson kinematic properties are given in Chap. 7.

Higgs boson quantum numbers In addition to the couplings measurement, probing the Higgs boson quantum numbers is essential to further unveil its properties. These measurements mainly probe the spin (J), charge conjugation (C) and parity (P) of the Higgs boson, or J^{PC}. The SM Higgs boson is predicted to have $J^{PC} = 0^{++}$, i.e.

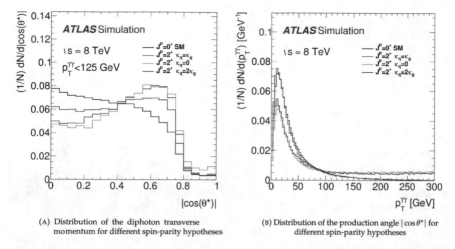

(A) Distribution of the diphoton transverse momentum for different spin-parity hypotheses

(B) Distribution of the production angle $|\cos\theta^*|$ for different spin-parity hypotheses

Fig. 1.18 Expected distributions of kinematic variables sensitive to the spin of the resonance considered in the $H \to \gamma\gamma$ analysis, **A** transverse momentum of the $\gamma\gamma$ system $p_T^{\gamma\gamma}$ and **B** the production angle of the two photons in the Collins-Soper frame $|\cos\theta^*|$, for a SM Higgs boson and for spin-2 particles with three different choices of the QCD couplings [61]

to be a CP-even scalar particle. The discovery of the Higgs boson in the diphoton channel excludes the spin-1 hypothesis following the Landau-Yang theorem that forbids such decays [59, 60], and therefore the observed Higgs boson can either be spin-0, spin-2, or a non-pure state with collimated photons that are detected as diphotons. To differentiate between these spin and parity hypotheses, several models of spin and parity are compared against the observed data: Spin-2 particle, CP-odd particle, and a mixture of SM CP-even and non-SM CP-odd [61]. Such comparisons rely on investigating the angular correlations in the Higgs boson decays. For example, for $H \to \gamma\gamma$ decays the different spin hypotheses are differentiated by means of the Higgs boson transverse momentum distribution or the production angle of the two photons in the Collins-Soper frame [62] as discriminating variables, as shown in Fig. 1.18. The latter is known as θ^* and is defined as:

$$|\cos\theta^*| = \frac{\sinh(\Delta\eta^{\gamma\gamma})}{\sqrt{1 + (p_T^{\gamma\gamma}/m_{\gamma\gamma})}} \frac{2p_T^{\gamma 1} p_T^{\gamma 2}}{m_{\gamma\gamma}^2} \tag{1.49}$$

Similarly looking at the angular distributions, in the Higgs boson decays to four leptons $H \to ZZ^* \to 4l$, to heavy fermions such as $H \to \tau\tau$, or $W^+W^- \to e\nu_e\mu\nu_\mu$, the different spin and parity hypotheses can be distinguished [61]. The observed data exclude all non-SM spin and parity models tested at more than 99.9% confidence level in favor of the SM $J^P = 0^+$ hypothesis. The results of these exclusion tests are shown in Fig. 1.19.

The charge conjugation quantum number C is multiplicative, and therefore given that the Higgs boson was observed in the $H \to \gamma\gamma$ channel with photons being C-

Fig. 1.19 Distributions of the test statistic q for the SM Higgs boson ($J^P = 0^+$) and for the J^P alternative hypotheses. They are obtained by combining the $H \to ZZ^* \to 4\ell$, $H \to WW^* \to e\nu\mu\nu$ and $H \to \gamma\gamma$ decay channels. The expected median (black dashed line) and the ± 1, ± 2 and $\pm 3\,\sigma$ regions for the SM Higgs boson (blue) and for the alternative J^P hypotheses (red) are shown for the signal strength fitted to data. The observed q values are indicated by the black points [61]

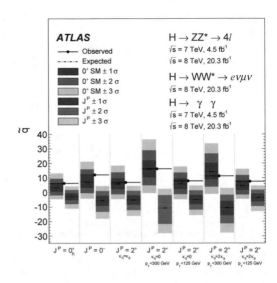

odd eigenstates, assuming C conservation, the observed neutral particle should be C-even. Therefore the observed Higgs boson quantum numbers are $J^{PC} = 0^{++}$, in agreement with the SM predictions [16, §11.V.1].

1.4 Beyond the Standard Model

Today the Standard Model represents the greatest achievement in particle physics. The SM explains all observed phenomena in collider physics, covering a wide range of energies [63]. The latest of these successes was the discovery of the Higgs boson, with properties matching those predicted by the SM as detailed in the previous section. Nevertheless, there are several experimental signatures, mostly from non-collider physics, that do not fit within the framework of the Standard Model. These signatures explicitly hint at some *new physics* that lies beyond the Standard Model (BSM).

The first and foremost of these phenomena is the fourth force of nature, *gravity*, which is not included in the SM.

A second one is that from a cosmological point of view, ordinary matter (i.e. SM constituents) compose only 4.9% of the observed universe [64]. The remaining contributions come from *dark matter* and *dark energy*. The prediction of dark matter resulted from astrophysical measurement on rotation curves of galaxies, which is the orbital velocity of a star in a galaxy as a function of the distance of this star to the center of this galaxy, that was very different from the predictions assuming only ordinary matter [65]. Therefore, dark matter was introduced as one of the possible solutions to this problem as an additional massive component. Dark matter has not been discovered yet and would contribute to a sizeable extent to the total mass of the

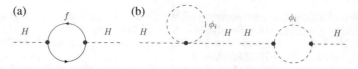

Fig. 1.20 Diagrams for the contributions of **a** fermions and **b** scalars to the Higgs boson mass [68]

galaxies, hence modifying their gravitational properties and the velocity curve. The SM does not include dark matter candidates. Therefore, various BSM models were developed with dark matter particle candidates [66].

Another experimental hint from cosmology is the matter/anti-matter asymmetry in the universe. This asymmetry is driven by CP-violation, which results in the prevalence of matter in the universe. Without this asymmetry, equal amounts of matter and anti-matter would have been produced in the universe. While the SM provides several sources of CP-violation in the weak sector, the amount is not large enough to explain the apparent asymmetry [16, 67] and hence hints at BSM physics.

In addition, there are a few theoretical arguments that concern the consistency of the electroweak theory. These arguments arise from the radiative corrections to the SM Higgs boson mass. The corrections are quadratically divergent in the cut-off scale Λ. For example, the correction to the Higgs boson mass from one-loop contributions (shown in Fig. 1.20) of a fermion f with number of fermion families N_f and Yukawa coupling $\lambda_f = \sqrt{2}m_f/v$ is [68]:

$$\Delta m_H^2 = N_f \frac{\lambda_f^2}{8\pi^2} \left[-\Lambda^2 + 6m_f^2 \log \frac{\Lambda}{m_f} - 2m_f^2 \right] + \mathcal{O}(1/\Lambda^2). \tag{1.50}$$

This shows the quadratic divergence as $\Delta m_H^2 \propto \Lambda^2$, and therefore for a cut-off scale equal to the Planck scale $M_p \sim 10^{18}$ GeV the corrections are huge. Nevertheless, the stability of the observed Higgs boson mass $m_H = 125$ GeV requires a high-level of fine tuning between the bare mass and radiative corrections [68, 69], which hints at new physics that should appear at a mass scale smaller than the Planck scale.

All these signatures hint at the existence of physics beyond the SM. In this thesis, we will use the Higgs sector to indirectly search for and constrain BSM physics via the *effective field theory* formulation. This formulation is detailed in Chap. 2. Such a search relies on precise measurements of the Higgs boson kinematics detailed in Chap. 7, and the results of the study are shown in Chap. 8.

References

1. Noether E (1918) Invariant variation problems. Gott Nachr 1918:235–257 [Transp. Theory Statist Phys 1,186 (1971)]. https://doi.org/10.1080/00411457108231446. arXiv:physics/0503066 [physics]
2. Peskin Michael E, Schroeder Daniel V (1995) An Introduction to quantum field theory. Addison-Wesley, Reading, USA. ISBN 978-0-201-50397-5

3. Weinberg S (2005) The quantum theory of fields. Vol. 1: foundations. Cambridge University Press, Cambridge. ISBN: 978-0-521-67053-1, 978-0-511-25204-4
4. Gell-Mann M (1964) A schematic model of baryons and mesons. Phys Lett 8:214–215. https://doi.org/10.1016/S0031-9163(64)92001-3
5. Zweig G (1964) An SU(3) model for strong interaction symmetry and its breaking. Version 2. In: Lichtenberg DB, Peter Rosen S (eds) Developments in the quark theory of hadrons, 1964–1978. Vol. 1, pp. 22–101
6. Glashow SL (1961) Partial-symmetries of weak interactions. Nucl Phys 22.4:579–588. ISSN: 0029-5582. https://doi.org/10.1016/0029-5582(61)90469-2. http://www.sciencedirect.com/science/article/pii/0029558261904692
7. Salam A (1969) The standard model. In: Svartholm N (ed) Elementary particle theory. Stockholm, p 367
8. Weinberg S (1967) A model of leptons. Phys Rev Lett 19:1264–1266. https://doi.org/10.1103/PhysRevLett.19.1264. https://link.aps.org/doi/10.1103/PhysRevLett.19.1264
9. Higgs PW (1964) Broken symmetries, massless particles and gauge fields. Phys Lett 12:132–133. https://doi.org/10.1016/0031-9163(64)91136-9
10. Englert F, Brout R (1964) Broken symmetry and the mass of gauge vector mesons. Phys Rev Lett 13:321–323. https://doi.org/10.1103/PhysRevLett.13.321
11. Guralnik GS, Hagen CR, Kibble TWB (1964) Global conservation laws and massless particles. Phys Rev Lett 13:585–587. https://doi.org/10.1103/PhysRevLett.13.585. https://link.aps.org/doi/10.1103/PhysRevLett.13.585
12. Wikipedia contributors (2018) Standard model of elementry particles - wikipedia, the free encyclopedia. https://en.wikipedia.org/wiki/StandardModel
13. Adrien Maurice Dirac P (1928) The quantum theory of the electron. In: Proc R Soc Lond 117. https://doi.org/10.1098/rspa.1928.0023
14. Djouadi A (2008) The anatomy of electro-weak symmetry breaking. I: the Higgs boson in the standard model. Phys Rept 457:1–216. https://doi.org/10.1016/j.physrep.2007.10.004. arXiv:hep-ph/0503172 [hep-ph]
15. Gell-Mann M (1962) Symmetries of baryons and mesons. Phys Rev 125:1067–1084. https://doi.org/10.1103/PhysRev.125.1067. https://link.aps.org/doi/10.1103/PhysRev.125.1067
16. Tanabashi M et al (2018) Review of particle physics. Phys Rev D 98:030001. https://doi.org/10.1103/PhysRevD.98.030001. https://link.aps.org/doi/10.1103/PhysRevD.98.030001
17. Bellm J et al (2016) Herwig 7.0/Herwig++ 3.0 release note. Eur Phys J C76.4:196. https://doi.org/10.1140/epjc/s10052-016-4018-8. arXiv:1512.01178 [hep-ph]
18. Sjöstrand T, Mrenna S, Skands PZ (2008) A brief introduction to PYTHIA 8.1. Comput Phys Commun 178:852–867. https://doi.org/10.1016/j.cpc.2008.01.036. arXiv:0710.3820 [hep-ph]
19. Lancaster T, Blundell SJ (2014) Quantum field theory for the gifted amateur. Oxford University Press, Oxford. Doi: 0199699321. https://cds.cern.ch/record/1629337
20. Goldstone J, Salam A, Weinberg S (1962) Broken symmetries. Phys Rev 127:965–970. https://doi.org/10.1103/PhysRev.127.965. https://link.aps.org/doi/10.1103/PhysRev.127.965
21. Rajasekaran G (2014) Fermi and the theory of weak interactions. Resonance J Sci Educ 19.1:18–44. https://doi.org/10.1007/s12045-014-0005-2. arXiv:1403.3309 [physics.hist-ph]
22. Salam GV (2009) QCD (for LHC) : parton distribution functions. https://gsalam.web.cern.ch/gsalam/repository/talks/2009-Bautzen-lecture2.pdf
23. Butterworth J et al (2016) PDF4LHC recommendations for LHC Run II. J Phys G 43:023001. https://doi.org/10.1088/0954-3899/43/2/023001. arXiv:1510.03865 [hep-ph]
24. Altarelli G, Parisi G (1977) Asymptotic freedom in parton language. Nucl Phys B126:298–318. https://doi.org/10.1016/0550-3213(77)90384-4
25. Dokshitzer YL (1977) Calculation of the structure functions for deep inelastic scattering and e+e- annihilation by perturbation theory in quantum chromodynamics. Sov Phys JETP 46:641–653 [Zh Eksp Teor Fiz 73, 1216 (1977)]
26. Gribov VN, Lipatov LN (1972) Deep inelastic e p scattering in perturbation theory. Sov J Nucl Phys 15:438–450 [Yad Fiz 15, 781 (1972)]

27. Dulat S et al (2016) New parton distribution functions from a global analysis of quantum chromodynamics. Phys Rev D93.3:033006. https://doi.org/10.1103/PhysRevD.93.033006. arXiv:1506.07443 [hep-ph]
28. Harland-Lang LA et al (2015) Parton distributions in the LHC era: MMHT 2014 PDFs. Eur Phys J C75.5:204. https://doi.org/10.1140/epjc/s10052-015-3397-6. arXiv: 1412.3989 [hep-ph]
29. Ball RD et al (2015) Parton distributions for the LHC Run II. JHEP 04:040. https://doi.org/10.1007/JHEP04(2015)040. arXiv:1410.8849 [hep-ph]
30. Huston J. K-factors. https://web.pa.msu.edu/people/huston/Kfactor/Kfactor.pdf
31. de Florian D et al (2016) Handbook of LHC Higgs cross sections: 4. Deciphering the nature of the Higgs sector. https://doi.org/10.23731/CYRM-2017-002. arXiv:1610.07922 [hep-ph]
32. Wackeroth D (2017) Electroweak corrections. https://indico.cern.ch/event/579660/contributions/2496119/attachments/1494651/2325090/Boost2017.pdf
33. Banfi A et al (2016) Jet-vetoed Higgs cross section in gluon fusion at N3LO+NNLL with small-R resummation. JHEP 04:049. https://doi.org/10.1007/JHEP04(2016)049. arXiv:1511.02886 [hep-ph]
34. Bizon W et al (2018) Momentum-space resummation for transverse observables and the Higgs $p_\perp$ at N^3LL+NNLO. JHEP 02:108. https://doi.org/10.1007/JHEP02(2018)108. arXiv:1705.09127 [hep-ph]
35. Hankele V, Klamke G, Zeppenfeld D (2006) Higgs + 2 jets as a probe for CP properties. In: Meeting on CP violation and non-standard higgs physics Geneva, Switzerland, December 2–3, 2004, pp 58-62. arXiv: hep-ph/0605117 [hep-ph]
36. Andersen JR et al (2013) Handbook of LHC Higgs cross sections: 3. Higgs properties. In: Heinemeyer S et al (eds). https://doi.org/10.5170/CERN-2013-004. arXiv:1307.1347 [hep-ph]
37. Dreyer FA, Karlberg A (2016) Vector-Boson fusion Higgs production at three loops in QCD. Phys Rev Lett 117.7:072001. https://doi.org/10.1103/PhysRevLett.117.072001. arXiv:1606.00840 [hep-ph]
38. Zhang Y et al (2014) QCD NLO and EW NLO corrections to $t\bar{t}H$ production with top quark decays at hadron collider. Phys Lett B738:1–5. https://doi.org/10.1016/j.physletb.2014.09.022. arXiv:1407.1110 [hep-ph]
39. Dawson S et al (2003) Associated Higgs production with top quarks at the large hadron collider: NLO QCD corrections. Phys Rev D68:034022. https://doi.org/10.1103/PhysRevD.68.034022. arXiv:hep-ph/0305087 [hep-ph]
40. Beenakker W et al (2003) NLO QCD corrections to t anti-t H production in hadron collisions. Nucl Phys B653:151–203. https://doi.org/10.1016/S0550-3213(03)00044-0. arXiv:hep-ph/0211352 [hep-ph]
41. Dawson S et al (2004) Exclusive Higgs boson production with bottom quarks at hadron colliders. Phys Rev D69:074027. https://doi.org/10.1103/PhysRevD.69.074027. arXiv:hep-ph/0311067 [hep-ph]
42. Dittmaier S, Krämer M, Spira M (2004) Higgs radiation off bottom quarks at the Tevatron and the CERN LHC. Phys Rev D70:074010. https://doi.org/10.1103/PhysRevD.70.074010. arXiv:hep-ph/0309204 [hep-ph]
43. Djouadi A, Kalinowski J, Spira M (1998) HDECAY: a program for Higgs boson decays in the standard model and its supersymmetric extension. Comput. Phys. Commun. 108:56–74. https://doi.org/10.1016/S0010-4655(97)00123-9. arXiv:hep-ph/9704448 [hep-ph]
44. Bredenstein A et al (2007) Precision calculations for H $\rightarrow$ WW/ZZ $\rightarrow$ 4 fermions with PROPHECY4f. In: Proceedings, international linear collider workshop (LCWS07 and ILC07): Hamburg , Germany, May 30–June 3, 2007, vol 1–2, pp 150–154. arXiv: 0708.4123 [hep-ph]
45. Search for the standard model Higgs boson decaying to charm quarks. Tech rep CMS-PAS-HIG-18-031. Geneva: CERN (2019). https://cds.cern.ch/record/2682638
46. A search for the dimuon decay of the standard model Higgs boson in pp collisions at ps = 13 TeV with the ATLAS Detector. Tech rep ATLAS-CONF-2019-028. Geneva: CERN (2019). https://cds.cern.ch/record/2682155

47. Quantum Diaries (2012) What to look for: the Higgs-to-gamma-gamma branching ratio. https://www.quantumdiaries.org/2012/07/03/what-to-look-for-the-higgs-to-gamma-gammabranching-ratio/
48. The ATLAS Collaboration (2008) The ATLAS experiment at the CERN large hadron collider. J Instrum 3.08:S08003–S08003. https://doi.org/10.1088/1748-0221/3/08/s08003. https://doi.org/10.1088%2F1748-0221%2F3%2F08%2Fs08003
49. The CMS Collaboration (2008) The CMS experiment at the CERN LHC. J Instrum 3.08:S08004–S08004. https://doi.org/10.1088/1748-0221/3/08/s08004. https://doi.org/10.1088%2F1748-0221%2F3%2F08%2Fs08004
50. Aad G et al (2012) Observation of a new particle in the search for the standard model Higgs boson with the ATLAS detector at the LHC. Phys Lett B716:1–29. https://doi.org/10.1016/j.physletb.2012.08.020. arXiv:1207.7214 [hep-ex]
51. Chatrchyan S et al (2012) Observation of a new boson at a mass of 125 GeV with the CMS experiment at the LHC. Phys Lett B716:30–61. https://doi.org/10.1016/j.physletb.2012.08.021. arXiv:1207.7235 [hep-ex]
52. Aaboud M et al (2018) Measurement of the Higgs boson mass in the H $\rightarrow$ ZZ* $\rightarrow$ 4ℓ and H $\rightarrow$ $\ell\ell$ channels with ps = 13 TeV pp collisions using the ATLAS detector. Phys Lett B784:345–366. https://doi.org/10.1016/j.physletb.2018.07.050. arXiv:1806.00242 [hep-ex]
53. Aad G et al (2015) Combined measurement of the Higgs boson mass in pp Collisions at ps = 7 and 8 TeV with the ATLAS and CMS experiments. Phys Rev Lett 114:191803. https://doi.org/10.1103/PhysRevLett.114.191803. arXiv: 1503.07589 [hep-ex]
54. Observation of H $\rightarrow$ b$\bar{b}$ AFb decays and VH production with the ATLAS detector. Tech rep ATLAS-CONF-2018-036. Geneva: CERN (2018). https://cds.cern.ch/record/2630338
55. Cross-section measurements of the Higgs boson decaying to a pair of tau leptons in proton-proton collisions at ps = 13 TeV with the ATLAS detector. Tech. rep. ATLAS-CONF-2018-021. Geneva: CERN (2018). https://cds.cern.ch/record/2621794
56. Aaboud M et al (2018) Observation of Higgs boson production in association with a top quark pair at the LHC with the ATLAS detector. Phys Lett B784:173–191. https://doi.org/10.1016/j.physletb.2018.07.035. arXiv: 1806.00425 [hep-ex]
57. David A et al (2012) LHC HXSWG interim recommendations to explore the coupling structure of a Higgs-like particle. arXiv:1209.0040 [hep-ph]
58. Combined measurements of Higgs boson production and decay using up to 80 fb^{-1}C0 1 of proton-proton collision data at ps = 13 TeV collected with the ATLAS experiment. Tech rep ATLAS-CONF-2018-031. Geneva: CERN (2018). https://cds.cern.ch/record/2629412
59. Yang CN (1950) Selection rules for the dematerialization of a particle into two photons. Phys Rev 77:242–245. https://doi.org/10.1103/PhysRev.77.242. https://link.aps.org/doi/10.1103/PhysRev.77.242
60. Landau LD (1948) On the angular momentum of a system of two photons. Dokl Akad Nauk Ser Fiz 60.2:207–209. https://doi.org/10.1016/B978-0-08-010586-4.50070-5
61. Aad G et al (2015) Study of the spin and parity of the Higgs boson in diboson decays with the ATLAS detector. Eur Phys J C75.10:476. https://doi.org/10.1140/epjc/s10052-015-3685-1,10.1140/epjc/s10052-016-3934-y. arXiv:1506.05669 [hep-ex]
62. Collins JC, Soper DE (1977) Angular distribution of dileptons in high-energy hadron collisions. Phys Rev D 16:2219–2225. https://doi.org/10.1103/PhysRevD.16.2219. https://link.aps.org/doi/10.1103/PhysRevD.16.2219
63. Standard Model Summary Plots Spring 2019. Tech rep ATL-PHYS-PUB-2019-010. Geneva: CERN (2019). https://cds.cern.ch/record/2668559
64. Aghanim N et al (2018) Planck 2018 results. VI. Cosmological parameters. arXiv: 1807.06209 [astro-ph.CO]
65. Rubin VC, Jr Ford WK, Thonnard N (1980) Rotational properties of 21 SC galaxies with a large range of luminosities and radii, from NGC 4605 (R=4kpc) to UGC 2885 (R=122kpc). APJ 238:471–487. https://doi.org/10.1086/158003

66. Lin T (2019) TASI lectures on dark matter models and direct detection. arXiv: 1904.07915 [hep-ph]
67. Rosner JL (2000) CP violation: a brief review. AIP Conf Proc 540.1:283–304. https://doi.org/10.1063/1.1328890. arXiv: hep-ph/0005258 [hep-ph]
68. Djouadi A (2008) The Anatomy of electro-weak symmetry breaking. II. The Higgs bosons in the minimal supersymmetric model. Phys Rept 459:1–241. https://doi.org/10.1016/j.physrep.2007.10.005. arXiv:hep-ph/0503173 [hep-ph]
69. Yorikiyo N (2014) Beyond the standard model of elementary particle physics. Wiley. https://doi.org/10.1002/9783527665020. https://onlinelibrary.wiley.com/doi/book/10.1002/9783527665020

Chapter 2
Beyond the Standard Model : **The Effective Field Theory Approach**

Since the 1970s, we have been witnessing the triumphs of the standard model of particle physics (SM) in describing the microscopic behavior of matter in terms of a few elementary constituents and fundamental interactions. The last of these triumphs was the discovery of the Higgs boson in 2012 by the ATLAS [1] and CMS [2] experiments. Furthermore, the measured properties of the Higgs boson show excellent agreement with the SM predictions, as detailed in Sect. 1.3.4. All these results seem to indicate that the SM is a valid theory at the electroweak energy scale [3].

Nevertheless, despite the successes of the SM in describing various phenomena up to the energy scales probed by the LHC, there are hints, as discussed in Sect. 1.4, that the SM is not valid to arbitrarily high energies, thus constituting only the low-energy effective limit of a complete unknown theory.

In this chapter, we will review the *effective field theory* approach as a tool to study and constrain the general features of the high-energy extension of the Standard Model. This approach will then be applied in Chap. 8 to the Higgs boson cross sections measured in the diphoton final state (Chap. 7) to constrain the anomalous Higgs boson couplings to gauge bosons that arise in such extensions of the SM.

2.1 Introduction to the EFT Approach

The effective field theory approach is based on the property of *decoupling*. Decoupling refers to the screening of high-energy phenomena for interactions at lower energy scales. For example, it is possible to describe lower energy phenomena (atomic or nuclear physics) without knowing the internal SM details. Similarly, in classical mechanics, it is possible to describe the motion of a macroscopic body without hav-

© The Editor(s) (if applicable) and The Author(s), under exclusive license
to Springer Nature Switzerland AG 2020
A. Tarek Abouelfadl Mohamed, *Measurement of Higgs Boson Production
Cross Sections in the Diphoton Channel*, Springer Theses,
https://doi.org/10.1007/978-3-030-59516-6_2

ing to deal with the motion of the microscopic elements (atoms and molecules) inside it. This is the result of degrees of freedom at a given energy scale being "integrated out" at lower energies. This idea is the core of the EFT approach in searching for deviations from the SM predictions.

To understand the nature of decoupling in quantum field theory, let us take a look at the propagator of a massive particle (with mass m) at low energies:

$$\frac{1}{p^2 - m^2} = \frac{-1}{m^2} \left(1 + \frac{p^2}{m^2} + \frac{p^4}{m^4} \cdots \right)$$

(2.1)

From this expansion, when $m^2 \gg p^2$, the propagation of the particle is suppressed. In other words, a non-local interaction via the exchange of a massive virtual particle can be approximated with a local (contact) interaction. This can be seen as well from the uncertainty principle. The virtual contribution of a heavy particle can violate the energy (momentum) conservation required for its production provided it takes place only over short times (distances). Therefore, at lower energies (momenta) the influence of heavy particles appears to be instantaneous, i.e. local in time (space) [4]. This means that one can construct an *effective* Lagrangian describing the interactions at lower energies only in terms of the light fields. This effect of heavy particles on low-energy processes is formally known as the Appelquist-Carazzone decoupling theorem (1975) [5]:

> The only role of the heavy fields in the low-momentum behavior of processes represented by Feynman diagrams without external heavy field lines is their contribution to coupling-constant and field-strength renormalization. The heavy fields effectively decouple, and the low-momentum behavior of the theory is described by a Lagrangian consisting of the massless fields only.

Using this picture, we can look at the SM as the low-energy renormalizable gauge theory (*effective theory*) embedded in a gauge theory at higher energy scales Λ_{NP}. This permits probing BSM effects, even if the BSM exists at energies much higher than what can be directly produced in the LHC, via their low-energy behavior (i.e. contribution to coupling constants and field-strength renormalizations). This view is commonly known as the *bottom-up* construction of the EFT, where an EFT at lower energies is used to infer more details of unknown theories at higher energies (shorter distances). This is opposed to the *top-down* EFT construction, in which the full theory is known in high energies, but it is used to describe lower-energy phenomena via its effective Lagrangian approximation.

The bottom-up EFT construction provides the basis for the so-called *indirect* searches for BSM physics. Indirect searches use precision measurements – and potential small deviations of these measurements from the corresponding SM predictions – to constrain BSM physics, as opposed to *direct* searches. Direct searches look for signatures of new physics in the final state such as resonances, not corresponding to known SM particles, in the invariant mass distributions of particles that could be produced by the decays of those resonances.

The bottom-up EFT construction simplifies the search for BSM physics substantially as follows. Given the absence of BSM signals, one can safely assume that BSM physics exists at an energy scale that is much larger than the EW symmetry breaking scale: $\Lambda_{NP} \gg \nu = 246$ GeV. Therefore, instead of guessing the nature of a complete theory with additional degrees of freedom (i.e. additional particles), we can use the Appelquist-Carazzone decoupling theorem, i.e. use the fact that at low energies the BSM heavy fields will be integrated out giving rise to additional higher order operators built exclusively of SM fields and suppressed with powers of $1/\Lambda_{NP}$:

$$\mathcal{L}_{EFT} = \mathcal{L}_{SM} + \mathcal{L}_{D=5} + \mathcal{L}_{D=6} \ldots, \tag{2.2}$$

where $\mathcal{L}_{SM}$ is the SM Lagrangian (detailed in Chap. 1), and D ($D \geq 5$) specifies the mass dimension of the lagrangian term $\mathcal{L}_D$, which is given by

$$\mathcal{L}_D = \sum_i \frac{c_i^{(D)}}{\Lambda_{NP}^{D-4}} \mathcal{O}_i^{(D)}. \tag{2.3}$$

The $c_i^{(D)}$ are dimensionless couplings, known as the *Wilson coefficients*, specifying the strength of the BSM interactions.

The SM operators are of dimension-4, which satisfies the renormalizability condition. The effective operators, on the other hand, are of dimension $D \geq 5$, and therefore are non-renormalizable, i.e. the effective theory is not valid to arbitrarily high energies but only up to energies much lower than the scale Λ_{NP}. The full effective Lagrangian is then an infinite series. However, the inverse of the high energy scale Λ_{NP} plays the role of an expansion parameter. Therefore, one can use dimensional analysis and the ratio between the energy scale of the experiment E and the energy scale Λ_{NP} (E/Λ_{NP}) to keep only the most relevant terms of the expansion. This procedure is known as *power counting* [6].

The $\mathcal{O}^{(D \geq 5)}$ operators are $SU(3) \times SU(2) \times U(1)$ invariant. These operators will modify the SM couplings, and lead to observables deviating from the SM predictions. This defines the strategy of searching for BSM effects using the EFT: measure (constrain) the Wilson coefficients of $D \geq 5$ operators. The measured values of the Wilson coefficients at a given value of the high energy scale (Λ_{NP}) can then be interpreted in a model-dependent manner as parameters of a UV-complete theory. This procedure is known as *matching* [7].

In the following section, we will use the Fermi theory of weak interactions as a concrete example of the different EFT concepts introduced above.

Case study: **The Fermi theory of weak-interactions** A famous example of the EFT paradigm is from the Fermi theory of weak interactions [8]. In this paradigm, the Fermi theory plays the role of an effective theory at lower energies for the SM which plays the role of a UV-complete theory at high energy scale $\Lambda_{NP} \sim m_W = 80.8$ GeV. In this case study, we will use the muon decay $\mu \to e\nu_\mu \bar{\nu}_e$ as an example. For this

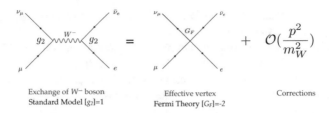

Exchange of W^- boson
Standard Model $[g_2]=1$ Effective vertex
Fermi Theory $[G_F]=-2$ Corrections

Fig. 2.1 An example of the muon decay using the EFT paradigm. The SM picture of exchanging W^- boson can be thought of as the UV complete theory of a lower-energy effective Fermi theory with the addition of $\mathcal{O}(\frac{1}{m_W^2})$ corrections. The SM coupling constant g_2 is dimensionless (*renormalizable* theory), whereas the Fermi constant G_F has dimensions $[G_F] = m^{-2}$ (*non-renormalizable* theory)

process, the SM energy scale is much higher than the energy scale of the interactions, $\mathcal{O}(m_\mu \sim 100 \text{ MeV})$.

In the SM, the muon decay is mediated by the exchange of a W boson, induced by the Lagrangian term:

$$\mathcal{L}_{SM} \supset \frac{g_2}{\sqrt{2}} \left[\bar{\nu}_\mu \gamma_\alpha (1 - \gamma_5)\mu + \bar{e}\gamma_\alpha(1 - \gamma_5)\nu_e \right] W_\alpha^+ + h.c. \ , \tag{2.4}$$

where g_2 is the dimensionless weak coupling constant.

The muon decay width can be computed from this Lagrangian and expanded as a series of the parameter p^2/m_W^2:

$$\frac{d\Gamma(\mu \to e\nu_\mu \bar{\nu}_e)}{dp} \approx \frac{g^4(m_\mu^2 - p^2)^2(m_\mu^2 + 2p^2)}{3072\pi^3 \, m_\mu^3 m_W^4}(1 + \frac{2p^2}{m_W^2} + \ldots). \tag{2.5}$$

where p is the momentum of the muon [9]. Therefore, given that $m_W^2 \gg m_\mu^2 \geq p^2$, the terms p^2/m_W^2 can be neglected to a good approximation. The expansion is sketched in Fig. 2.1, where the first order of the fundamental interaction is the effective vertex and the sub-leading terms are of the order $\frac{p^2}{m_W^2}$.

On the other hand, in the Fermi effective theory, weak interactions are described by contact 4-fermion vertices with a coupling $\frac{c}{\Lambda^2}$, where c is the Wilson coefficient and Λ is the energy scale of the EFT, via the effective Lagrangian:

$$\mathcal{L}_{EFT} \supset \frac{c}{\Lambda^2} \left[\bar{\nu}_\mu \gamma^\alpha (1 - \gamma_5)\mu \right] \left[\bar{e}\gamma_\alpha(1 - \gamma_5)\nu_e \right] + h.c. \tag{2.6}$$

The coupling $\frac{c}{\Lambda^2}$ has dimension m^{-2}. From this Lagrangian, the decay width can be obtained as:

$$\frac{d\Gamma(\mu \to e\nu_\mu \bar{\nu}_e)}{dp} = \frac{c^2(m_\mu^2 - p^2)^2(m_\mu^2 + 2p^2)}{768\pi^3 \, m_\mu^3 \Lambda^4} \tag{2.7}$$

Setting $\Lambda = m_W$ and $c = \frac{g_2^2}{2}$ in Eq. (2.7), we can match the decay width from the SM with that from the Fermi theory. The coupling of the EFT Lagrangian that appears in the 4-fermion contact vertex, c/Λ^2, is thus related to the Fermi constant G_F by $G_F = \frac{\sqrt{2}c}{4\Lambda^2} = \frac{\sqrt{2}g_2^2}{8m_W^2}$.

This example illustrates the power of the EFT approach, as the Fermi constant, $G_F = 1.16 \times 10^{-5}$ GeV^{-2}, was measured from the muon lifetime much earlier than the discovery of W boson, and could thus be used to infer constraints on the W boson mass, even without knowing the underlying UV complete theory. However, for this bottom-up approach to work, the matching of the couplings in the EFT (i.e G_F) to the UV complete parameter (i.e. m_W) requires an assumption on the value of the coupling in the UV complete theory (g_2). Hence, determining the validity of the EFT requires a degree of model-dependence. For example, for very small values of $g_2 = \mathcal{O}(10^{-4})$, the bound on the W-boson mass can be even smaller than the muon mass which contradicts the initial assumption ($m_W \gg m_\mu$). The self-consistency of this approach thus requires $g_2 \gg 10^{-2}$. In addition, an upper-limit on the mass $m_W \leq 1.5$ TeV can be obtained by setting g_2 to its maximally strongly-coupled limit $g_2 \sim 4\pi$.

The description of the muon decay using the Fermi (effective) theory can be improved by considering more terms, of order $c^{(D=8)}/\Lambda^4$, in the expansion of the EFT Lagrangian. The range of validity of the Fermi (effective) theory can be deduced directly from Eq. (2.5), as p approaches the W boson mass m_W the approximation breaks down.

2.2 The EFT Expansion

As detailed in the previous section, the EFT framework provides a systematic expansion of the SM Lagrangian with additional operators suppressed by the mass scale, Λ_{NP}. Therefore, one can construct the most general EFT Lagrangian consistent with symmetry principles. The result of such Lagrangian will be the most general S-matrix consistent with the assumed symmetry properties [10].

There are two main classes of effective field theories describing BSM physics:

- *linear* EFT, in which the Higgs boson h is included in a $SU(2)$ doublet H. In this linearly-realized EFT the doublet H transforms linearly under $SU(2)$.
- *non-linear* EFT, in which general anomalous couplings are introduced for the physical Higgs boson h [7]. The EFT, in this case, is known as the electroweak chiral Lagrangian.

In our analysis, we will study only linearly-realized EFT, i.e. the Higgs boson transforms under $SU(2)$ generating couplings proportional to $h + v$, where v is the vacuum expectation value.

In this section, we will review the different operators resulting from the EFT expansion at different orders.

2.2.1 Dimension-5 (Weinberg) Operator

The first BSM operators in the EFT expansion of Eq. (2.3) are dimension-5 terms. There exists only a single dimension-5 operator built with SM fields that respects $SU(3) \times SU(2) \times U(1)$ gauge invariance, and is known as the Weinberg operator [11]. This operator violates the conservation of lepton number, and hence its effects cannot be probed in the LHC due to strong constraints on lepton number violation. Therefore, it is not included in our study. Nevertheless, it is of importance to neutrino physics, as the dimension-5 operator can give rise to Majorana neutrino masses via EWSB [12].

2.2.2 Dimension-6 Operators

Dimension six operators are the first BSM operators in the EFT expansion that can be probed using LHC data. The first attempts to classify dimension-6 operators excluding lepton and baryon number violating terms date back to 1986 [13]. Using this classification, one finds 80 operators for each flavor. It was found that there are several ways of constructing the operators, as it is possible to transform one set of operators to another set. Therefore, it is needed to define a *basis* of operators that should not be redundant when using the equations of motion, integration by parts, field re-definitions, and Fierz transformations. There exist many bases of dimension-6 operators. In general, the different bases are equivalent, meaning that any complete basis will lead to the same BSM effects [7]. In addition, it is possible to translate between the different bases using, for example, the ROSETTA tool [14].

In this section, we will review the bases that are studied in this thesis, focusing on their $H \rightarrow \gamma\gamma$ phenomenology. The Higgs boson production in the Standard Model is dominated by gluon fusion (induced by the effective gluon coupling arising from top-quark mediated loop diagrams) and vector-boson fusion and associated VH ($V = W, Z$) production, arising from HVV vertices. The Higgs boson decay to two photons is dominated by W-mediated loop diagrams induced by the HWW vertex, in addition to contributions from fermion-loop induced diagrams that are absorbed together with possible loop diagrams induced by heavier particles in an effective $H\gamma\gamma$ coupling. Therefore, in our analysis we only study operators associated with three-point interactions between the Higgs boson and the gauge bosons. The different bases that we study also include CP-violating operators allowing us to probe CP-violation in the Higgs sector.

2.2.2.1 The Warsaw Basis (SMEFT)

The Warsaw basis [15] is the first complete and non-redundant set of dimension-6 operators, proposed in 2010. The Warsaw basis includes 59 non-redundant operators (conserving lepton and baryon numbers). The Warsaw basis is self-consistent at one loop as it has been completely renormalized [16]. In our analysis, we studied the Warsaw basis as implemented in the Standard Model Effective Field Theory or *SMEFT* [17]. The SMEFT is a consistent EFT generalization of the SM constructed out of a series of $SU_C(3) \times SU_L(2) \times U_Y(1)$ invariant higher-dimensional local contact operators, built using the SM fields including the Higgs doublet.

A full list of the baryon-number-conserving dimension-6 SMEFT operators can be found in Ref. [17]. Among them, we are interested in the operators that can alter the $pp \to H \to \gamma\gamma$ cross sections. These include the following CP-even operators:

$$\mathcal{L}_{\text{SMEFT}} \supset \frac{\overline{C}_{HG}}{v^2} H^\dagger H G_{\mu\nu}^A G^{\mu\nu A} + \frac{\overline{C}_{HW}}{v^2} H^\dagger H W_{\mu\nu}^I W^{\mu\nu I} + \frac{\overline{C}_{HB}}{v^2} H^\dagger H B_{\mu\nu} B^{\mu\nu}$$
$$+ \frac{\overline{C}_{HWB}}{v^2} H^\dagger \sigma^I H W_{\mu\nu}^I B^{\mu\nu}, \tag{2.8}$$

where H is the Higgs boson doublet, $G_{\mu\nu}$, $W_{\mu\nu}$ and $B_{\mu\nu}$ are the gauge field strength tensors for the $SU(3)_c$, $SU(2)_L$ and $U(1)_Y$ generators and σ^I are the Pauli matrices. $\overline{C}_{HG}$, $\overline{C}_{HW}$, $\overline{C}_{HB}$ and $\overline{C}_{HWB}$ are the different Wilson coefficients specifying the strength of the coupling. For these coefficients we have absorbed the scale Λ_{NP} in the definition of the Wilson coefficient, $\overline{C} \equiv \frac{v^2}{\Lambda^2} C$. In addition, there are the following CP-odd operators:

$$\mathcal{L}_{\text{SMEFT}}^{\text{CP-odd}} \supset \frac{\tilde{C}_{HG}}{v^2} H^\dagger H \tilde{G}_{\mu\nu}^A G^{\mu\nu A} + \frac{\tilde{C}_{HW}}{v^2} H^\dagger H \tilde{W}_{\mu\nu}^I W^{\mu\nu I} + \frac{\tilde{C}_{HB}}{v^2} H^\dagger H \tilde{B}_{\mu\nu} B^{\mu\nu}$$
$$+ \frac{\tilde{C}_{HWB}}{v^2} H^\dagger \sigma^I H \tilde{W}_{\mu\nu}^I B^{\mu\nu}, \tag{2.9}$$

where $\tilde{G}_{\mu\nu}$, $\tilde{W}_{\mu\nu}$ and $\tilde{B}_{\mu\nu}$ are the dual field strength tensors, defined as $\tilde{X}_{\mu\nu} = \frac{1}{2} \epsilon_{\mu\nu\rho\sigma} X^{\rho\sigma}$. Similarly to the CP-even case, the energy scale Λ_{NP} is absorbed in the definition of the $\tilde{C}$ coefficients.

These operators contain all the possible contractions of two field-strength tensors that form singlets or triplets of $SU(2)_L$, and singlets of $SU(3)_C$. The operators appear in the form $H^\dagger H$ or $H^\dagger \sigma^I H$ due to electroweak constraints [15]. A complete dictionary of SM extensions in the Warsaw basis was given in Ref. [18] matching the different extensions to dimension-6 operators. The LO modifications to the SM interactions in the SMEFT are derived by expanding around the vacuum expectation value in the unitary gauge and rotating to mass eigenstate fields [7]. The non-zero values of the Wilson coefficients modify the SM coupling constants as follows:

$$\bar{g}_s = g_s(1 + C_{HG} v_T^2), \quad \bar{g}_2 = g_2(1 + C_{HW} v_T^2), \quad \bar{g}_1 = g_1(1 + C_{HB} v_T^2), \tag{2.10}$$

where $\bar{g}$ denotes the modified coupling constants and v_T is the modified VEV in SMEFT. Using the modified couplings, one can derive the modifications to the Higgs boson interactions with gauge bosons. These modifications will affect both the production and decay of the Higgs boson in the $pp \to H \to \gamma\gamma$ process.

The operators of the Warsaw basis are of the form $|H|^2 A_{\mu\nu} A^{\mu\nu}$, which is the same as the one stemming from the SM loop contribution. This results in an overall rescaling of the bosonic partial decay widths, for example for $H \rightarrow \gamma\gamma$ the following analytical expression from Ref. [17] was derived giving the correction to $H \rightarrow \gamma\gamma$ partial width due to non-zero Wilson coefficients :

$$\frac{\Gamma(H \rightarrow \gamma\gamma)}{\Gamma^{SM}(H \rightarrow \gamma\gamma)} \simeq \mid 1 + \frac{8\pi^2 \bar{v}_T^2}{I^\gamma} \mathscr{C}_{\gamma\gamma} \mid^2 + \mid \frac{8\pi^2 \bar{v}_T^2}{I^\gamma} \tilde{\mathscr{C}}_{\gamma\gamma} \mid^2 \qquad (2.11)$$

where

$$\mathscr{C}_{\gamma\gamma} = \frac{1}{\bar{g}_2^2} \overline{C}_{HW} + \frac{1}{\bar{g}_1^2} \overline{C}_{HB} - \frac{1}{\bar{g}_1 \bar{g}_2} \overline{C}_{HWB} \qquad (2.12)$$

$$\tilde{\mathscr{C}}_{\gamma\gamma} = \frac{1}{\bar{g}_2^2} \tilde{C}_{HW} + \frac{1}{\bar{g}_1^2} \tilde{C}_{HB} - \frac{1}{\bar{g}_1 \bar{g}_2} \tilde{C}_{HWB} \qquad (2.13)$$

The Wilson coefficients in Eqs. (2.8) and (2.9) have the following effects on the main Higgs boson production modes:

- $\overline{C}_{HG}$ and $\tilde{C}_{HG}$ affect the gluon-fusion production mode. The gluon-fusion production in the SM is induced by quark-loop diagrams (as detailed in Chap. 1), where the leading contribution arises from top quark loops. In the limit $m_t \rightarrow \infty$ the contact term $HG_{\mu\nu}^A G^{A\mu\nu}$ can be a good approximation. In SMEFT the gluon-fusion production cross section is modified via $\overline{C}_{HG}$ and $\tilde{C}_{HG}$ as follows:

$$\frac{\sigma(gg \rightarrow H)}{\sigma^{SM}(gg \rightarrow H)} \simeq \mid 1 + \frac{16\pi^2 v^2}{I^g \bar{g}_3^2} \overline{C}_{HG} \mid^2 + \mid \frac{16\pi^2 v^2}{I^g \bar{g}_3^2} \tilde{C}_{HG} \mid^2, \qquad (2.14)$$

where I^g is a Feynman integral accounting for the top-quark loop contribution. The change in the Higgs boson decay width to two gluons is also modeled with approximately the same corrections in Eq. (2.14) [17]. This can affect the $h \rightarrow \gamma\gamma$ decay by modifying the Higgs boson total width and hence affect the branching ratio of $h \rightarrow \gamma\gamma$. However, the effect of this variation on the total Higgs width is tiny since the SM width to gluon pairs is very small (SM branching ratio = 8.18×10^{-2} [19]).

- $\overline{C}_{HW}$, $\overline{C}_{HB}$ and $\overline{C}_{HWB}$ (and the corresponding CP-odd terms) affect the cross sections of vector boson fusion (VBF) and of associated production with a vector boson (VH). As an example, the following equation gives the analytical expression of the change in the VH production cross sections [17]:

$$\frac{\sigma(\psi\bar{\psi} \rightarrow VH)}{\sigma^{SM}(\psi\bar{\psi} \rightarrow VH)} \supset \frac{\hat{m}_V^2 q^2}{|\vec{p_H}|^2 + 3\hat{m}_V^2} \left[|f_3^V(q^2)|^2 (3\hat{m}_V^2 + 2|\vec{p_H}|^2) + 2|\vec{p_H}|^2 |f_4^V(q^2)| \right], \quad (2.15)$$

where $f_i^V(q^2)$ are form factors that can be decomposed into $f_i^V(q^2) = f_i^{V,SM}(q^2) + \delta f_i^V(q^2)$ with $\delta f_i^V(q^2)$ the correction due to non-zero Wilson coefficients, q^2 is the four momentum of the fermion pair (from the proton) and p_h is the transverse momentum of the Higgs boson. Therefore, these variations have a dependence on the Higgs boson kinematics from the momentum structure of the operators. The corrections $\delta f_i^V(q^2)$ to the SM form factors are functions of the Wilson coefficients, for example for $V = W^+$ the variation will be $\delta f_i^W(q^2) \propto \overline{C}_{HW}/\hat{m}_W^2$. In addition, these coefficients modify the Higgs boson decay width to photons following Eq. (2.12). The effect of these coefficients on modifying the Higgs boson decay to photons is much larger than their effect in increasing the $VBF + VH$ cross sections. In addition, these coefficients modify the Higgs boson decay width in the channels $H \to ZZ^*$ and $H \to W^+W^-$ in their different decay modes detailed in Chap. 1. Therefore, these Wilson coefficients modify the Higgs boson branching ratio to two photons through the change in the total Higgs boson width, in addition to their effect on the two-photon width as in Eq. (2.12).

2.2.2.2 The SILH Basis (HEL)

An additional operator basis that we studied in our analysis is the Strongly Interacting Light Higgs (*SILH*) basis [20]. This basis was originally built for composite Higgs models in which Λ_{NP} can be related to the compositeness scale. We study the SILH basis as implemented in the Higgs effective lagrangian (*HEL*) [21] as follows:

$$\mathcal{L}_{\text{HEL}} = \mathcal{L}_{\text{SM}} + \mathcal{L}_{\text{SILH}} + \mathcal{L}_{CP} + \mathcal{L}_{F_1} + \mathcal{L}_{F_2} + \mathcal{L}_G , \qquad (2.16)$$

where the dimension-six operators have been grouped in the following way:

- $\mathcal{L}_{\text{SILH}}$ contains all CP-even three-point interactions between a single Higgs boson and either a pair of gauge bosons or a fermion-antifermion pair;
- $\mathcal{L}_{CP}$ contains the operators describing the corresponding CP-odd interactions;
- $\mathcal{L}_{F_1}, \mathcal{L}_{F_2}, \mathcal{L}_G$ contain operators that induce anomalous triple and quartic gauge boson interactions, or four-point interactions between a fermion-antifermion pair, a Higgs boson and a gauge boson or a second Higgs boson.

The Wilson coefficients in $\mathcal{L}_{\text{SILH}}$ associated with interactions between a Higgs boson and a fermion-antifermion pair are not probed. The $\mathcal{L}_{\text{SILH}}$ part of the HEL Lagrangian is:

$$\mathcal{L}_{\text{SILH}} = \frac{\overline{c}_H}{v^2}\frac{1}{2}\partial^\mu[H^\dagger H]\partial_\mu[H^\dagger H] + \frac{\overline{c}_T}{v^2}\frac{1}{2}[H^\dagger \overset{\leftrightarrow}{D}{}^\mu H][H^\dagger \overset{\leftrightarrow}{D}_\mu H] - \frac{\overline{c}_6 \lambda}{v^2}[H^\dagger H]^3$$

$$- \left[\frac{\overline{c}_u}{v^2} y_u H^\dagger H H^\dagger \cdot Q_L u_R + \frac{\overline{c}_d}{v^2} y_d H^\dagger H Q_L d_R + \frac{\overline{c}_l}{v^2} y_\ell H^\dagger H H L_L e_R + \text{h.c.} \right]$$

$$+ \frac{g \, \overline{c}_W}{m_W^2}\frac{i}{2}[H^\dagger \sigma_k \overset{\leftrightarrow}{D}{}^\mu H]D^\nu W_{\mu\nu}^k + \frac{g' \, \overline{c}_B}{m_W^2}\frac{i}{2}[H^\dagger \overset{\leftrightarrow}{D}{}^\mu H]\partial^\nu B_{\mu\nu}$$

$$+ \frac{g \, \overline{c}_{HW}}{m_W^2}i[D^\mu H^\dagger \sigma_k D^\nu H]W_{\mu\nu}^k + \frac{g' \, \overline{c}_{HB}}{m_W^2}i[D^\mu H^\dagger D^\nu H]B_{\mu\nu}$$

$$+ \frac{g'^2 \, \overline{c}_\gamma}{m_W^2}H^\dagger H B_{\mu\nu}B^{\mu\nu} + \frac{g_s^2 \, \overline{c}_g}{m_W^2}H^\dagger H G_{\mu\nu}^a G_a^{\mu\nu} ,$$

$$(2.17)$$

where:

- H is the Higgs doublet;
- Q_L, L_L are the left-handed quark and lepton doublets;
- u_R, d_R, e_R are the right-handed up-type quark, down-type quark and charged lepton fields;
- $G_{\mu\nu}$, $W_{\mu\nu}$ and $B_{\mu\nu}$ are the gauge field strength tensors for the $SU(3)_c$, $SU(2)_L$ and $U(1)_Y$ generators;
- λ is the Higgs quartic coupling;
- y_u, y_d and y_ℓ are the 3×3 Yukawa coupling matrices in flavour space;
- g', g and g_s are the $U(1)_Y$, $SU(2)_L$ and $SU(3)_c$ coupling constants, respectively;
- σ_k are the Pauli matrices.

The $\bar{c}$ coefficients are related to the Wilson coefficients by:

$$\bar{c}_H \equiv c_H \frac{v^2}{\Lambda^2} \quad \bar{c}_T \equiv c_T \frac{v^2}{\Lambda^2} \quad \bar{c}_6 \equiv c_6 \frac{v^2}{\lambda\Lambda^2} \tag{2.18}$$

$$\bar{c}_{u,d,\ell} \equiv c_{u,d,\ell} \frac{v^2}{y_{u,d,\ell}\Lambda^2} \tag{2.19}$$

$$\bar{c}_W \equiv c_W \frac{m_W^2}{g\Lambda^2} \quad \bar{c}_B \equiv c_B \frac{m_W^2}{g'\Lambda^2} \quad \bar{c}_{HW} \equiv c_{HW} \frac{m_W^2}{g\Lambda^2} \quad \bar{c}_{HB} \equiv c_{HB} \frac{m_W^2}{g'\Lambda^2} \tag{2.20}$$

$$\bar{c}_\gamma \equiv c_\gamma \frac{m_W^2}{g'\Lambda^2} \quad \bar{c}_g \equiv c_g \frac{m_W^2}{g_s\Lambda^2} \tag{2.21}$$

The corresponding CP-odd interactions are contained in $\mathcal{L}_{CP}$, defined as:

$$\mathcal{L}_{CP} = \frac{g\,\tilde{c}_{HW}}{m_W^2}\frac{i}{2}D^\mu H^\dagger \sigma_k D^\nu H \widetilde{W}_{\mu\nu}^k + \frac{g'\,\tilde{c}_{HB}}{m_W^2}iD^\mu H^\dagger D^\nu H \widetilde{B}_{\mu\nu} + \frac{g'^2\,\tilde{c}_\gamma}{m_W^2}H^\dagger H B_{\mu\nu}\widetilde{B}^{\mu\nu}$$
$$+ \frac{g_s^2\,\tilde{c}_g}{m_W^2}H^\dagger H G_{\mu\nu}^a\widetilde{G}_a^{\mu\nu} + \frac{g^3\,\tilde{c}_{3W}}{m_W^2}\epsilon_{ijk}W_{\mu\nu}^i W^{\nu\,j}_{\rho}\widetilde{W}^{\rho\mu k} + \frac{g_s^3\,\tilde{c}_{3G}}{m_W^2}f_{abc}G_{\mu\nu}^a G^{\nu b}_{\rho}\widetilde{G}^{\rho\mu c} , \tag{2.22}$$

where the $\widetilde{G}_{\mu\nu}$, $\widetilde{W}_{\mu\nu}$ and $\widetilde{B}_{\mu\nu}$ are the dual field strength tensors, defined as $\widetilde{X}_{\mu\nu} = \frac{1}{2}\epsilon_{\mu\nu\rho\sigma}X^{\rho\sigma}$. Again, the $\tilde{c}$ coefficients are related to the Wilson coefficients by relations similar to those written before for the $\bar{c}$ ones.

The operators that can modify the $pp \to h \to \gamma\gamma$ cross sections correspond to the following 12 coefficients:

$$\bar{c}_\gamma, \ \tilde{c}_\gamma, \ \bar{c}_g, \ \tilde{c}_g, \ \bar{c}_T, \ \bar{c}_B, \ \bar{c}_H, \ \bar{c}_W, \ \bar{c}_{HW}, \ \tilde{c}_{HW}, \ \bar{c}_{HB} \quad \text{and} \quad \tilde{c}_{HB} . \tag{2.23}$$

The $\bar{c}_T$ coefficient has been constrained to be $-0.0015 < \bar{c}_T < 0.0022$ [22] at 95% confidence level (CL) by precision electroweak data from LEP. For these values, the effect on the measured $H \rightarrow \gamma\gamma$ cross sections would be negligible and thus the $\bar{c}_T$ coefficient is set to zero in this analysis. The sum of $\bar{c}_W$ and $\bar{c}_B$ has been constrained by LEP data to be $-0.0014 < \bar{c}_W + \bar{c}_B < 0.0019$ at 95% CL. As both the $\bar{c}_B$ and $\bar{c}_W$ coefficients are found to have a small impact on the normalization and shapes of the distributions, they are set to zero and not further investigated. Finally, the $\bar{c}_H$ parameter is also observed to have a very small effect on the differential $H \rightarrow \gamma\gamma$ cross sections ($<0.1\%$ for $\bar{c}_H = 1$). It is therefore also set to zero.

The remaining Wilson coefficients belong to two categories:

i. Coefficients of operators that pre-dominantly alter the total cross section or change the $H \rightarrow \gamma\gamma$ branching fraction
ii. Coefficients of operators that alter the Higgs boson kinematics, the amount of radiation or the angular correlations of the radiation.

The coefficients $\bar{c}_\gamma$ and $\bar{c}_g$ belong to the first category. The coefficient $\bar{c}_\gamma$ corresponds to an operator that interferes with the SM $H \rightarrow \gamma\gamma$ diagrams, affecting the decay but not the properties of the production. The interference can be either constructive or destructive. In addition, that operator can introduce a small $\gamma\gamma \rightarrow H$ production cross section that would change the shape of the observed production kinematics. However, the overall effect is completely negligible with respect to the amount of total cross section change that will drive any limit. The coefficient $\bar{c}_g$ corresponds to an operator that has the same structure as the SM $gg \rightarrow H$ diagram and can interfere with it, leading to a change in the normalization of the gluon fusion production cross section, but not in the kinematic distributions. The corresponding CP conjugate operators, that are multiplied by the $\tilde{c}_\gamma$ and $\tilde{c}_g$ Wilson coefficients, also modify the $H \rightarrow \gamma\gamma$ decay and $gg \rightarrow H$ production, though they cannot interfere with the SM amplitudes.

The coefficients $\bar{c}_{HW}$ and $\bar{c}_{HB}$, as well as their CP conjugate partners belong to the second category: they induce large shape changes in isolated parts of phase space. The corresponding operators are not proportional to SM contributions, thus no pure interference effects leading to simple changes in normalization are possible. Instead, they induce changes in the kinematic distribution of the Higgs boson, e.g. produce a Higgs boson with higher transverse momenta.

2.2.2.3 Other Bases

In addition to the previous bases, there exist other dimension-6 EFT bases that were not included in our study, such as the Higgs basis. This basis was proposed by the LHC cross section working group [7] to separate the Wilson coefficients that are strongly constrained by the electroweak precision tests (EWPT) and the Higgs studies. This is done by rotating other dimension-6 bases such that one can isolate the linear combination that is affecting only the Higgs sector. More details can be found in Ref. [23].

More details on other different bases can be found in Refs. [7, 24]. As mentioned before, complete bases are equivalent, describing the same effects, and there exist tools to translate constraints on the coefficients of the operators in one basis to constraints on the operators of an alternative basis.

2.2.3 Beyond Dimension-6 Operators

The next terms in the EFT expansion of Eq. (2.3), beyond dimension-6 operators, are the dimension-7 terms. These operators violate lepton number conservation, and some of them also violate baryon number conservation [25]. Therefore, the effects stemming from these operators can not be probed at the LHC [7], and hence, they are not considered in this thesis.

Operators with dimension ≥ 8 are suppressed by at least $1/\Lambda_{NP}^4$, making their effect negligible with respect to dimension-6 operators from a simple power counting procedure. The relative size of the higher dimensional terms (e.g. dimension-8) with respect to the dimension-6 terms is controlled by $C^{(8)}/C^{(6)}(E_{exp}^2/\Lambda_{NP})$ [7], where $C^{(8)}$ and $C^{(6)}$ are generic dimension-8 and dimension-6 Wilson coefficients, and E_{exp} is the energy scale of the process. Nevertheless, a complete set of dimension-8 operators were derived in Ref. [26]. There exist around 900 baryon-number-conserving dimension-8 operators. Therefore, the inclusion of full dimension-8 operators will over-complicate setting limits on the EFT models, as the current status of data does not allow lifting the degeneracy between dimension-6 and dimension-8 [6]. Therefore, for this thesis we will truncate the EFT expansion to dimension-6 operators. The truncation of the EFT series assumes that the EFT expansion is still valid to describe the UV-complete theory at higher energy scales Λ_{NP}. The full assessment of this statement can not be done in a fully model-independent manner as the same values of the Wilson coefficient can be matched to different energy scales and the UV-complete coupling parameter as was shown in the Fermi theory case. This degeneracy can be lifted by inserting assumptions on the UV complete model. A concrete example of this procedure is shown in Ref. [6].

2.2.3.1 Higher Order Correction

Another modification that can affect dimension-6 operators is due to higher-order (NLO) QCD corrections. Tree-level EFT predictions can be extended with higher-order loop corrections. The NLO corrections can change the cross section value or modify the differential distributions. The one-loop effects of dimension-6 operators are suppressed by $\mathcal{O}(g_{SM}^2/16\pi^2)$, with g_{SM} denoting a SM coupling, thus they are in general sub-leading [6]. Therefore, the inclusion of the loop corrections for the EFT modification is less crucial. This is not the case for the nominal SM predictions, in which we use the best known high order corrections in QCD and EW in order to describe the data with the highest accuracy. More details are given in Sect. 7.2.

References

1. Aad G et al (2012) Observation of a new particle in the search for the standard model Higgs boson with the ATLAS detector at the LHC. Phys. Lett. B716:1–29. https://doi.org/10.1016/j.physletb.2012.08.020. arXiv:1207.7214 [hep-ex]
2. Chatrchyan S et al (2012) Observation of a new boson at a mass of 125 GeV with the CMS experiment at the LHC. Phys Lett B716:30–61. https://doi.org/10.1016/j.physletb.2012.08.021. arXiv: 1207.7235 [hep-ex]
3. Dawson S (2017) Electroweak symmetry breaking and effective field theory. In: Proceedings, theoretical advanced study institute in elementary particle physics : anticipating the next discoveries in particle physics (TASI 2016): Boulder, CO, USA, June 6-July 1, 2016, pp 1–63. https://doi.org/10.1142/9789813233348_0001. arXiv: 1712.07232 [hep-ph]
4. Burgess CP (2007) Introduction to effective field theory. Ann Rev Nucl Part Sci 57:329–362. https://doi.org/10.1146/annurev.nucl.56.080805.140508. arXiv: hep-th/0701053 [hep-th]
5. Appelquist T, Carazzone J (1975) Infrared singularities and massive fields. Phys Rev D11:2856. https://doi.org/10.1103/PhysRevD.11.2856
6. Contino R et al (2016) On the validity of the effective field theory approach to SM precision tests. JHEP 07:144. https://doi.org/10.1007/JHEP07(2016)144. arXiv: 1604.06444 [hep-ph]
7. de Florian D et al (2016) Handbook of LHC Higgs cross sections: 4. deciphering the nature of the Higgs sector. https://doi.org/10.23731/CYRM-2017-002. arXiv: 1610.07922 [hep-ph]
8. Rajasekaran G (2014) Fermi and the theory of weak interactions. Resonance J Sci Educ 19.1:18–44. https://doi.org/10.1007/s12045-014-0005-2. arXiv: 1403.3309 [physics.hist-ph]
9. Griffiths DJ (2008) Introduction to elementary particles; 2nd rev. version. Physics textbook. Wiley, New York. https://cds.cern.ch/record/111880
10. Weinberg S (2009) Effective field theory, past and future. In: PoS CD09, p 001. https://doi.org/10.22323/1.086.0001. arXiv: 0908.1964 [hep-th]
11. Weinberg S (1979) Baryon- and Lepton-nonconserving processes. In: Phys Rev Lett 43:1566–1570. https://doi.org/10.1103/PhysRevLett.43.1566. https://link.aps.org/doi/10.1103/PhysRevLett.43.1566
12. Jenkins EE, Manohar AV, Stoffer P (2018) Low-energy effective field theory below the electroweak scale: operators and matching. HEP 03:016. https://doi.org/10.1007/JHEP03(2018)016. arXiv:1709.04486 [hep-ph]
13. Buchmuller W, Wyler D (1986) Effective Lagrangian analysis of new interactions and flavor conservation. Nucl. Phys. B268:621–653. https://doi.org/10.1016/0550-3213(86)90262-2
14. Falkowski A et al (2015) Rosetta: an operator basis translator for standard model effective field theory. Eur Phys J C75.12:583. https://doi.org/10.1140/epjc/s10052-015-3806-x. arXiv: 1508.05895 [hep-ph]
15. Grzadkowski B et al (2010) Dimension-six terms in the standard model lagrangian. JHEP 10:085. https://doi.org/10.1007/JHEP10(2010)085. arXiv:1008.4884 [hep-ph]
16. Grojean C et al (2013) Renormalization group scaling of higgs operators and $G(h - > \ell\ell)$. JHEP 1304:016. https://doi.org/10.1007/JHEP04(2013)016. arXiv:1301.2588 [hep-ph]
17. Brivio I, Trott M (2017) The standard model as an effective field theory. https://doi.org/10.1016/j.physrep.2018.11.002. arXiv:1706.08945 [hep-ph]
18. de Blas J et al (2018) Effective description of general extensions of the standard model: the complete tree-level dictionary. JHEP 03:109. https://doi.org/10.1007/JHEP03(2018)109. arXiv:1711.10391 [hep-ph]
19. Andersen JR et al (2013) Handbook of LHC Higgs cross sections: 3. Higgs properties. In: Heinemeyer S et al (eds). https://doi.org/10.5170/CERN-2013-004. arXiv:1307.1347 [hep-ph]
20. Giudice GF et al (2007) The strongly-interacting light Higgs. JHEP 06:045. https://doi.org/10.1088/1126-6708/2007/06/045. arXiv: hep-ph/0703164 [hep-ph]
21. Contino R et al (2013) Effective Lagrangian for a light Higgs-like scalar. JHEP 07:035. https://doi.org/10.1007/JHEP07(2013)035. arXiv:1303.3876 [hep-ph]

22. Alloul A, Fuks B, Sanz V (2014) Phenomenology of the Higgs effective Lagrangian via FEYN-RULES. JHEP 04:110. https://doi.org/10.1007/JHEP04(2014)110. arXiv:1310.5150 [hep-ph]
23. Falkowski A (2015) Higgs basis: proposal for an EFT basis choice for LHC HXSWG. https://cds.cern.ch/record/2001958
24. Willenbrock S, Zhang C (2014) Effective field theory beyond the standard model. Ann Rev Nucl Part Sci 64:83–100. https://doi.org/10.1146/annurev-nucl-102313-025623. arXiv:1401.0470 [hep-ph]
25. Lehman L (2014) Extending the standard model effective field theory with the complete set of dimension-7 operators. Phys Rev D90.12:125023. https://doi.org/10.1103/PhysRevD.90. 125023. arXiv:1410.4193 [hep-ph]
26. Henning B et al (2017) 2, 84, 30, 993, 560, 15456, 11962, 261485, ...: higher dimension operators in the SM EFT. JHEP 08:016. https://doi.org/10.1007/JHEP08(2017)016. arXiv:1512.03433 [hep-ph]

Chapter 3
Interlude A. Statistical Analysis

> It is often said that the language of
> science is mathematics. It could well
> be said that the language of
> experimental science is statistics.

Kyle Cranmer

Basic Principles

The main challenge in experimental particle physics, after performing the measurements, is the interpretation of the measured observables in terms of the parameters of a theory. Statistical methods come to the rescue providing tools and concepts for quantifying the correspondence between the measured observables and model parameters. One of the goals of statistical analysis is the estimation of parameters of interest for a given model from a measured dataset and quantifying the uncertainty of such estimates. In our analysis, we aim at estimating the number of Higgs boson signal events from a dataset containing both Higgs boson signal and background events. The estimated Higgs boson signal yield and the theoretically predicted yield from the SM can be used to assess the agreement between our data and the SM. Similarly, one can assess the agreement of BSM scenarios with the data, and estimate the values of BSM couplings. In this interlude, we will briefly review the statistical tools that are used in this analysis.

One of the most fundamental concepts in statistics is that of random variables, or in the particle physics context that of uncertainty. Uncertainties in particle physics measurements can arise from different factors such as the lack of knowledge about the experiment, or the intrinsically random nature of quantum mechanics. A variable is said to be random if it cannot be predicted with complete certainty [1]. As this

© The Editor(s) (if applicable) and The Author(s), under exclusive license
to Springer Nature Switzerland AG 2020
A. Tarek Abouelfadl Mohamed, *Measurement of Higgs Boson Production
Cross Sections in the Diphoton Channel*, Springer Theses,
https://doi.org/10.1007/978-3-030-59516-6_3

is generally the case in the results of particle physics measurements, the tools of statistical probability are used to quantify the uncertainties on a given estimate. The mathematical definition of probability was given by Kolmogorov in a few axioms based on set theory. The axioms give the probability of a subset A in a sample space S, $P(A)$. Nevertheless, the interpretation of the probability $P(A)$ can take one of two forms:

- the **frequentist** interpretation views the probability of an event A as the limit frequency of obtaining A when the experiment is repeated an infinite amount of times.

$$P(A) = \lim_{N \to \infty} \frac{N(A)}{N} \tag{3.1}$$

- the **Bayesian** interpretation, on the other hand, considers $P(A)$ as the degree of belief that A is true.

Both interpretations follow the Kolmogorov axioms. In our analysis, we use the frequentist interpretation, i.e. probabilities are only associated with the outcome of repeatable observations instead of hypothetical statements that do not change with the repetition of the experiment. Frequentist interpretation tools can also be used to make statements about the different hypotheses.

Parameter Estimation

One of our goals is to estimate a set of parameters of interest (POI) $\vec{\mu}$ by measuring a set of observables $\vec{x}$ whose probability depends on $\vec{\mu}$. For simplicity of notation, we will consider the case of a single parameter of interest, μ. The definition in Eq. (3.1) can be used for cases where A takes discrete values (e.g. event counting); however, we are generally interested in continuous variables, such as the diphoton invariant mass for a given event. For a continuous variable x, the probability can be defined using a probability density function (PDF), $f(x)$, giving the probability of observing x within an infinitesimal interval $[x, x + dx]$ as follows:

$$P(x \in [x, x + dx]) = f(x)dx, \tag{3.2}$$

where the function $f(x)$ satisfies the normalization $\int_{-\infty}^{\infty} f(x)dx = 1$. The function $f(x)$ should be varying with our parameter of interest μ, $f(x|\mu)$. The analytical form of $f(x|\mu)$ can be obtained using simulations (such as the modeling of the Higgs boson signal shape) or from data control regions (such as the modeling of the background shape). In the case of multiple events n, the random variable x becomes the vector $\vec{x} = x_1, \ldots x_n$, and the combined PDF for all events will be the product of the PDF for each event, also known as *likelihood function*, $\mathcal{L}(\vec{x}|\alpha) = \prod_{i=1}^{n} f(x_i|\mu)$. In particle physics counting experiments (such as our analysis), the number of selected events is also a random variable: the observed number of events n is fluctuating around an expected number $\nu(\mu)$ according to a Poisson distribution $\text{Poisson}(n|\nu) = \frac{\nu^n e^{-\nu}}{n!}$. The likelihood function in this case is known as the *extended likelihood*:

$$\mathcal{L}(\overrightarrow{x}\,|\mu) = \text{Poisson}(n|\nu) \prod_{i=1}^{n} f(x_i|\mu), \qquad (3.3)$$

Using this likelihood function, we would like to estimate the parameter of interest given the dataset we collected $\overrightarrow{x} = x_1, \ldots x_n$. A good estimator $\hat{\mu}(\overrightarrow{x})$ of the true value of μ must satisfy the following conditions:

- *Consistency.* The estimator should converge to the true value in the limit of infinite statistics $\lim_{n\to\infty} \hat{\mu} = \mu$.
- The difference between the expectation value of the estimator $E(\hat{\mu})$ and the true value μ, known as the *bias*, should be minimal with respect to other estimators.
- The *variance*, defined as $var[\hat{\mu}] = E((\mu - E(\hat{\mu}))^2)$, should be minimal as well.

In general, there is a trade-off between the bias and the variance, and even for unbiased estimators, there is a well-defined minimum variance bound [2]. One of the most widely used estimators is the maximum likelihood estimator (ML). It is defined as the value of μ that maximizes the likelihood function $\mathcal{L}(\overrightarrow{x}\,|\mu)$ of Eq. (3.3). In practice, this is done by minimizing the negative logarithm of $\mathcal{L}$, $-\log \mathcal{L}(\overrightarrow{x}\,|\mu)$, as it is computationally more efficient, in a numerical procedure known as likelihood fitting. The uncertainty on the ML estimate of the POI can be obtained using several methods [1]. One of these is a graphical method where the $\pm 1\sigma$ uncertainty on the estimator corresponds to the width of the negative log-likelihood curve at an increase of 0.5 from the $-\log \mathcal{L}_{min}$. A schematic overview of the ML fit is shown in Fig. 3.1.

Systematic Uncertainties

In practice, the likelihood function that is used to estimate the parameter of interest, the number of Higgs boson signal events in our case, will not be only a function

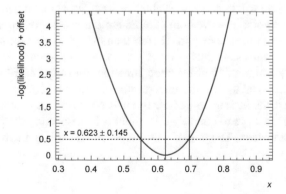

Fig. 3.1 An illustration of the maximum likelihood fit, minimizing the negative log of the likelihood function. The negative log-likelihood function is shown in blue with the minimum denoting the best fit estimate for $\hat{x}$ (vertical red line). The function has been offset to have minimum at zero for clarity. The uncertainty on the estimate of $\hat{x}$ is given by the intersection of $-\log \mathcal{L}_{min} + 0.5$ (horizontal dashed line) with the likelihood curve

of the parameter of interest μ. This is a result of our imperfect knowledge of other parameters that also affect the distributions of our observables. These parameters are known as *nuisance parameters* $\vec{\theta}$. Examples of nuisance parameters in our analysis are the number of background events, the different parameters that describe the signal and background shapes, and the different systematic uncertainties affecting them. These different parameters affect our estimate of the Higgs boson signal yield and hence have to be included to describe our dataset fully. The inclusion of nuisance parameters in the likelihood function will increase the number of degrees of freedom, as the fit tries to accommodate the observed data, resulting in a larger uncertainty on the estimate of the parameter of interest.

The number of nuisance parameters can be quite large for our dataset to be able to constrain all of them at the same time. One can overcome this limitation with the help of *auxiliary measurements*. An auxiliary measurement, performed in some data control region, provides a best estimate for a certain nuisance parameter $\tilde{\theta}$ and a measure of its uncertainty σ_θ. From the auxiliary measurement, we can build a variational estimate $\theta^\pm$ corresponding to $\pm 1\sigma$ variations in θ. Then, a constraint term $f(\theta|\tilde{\theta}, \sigma_\theta)$ can be multiplied by our full measurement likelihood function, leading to:

$$\mathcal{L}(\vec{x}|\mu) = \text{Poisson}(n|\nu) \prod_{i=1}^{n} f(x_i|\mu) \cdot f(\theta|\tilde{\theta}, \sigma_\theta). \tag{3.4}$$

A common way to implement the constraint is via a Gaussian penalty term parameterized such that $\theta = 0$ is the nominal value of the parameter, and $\theta = \pm 1$ are the $\pm 1\sigma$ variations. In some cases, however, a Gaussian constraint term is not appropriate, such as when the parameter can only be positive. The signal resolution is an example and is typically included via a log-normal constraint term. A concrete example of adding constraints to the likelihood function is shown in Sect. 7.6.

The fit procedure will then find the values of $\hat{\mu}$ and $\hat{\theta}$ that maximize the likelihood. From a statistical point of view, both parameters μ and θ are on equal footing, and in some cases, the fit may constrain θ better than the auxiliary measurement. While this is "normal" from a statistical point of view, from an experimental point of view this is generally a sign of the likelihood function not fully describing the dataset (i.e. missing additional systematic uncertainties). This problem is known as *over-constraining* and might lead to underestimated uncertainties on the parameter of interest. One of the ways to quantify this effect is by inspecting the post-fit *pulls* of the nuisance parameters. The pull of a nuisance parameter is defined as

$$\text{pull}(\theta) = \frac{\hat{\theta} - \tilde{\theta}}{\hat{\sigma}_\theta}, \tag{3.5}$$

i.e. the pulls quantifies how far our estimate $\hat{\theta}$ deviates (is pulled) from the expected value $\tilde{\theta}$ given the post-fit uncertainty $\hat{\sigma}_\theta$. Using this definition, the nominal or the pre-fit systematic uncertainties nuisance parameters are centered at zero with an uncertainty of 1. One can then examine over-constraining by observing the post-fit pulls. Deviations from a central value of zero indicate that some data features are absorbed by this non-zero central value of the pull, while a deviation from a standard deviation of 1 indicates that the data is constraining the systematic uncertainty which might require additional investigations. The *impact* of a nuisance parameter on the parameter of interest is defined as follows:

$$\text{Impact}(\theta) = \Delta\mu^{\pm} = \hat{\hat{\mu}}_{\hat{\theta}\pm\sigma_\theta} - \hat{\mu}, \tag{3.6}$$

where $\hat{\hat{\mu}}_{\hat{\theta}\pm\sigma_\theta}$ is the fitted value of μ when the nuisance parameter is fixed to its expectation value plus or minus one standard deviation for the pre-fit impacts, and fixed to their observed values for the post-fit impacts. Examples of the nuisance parameters pulls and impacts are shown in Sect. 7.6.1.

Asimov Datasets

In a particle physics analysis, such as this thesis, we are interested in estimating the sensitivity and the expected uncertainties before performing the measurement. This requires the knowledge of the distribution of the likelihood function. This can be obtained via an *Asimov* dataset, where the parameters of interest are fixed, and the nuisance parameters are fixed to their best fit values from data resulting in a post-fit Asimov dataset or by fixing nuisance parameters to their expected values resulting in a pre-fit Asimov dataset. This results in a distribution (histogram) that matches perfectly the model. Such histogram is built setting the number of events at each bin $n_i^{\text{Asimov}} = E[n_i](\mu, \overrightarrow{\hat{\theta}})$, thus eliminating all statistical fluctuations. More details on the construction of the Asimov dataset are given in Ref. [3]. An illustrative example of an Asimov dataset is shown in Fig. 3.2 for a simple signal plus background model.

Hypothesis Testing and Confidence Intervals

Another statistical tool that we will be using in our analysis is that of hypothesis testing. It is a statistical procedure used to make statements regarding the compatibility of the data collected with two alternative hypotheses by rejecting one of the

Fig. 3.2 An example of an Asimov dataset (black points) built from a simple signal plus background model (blue line)

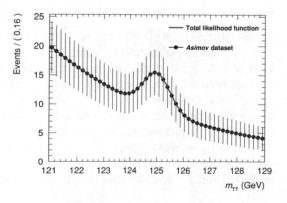

hypotheses if the other one is in much better agreement with the data. As an example, when the Higgs boson was discovered the two hypotheses under test were the SM with and without a Higgs boson.

In our analysis, we expect to observe a signal (the Higgs boson signal) on top of a background from known processes. In order to estimate the significance of the Higgs boson signal yield, we can then define two hypotheses: a background-only hypothesis (H_b) and a signal-plus-background hypothesis (H_{s+b}). To distinguish between the two hypotheses, a *test statistics* $t(\vec{x})$ is built in a way such that the value of t will be different between the two hypotheses. This is done via the definition of an *acceptance region* such that if $t(\vec{x}) < k$, we accept the background-only hypothesis. The test-statistics should minimize the probabilities of type-I errors (rejecting the background-only hypothesis when it is true), quantified by the *size* α of the test, and type-II errors (accepting the background-only hypothesis when the alternative is true), quantified with the *power* of the test $1 - \beta$.

One of the ways we can use to quantify if the discrepancy between data and one of the alternative hypotheses is large enough to reject it is the *significance*. The significance of an observation is related to the *p-value* under the alternative hypothesis (background-only hypothesis in our case). The p-value is the fraction of times one obtains a dataset that is as compatible (or more) with the alternative (background-only) hypothesis. For example, the famous 5σ significance that is used in particle physics to claim discoveries is equivalent to a p-value of 2.87×10^{-7}. It also quantifies the rate of type-I error $\alpha = 2.87 \times 10^{-7}$.

In general, there are various methods to construct test-statistics for hypothesis testing. The most powerful one is the likelihood ratio test $t(\vec{x}) = \mathcal{L}(\vec{x} | H_{S+B})/ \mathcal{L}(\vec{x} | H_B)$ according to the Neyman-Pearson lemma [1, 4]. A generalization of the likelihood ratio is called the *profile likelihood ratio*, in which the likelihood functions of the two hypotheses are maximized for all parameters except for the parameter of interest:

$$\lambda(\vec{\mu}) = \frac{\mathcal{L}(\vec{x} | \mu, \hat{\hat{\vec{\theta}}}_{\mu})}{\mathcal{L}(\vec{x} | \hat{\mu}, \hat{\vec{\theta}})} \tag{3.7}$$

where μ is the parameter of interest labeling the hypothesis, $\hat{\hat{\vec{\theta}}}_{\mu}$ maximizes the likelihood function for a given μ, and $\hat{\mu}$, $\hat{\vec{\theta}}$ maximize the likelihood function.

One of the uses of hypothesis testing is to find confidence intervals. A confidence interval is a region of allowed values for a given model parameter. A confidence interval is defined based on a *confidence level* between 0 and 1. Typical examples are 95% and 68% confidence levels. For example, a 95% confidence interval for a parameter means that the interval determined with this procedure will contain the true value of the parameter in 95% of identical experiments that could be performed. This property is called *coverage*. The procedure for building confidence intervals is called the Neyman Construction. It is based on inverting a series of *hypothesis tests*, as shown in Fig. 3.3. This means that for each value of μ in the parameter space, we

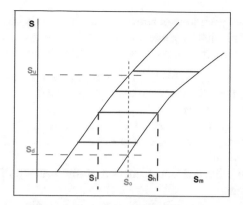

Fig. 3.3 An illustration of the Neyman construction. The x-axis shows the measurement parameter space S_m and the y-axis shows the true parameter space S. The horizontal lines are the acceptance intervals in the measured space for a given true value s, $[S_l, S_h](S)$. Given an observation S_0 one can construct a confidence interval $[S_d, S_u]$ via inversion [4]

perform a hypothesis test. A confidence interval $I(\overrightarrow{x})$ is constructed as follows

$$I(\overrightarrow{x}) = \{\mu | P(\lambda > k | \mu) < \alpha\}, \tag{3.8}$$

where α refers to the size of the test, i.e. $\alpha = 5\%$ for a 95% confidence interval. In practice, the confidence interval can be determined using two alternative methods [5]:

- **The asymptotic formula**. This approximation is based on the theorems by Wilks and Wald [3], which show that for a sufficiently large data sample the distributions of the likelihood-ratio-based test statistics follows a Gaussian distribution. This means that $-2 \log \mathcal{L}(\mu)$ will follow a χ^2 distribution with $n_{dof} = 1$.

$$\chi^2(\mu) = -2 \log \mathcal{L}(\mu) \tag{3.9}$$

The determination of the confidence interval begins by finding all possible global minima (as multiple solutions may exist). At each minimum $\chi^2 = \chi^2_{\min}$. The confidence level for a given value of μ (e.g. μ_0) is computed from the minimum at μ_0 with respect to the nuisance parameters of the model $\chi^2_{\min}(\mu_0)$. The difference in χ^2 between the the global minimum and the minimum at μ_0 will satisfy $\Delta\chi^2 = \chi^2_{\min}(\mu_0) - \chi^2_{\min} \geq 0$. The p-value $P \equiv 1 - \mathrm{CL}$ will be obtained from the χ^2 distribution with one degree of freedom:

$$1 - \mathrm{CL} = \mathrm{Prob}(\Delta\chi^2, n_{\mathrm{dof}} = 1) \text{ where } \mathrm{Prob}(\Delta\chi^2, n_{\mathrm{dof}} = 1) = \frac{1}{\sqrt{2}\Gamma(1/2)} \int_{\Delta\chi^2}^{\infty} e^{-t/2} t^{-1/2} dt \tag{3.10}$$

The 95% confidence intervals, for example, will be computed from the intersection of the $1 - \mathrm{CL} = 0.05$. The method is commonly known as the PROB method [6], [§39.3.2.3].

- **Distribution from pseudo-experiments**. This method is based on obtaining the test statistics distribution by generating pseudo datasets, and can be summarized in the following steps, for a given value (e.g. μ_0) of the parameter of interest:

  1. Calculate $\Delta\chi^2 = \chi^2_{\min}(\mu_0) - \chi^2_{\min}$, where $\chi^2_{\min}(\mu_0)$ is the χ^2 minimum at μ_0 with respect to the nuisance parameters of the model.
  2. Generate a pseudo dataset $\vec{x}_{\text{toy}}$ at $\mu = \mu_0$.
  3. Calculate $\Delta\chi^{2\prime}$ using $\vec{x}_{\text{toy}}$ instead of $\vec{x}_{\text{observed}}$, similar to step (1), where $\Delta\chi^{2\prime}$ is computed from the difference of the χ^2 minima once with μ fixed to μ_0, and once with μ floating.
  4. $1 - \text{CL}$ can be estimated as the fraction of toy results performing worse than the measured data

  $$1 - \text{CL} = N(\Delta\chi^2 < \Delta\chi^{2\prime})/N_{\text{toy}} \tag{3.11}$$

In general, using pseudo data to estimate the confidence limits results in better coverage, though not necessarily a perfect one [5].

References

1. Cowan G (1998) Statistical data analysis. Oxford science publications, Clarendon Press. ISBN 9780198501558. https://books.google.ch/books?id=ff8ZyW0nlJAC
2. Cranmer K (2015) Practical statistics for the LHC. In: Proceedings, 2011 European school of high-energy physics (ESHEP 2011): Cheile Gradistei, Romania, September 7-20, 2011. [247(2015)], pp 267–308. https://doi.org/10.5170/CERN-2015-001.247,10.5170/CERN-2014-003.267. arXiv: 1503.07622 [physics.data-an]
3. Cowan G et al (2011) Asymptotic formulae for likelihood-based tests of new physics. Eur Phys J C71:1554. https://doi.org/10.1140/epjc/s10052-011-1554-0,10.1140/epjc/s10052-013-2501-z. arXiv: 1007.1727 [physics.data-an]
4. Gross E (2017) Practical statistics for high energy physics. In: CERN yellow reports: school proceedings 4.0, p 165. ISSN: 2519-805X. https://e-publishing.cern.ch/index.php/CYRSP/article/view/303
5. GammaCombo User Manual. Tech. rep. (2017) https://gammacombo.github.io/manual.pdf
6. Tanabashi M et al (2018) Review of particle physics. In: Phys. Rev. D 98:030001. https://doi.org/10.1103/PhysRevD.98.030001. https://link.aps.org/doi/10.1103/PhysRevD.98.030001

Part II
Experiment

Chapter 4
The LHC and the ATLAS Experiment

The results in this thesis are based on pp collision data from the ATLAS experiment at the Large Hadron Collider (LHC). In this chapter, a brief introduction of the LHC and the ATLAS experiment is given with an emphasis on the detector systems most relevant for the analysis.

4.1 The Large Hadron Collider

The Large Hadron Collider (LHC) [1] is a circular pp and heavy ion collider at the European Organization for Nuclear Research (CERN). The LHC lies in a tunnel with approximately 27 km of circumference, 100 m beneath the French-Swiss border near Geneva, Switzerland. The construction of the LHC relies on the successful operation of various generations of accelerators [2]. The official LHC proposal took place in Lausanne in March 1984 during the first LHC workshop [3]. Twenty-five years later, in November 2009, the LHC started its successful operation.

4.1.1 The LHC Acceleration Chain

The acceleration process at the LHC is done through various phases that successively increase the energy of the colliding beams via the CERN accelerator complex sketched in Fig. 4.1. The acceleration chain is based on a series of lower energy accelerators that inject their accelerated beams in higher energy accelerators reaching the target beam energy. Protons are collected from a Hydrogen container, where the Hydrogen molecules are submitted to an intense electric field, breaking them

© The Editor(s) (if applicable) and The Author(s), under exclusive license to Springer Nature Switzerland AG 2020
A. Tarek Abouelfadl Mohamed, *Measurement of Higgs Boson Production Cross Sections in the Diphoton Channel*, Springer Theses,
https://doi.org/10.1007/978-3-030-59516-6_4

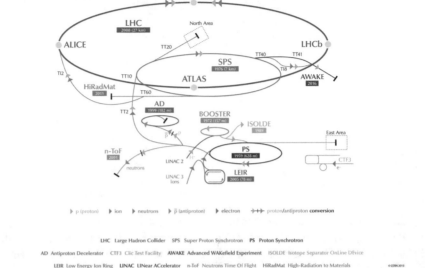

Fig. 4.1 Sketch of the acceleration chain for the LHC. The chain relies on the different accelerators at CERN [7]

into protons and electrons. The protons are first accelerated in the LINAC 2 linear accelerator (which started operation in 1978 and will be replaced by the new LINAC 4 [4] after 2020) reaching an energy of 50 MeV. They are then injected into the proton synchrotron booster (PSB), which accelerates them to an energy of 1.4 GeV, before injecting them into the Proton Synchrotron (PS). The PS is CERN's first synchrotron (it started operation in 1959) that accelerates protons to an energy of 26 GeV. After the PS, protons are injected in the 7 km long Super Proton Synchrotron (SPS) and accelerated to an energy of 450 GeV. The SPS is one of the pillars of the CERN accelerators complex, providing high-energy proton (and anti-proton) beams since 1976 that led to several breakthroughs such as the discovery of the W [5] and Z [6] bosons. At this stage, the protons injected in the LHC where they get accelerated to multi-TeV energies. The LHC is designed to collide proton beams with a center-of-mass energy of $\sqrt{s} = 14$ TeV, this requires accelerating the beam in opposite directions at an energy $\sqrt{s}/2 = 7$ TeV. Due to technical problems the operational center-of-mass energy was lowered to 7 and 8 TeV in the first run of the LHC (2011–2012). The energy was then increased in the second run of the LHC (2015–2018) reaching $\sqrt{s} = 13$ TeV.

The acceleration at the LHC is based on superconducting radio-frequency (RF) cavities for the beam acceleration and superconducting dipole and quadrupole magnets for beam bending and focusing. The LHC uses 8 RF cavities per beam (operating at 4.5 K), providing an accelerating field of 5 MV/m at 400MHz [8]. This results in an acceleration time of around 20 min to reach 7 TeV. The magnets of the LHC are designed to reach a magnetic field up to 8.3 T, providing the necessary curvature matching the tunnel radius at a nominal beam energy of 7 TeV. The magnet system consists of 1232 dipoles to bend the beam and 392 quadrupoles to focus the beam. The magnets are made of superconducting coils of Niobium-Titanium (NbTi) alloy that are used to generate the required high magnetic field and are cooled via super-fluid Helium operating at 1.4 K [9]. Along the LHC ring, different particle physics detectors are built around the beam collision points. There are two big general-purpose experiments: ATLAS [10] and CMS [11]. These two experiments have quite similar layouts, though they are based on different detector technologies. They were designed to be sensitive to a broad spectrum of experimental signatures for new physics searches, in addition to performing Standard Model precision measurements. Also, there are two other major experiments with more focused aims: the ALICE experiment [12], which is mainly used for heavy-ion collisions to study the quark-gluon plasma, and the LHCb experiment [13], detecting bottom and charm quarks mesons produced in the forward direction to study flavor physics and CP-violation.

4.1.2 Beam Structure and Luminosity

The beam in the LHC is not continuous but rather divided in *bunches*. A bunch is a collection of particles that get clumped around the synchronous particle which is the particle that is exactly synchronized with the RF frequency. The LHC beam at full intensity nominally consists of 2808 bunches, with each bunch containing 1.15×10^{11} protons and spaced by 25 ns [14]. A scheme of the LHC bunch structure is shown in Fig. 4.2. The missing (empty) bunches in that scheme provide the necessary time for various procedures such as: the beam dumps, the injection from the SPS to LHC and the rise of the magnetic field of the kicker magnets.

The event rate of a given process with cross section σ_i is given by:

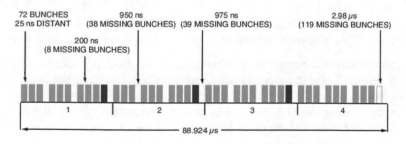

Fig. 4.2 Sketch of the nominal bunch structure of the LHC beam [15]

$$\frac{dN}{dt} = \mathcal{L}_i \sigma_i, \tag{4.1}$$

where the quantity $\mathcal{L}_i$ is known as the instantaneous luminosity. The instantaneous luminosity is a characteristic of the accelerator given by the following formula:

$$\mathcal{L}_i = \frac{N_b^2 k_b f \gamma}{4\pi \sigma_x \sigma_y} F \tag{4.2}$$

where:

- N_b is the number of particles per bunch
- k_b is the number of bunches
- γ is the relativistic factor of the accelerated particles
- f is the revolution frequency of the accelerator. It is 11.2 kHz for the LHC
- σ_x and σ_y are the horizontal and vertical beam size. They are typically around 2.5 μm at the LHC
- F is a geometrical correction factor from the crossing-angle of the two beams at the interaction point (IP). The angle is typically around 150–200 μrad at the LHC

The instantaneous luminosity is measured in units of $cm^{-2}s^{-1}$. The LHC has a design maximum instantaneous luminosity of 10^{34} $cm^{-2}s^{-1}$. The peak instantaneous luminosity of the LHC increased gradually between Run-1 and Run-2 even exceeding the design peak instantaneous luminosity, reaching 2.1×10^{34} $cm^{-2}s^{-1}$ in 2018. From the instantaneous luminosity, one can define the *integrated luminosity* $\mathcal{L} = \int \mathcal{L}_i dt$. The integrated luminosity is related to the number N_i of produced events for a process of a given cross section σ_i by $N_i = \mathcal{L}\sigma_i$. The integrated luminosity is measured in units of [area^{-2}] and typically expressed in inverse femto-barn fb($^{-1}$), where $1 b = 10^{-28}$ m^2.

Another quantity that is related to the luminosity is the *pileup*. Pileup refers to additional low transverse momentum pp inelastic collisions accompanying the hard scattering pp interactions. One can distinguish between two kinds of pileup:

- *In-time* pileup resulting from additional collisions occurring within the same bunch as that of the hard scatter.
- *Out-of-time* pileup occurring in the previous or the following bunch crossings relative to that of the hard scatter.

Pileup represents a challenge to the physics data analysis as it results in additional energy deposits in the detector and hence complicating the identification of the different physics objects. Pileup is quantified via the average number of interactions per bunch crossing, $\langle \mu \rangle$. The average number of interactions per bunch crossing is directly related to the instantaneous luminosity via the relation:

$$\mu = \frac{\mathcal{L}_{bunch} \sigma_{inelastic}}{f}, \tag{4.3}$$

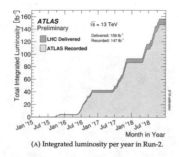
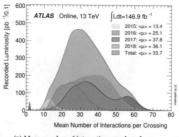

(A) Integrated luminosity per year in Run-2.

(B) Mean number of interactions per bunch crossing per year in Run-2.

Fig. 4.3 **A** Integrated luminosity versus time delivered to ATLAS (green) and recorded by ATLAS (yellow) during stable beams for pp collisions at 13 TeV center-of-mass energy in the LHC Run 2. **B** Mean number of interactions per bunch crossing $\langle \mu \rangle$ per year in Run-2 [16]

where $\mathcal{L}_{bunch}$ is the per-bunch instantaneous luminosity, $\sigma_{inelastic}$ is the pp inelastic cross section ($\sigma_{inelastic} = 80$ mb at $\sqrt{s} = 13$ TeV) and f is the revolution frequency of the LHC.

4.1.3 LHC Run-2 Performance

The dataset used in this thesis comprises the events recorded by the ATLAS detector during the LHC Run-2 at $\sqrt{s} = 13$ TeV. The LHC delivered 156 fb($^{-1}$) of pp collisions, among which 139 fb($^{-1}$) of data were recorded with stable beams and detector conditions providing data that can be used for physics analysis in ATLAS. The integrated luminosity collected in Run-2 as a function of time is shown in Fig. 4.3a. Run-2 witnessed an outstanding performance of the LHC, exceeding the design instantaneous luminosity and providing more integrated luminosity than what was predicted. The increase of the instantaneous luminosity also resulted in a higher number of interactions per bunch crossing across the years with an average number of $\langle \mu \rangle = 33.7$ for all of Run-2, compared to $\langle \mu \rangle \approx 20$ for Run-1. The $\langle \mu \rangle$ distribution for each year in Run-2 is shown in Fig. 4.3b. The resulting higher pileup conditions require additional procedures to mitigate its effects on the calibration and identification of the physics objects, as will be detailed in the next chapters.

4.2 The ATLAS Experiment

ATLAS (A Toroidal LHC ApparatuS) [10] is a general-purpose detector located in an experimental cavern at point 1 at the LHC. ATLAS is the biggest experiment on the LHC ring. It has a cylindrical shape extending for 46 m, with a diameter of 25 m and weighing about 7000 tonnes. A scheme of the ATLAS detector is shown in

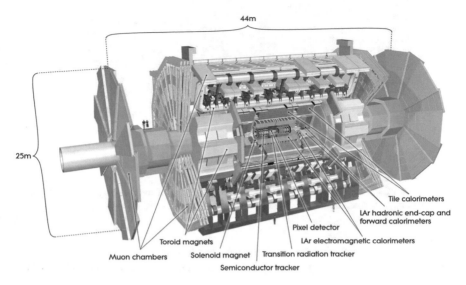

44m

25m

Tile calorimeters

LAr hadronic end-cap and
forward calorimeters

Pixel detector

Toroid magnets   LAr electromagnetic calorimeters

Muon chambers   Solenoid magnet │ Transition radiation tracker

Semiconductor tracker

Fig. 4.4 A general scheme of the ATLAS detector showing its various sub-detectors [10]

Fig. 4.4. The ATLAS detector is made of different sub-detectors, similar to previous generations of detectors, that are sensitive to different groups of particles. The different sub-detectors are built as different coaxial layers around the beam pipe, with the collision point at the center of the detector. The cylindrical shape of the detector is divided into two parts: *barrel* and *endcap*. The sub-detectors start (from the beam) with a tracking detector built from silicon and transition-radiation gas detectors. The tracking detector lies inside a 2 T magnet, which allows the reconstruction of charged particles and the measurement of their momentum through their curved tracks. Going further away from the beam, there are the electromagnetic and hadronic calorimeters measuring the energies of incident particles via their electromagnetic and hadronic showers of secondary particles. The outermost layer is a muon spectrometer, where the momentum of muons, which are the only known charged particles that can cross the previous layers, is measured, thanks to a toroidal magnetic field of 0.5 T (1 T) in the barrel (endcap) that curves their trajectories. The ATLAS detector is forward-backward symmetric with respect to the interaction point. The cylindrical geometry allows nearly 4π coverage for the detector.

ATLAS is a general-purpose detector, meaning that it was designed to search for various signatures of new physics in addition to performing very precise Standard Model measurements, on top of which was the search and discovery of the Higgs boson in 2012 [17]. In this section, we will review the ATLAS detector with particular emphasis on the sub-detectors systems used in this analysis.

Fig. 4.5 The ATLAS
coordinate system. Scheme
based on Ref. [10]

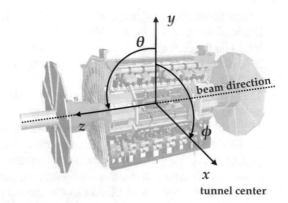

4.2.1 ATLAS Coordinate System

ATLAS uses a right-handed coordinate system, sketched in Fig. 4.5, with the interaction point (IP) defining the origin. The x axis is in the plane defined by the LHC
ring, oriented from the detector center towards the center of the LHC. The x-y plane
defines the transverse plane to the beam axis. The beam direction defines the z-axis of
the Cartesian coordinates. In particle physics, we generally use the azimuthal angle
ϕ and the pseudorapidity η defined as a function of the polar angle θ:

$$\eta = -\log \tan \left(\frac{\theta}{2} \right) \tag{4.4}$$

The change in pseudorapidity $\Delta \eta$ is invariant under boosts along the beam axis. The
rapidity $y = \frac{1}{2} \log \left[(E + p_z)(E - p_z) \right]$ is used when dealing with massive particles,
where E is the energy and p_z is the z-component of the momentum. The angular
distance between two objects in the detector may be determined using the variable
ΔR, which is computed as $\Delta R^2 = \Delta \eta^2 + \Delta \phi^2$ which is also invariant under a boost
along the z axis.

4.2.2 ATLAS Magnets

The ATLAS detector has a hybrid system of four large superconducting magnetic
systems :

- One solenoid magnet is responsible for curving the charged particles trajectories in
  the inner tracker. The solenoid provides an axial magnetic field of 2 T. The solenoid
  design is optimized to keep the material thickness in front of the calorimeter as
  low as possible to reduce the probability of particle interactions. The solenoid
  thickness is about 0.66 radiation lengths [18]. The magnetic field is generated via

a superconducting magnet, made out of a Niobium-Titanium alloy, and cooled by liquid Helium at 4.5 K. The magnetic field is precisely measured using probes based on the Hall effect. This results in an accurate measurement of the magnetic field in the inner detector (accuracy of around 0.01 mT) which is necessary for the tracking.

- Three toroidal magnets, one in the barrel region and two in the endcap regions provide the field in the muon spectrometer. The generated magnetic fields are orthogonal to one another, and together with the inner tracker, they allow for independent measurements of the muon momentum in the innermost and outermost part of the ATLAS detector. The barrel system is constituted of 8 large barrel coils, that are arranged in a star configuration, and of two endcap systems that are also made of 8 coils each. Similar to the solenoid, they use liquid-Helium cooled superconducting magnets based on a Niobium-Titanium alloy. The magnetic field provided by these magnets is 0.5 T for the barrel and 1 T for the endcaps.

4.2.3 The Inner Detector (ID)

The inner detector aims at providing a precise measurement of the parameters of charged particles trajectories, namely an excellent vertex and momentum resolution. It is composed of four separate parts based on different technologies. The first three sub-detectors are based on sensors made of silicon semiconductors, while the last one uses the transition radiation technique. The inner detector extends up to $|\eta| = 2.5$. A global description of the installation and the composition of the inner detector is schemed in Fig. 4.6.

The inner detector encompasses the following four sub-detectors:

The Insertable B-Layer (IBL) The IBL [19] is the innermost component of the inner detector. The IBL was added during the shutdown between Run-1 and Run-2

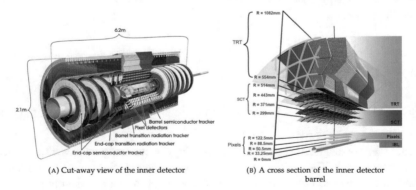

(A) Cut-away view of the inner detector

(B) A cross section of the inner detector barrel

Fig. 4.6 **A** General scheme of the ATLAS inner detector and its components. **B** Drawing showing the sensors and structural elements traversed by a charged track of $p_T = 10$ GeVin the barrel inner detector [10]

with the aim of providing measurements closest to the interaction point. The IBL has particular importance for the identification of b-jets, that relies on precise vertex reconstruction to identify the displaced vertex of the B hadron decays. The IBL was added as the original pixel detector suffered from radiation damage. The IBL has a barrel structure with a radius of 3.2 cm composed of 14 carbon fiber staves with 32 or 16 modules in each stave. The IBL modules include two different kinds of pixel sensors: ATLAS pixel planar sensors and 3D sensors, with a pixel size of 50×250 μm^2 providing a resolution on the longitudinal impact parameter, z_0, of approximately $200 - 80$ μm for $0.4 < p_T < 20$ GeV.

The Pixel detector The pixel detector [20] consists of three coaxial cylinders and three disks perpendicular to the beam. The pixel detector is designed to provide high-resolution track and vertex reconstruction. The detector consists of 1744 pixel sensors. Each of the pixels corresponds to a semiconductor sensor made from silicon with a size of 50×400 μm^2. The complete pixel detector contains approximately 80 million read-out channels, each one corresponding to a pixel. The resolution of the charged particle positions is of 10 μm in $(R - \phi)$ and 115 μm in z.

The Semiconductor Tracker (SCT) The next part of the inner tracker is the Semiconductor Tracker (SCT) [21]. The SCT comprises 61 m^2 of silicon microstrip sensors arranged into four concentric cylinders and nine disks at each end. It is 5.6 m long and extends to 0.7 m in radius at the disks, providing four additional space points for each track. This results in a hit resolution of $\sim$16 μm in the R-ϕ plane and $\sim$580 μm along the z direction.

The Transition Radiation Tracker (TRT) The TRT is the outermost part of the inner tracker. It has different detector technology from that of the Pixel and SCT detectors. The TRT is a transition radiation detector that uses gas ionization for particle detection [22]. The TRT is composed of straw-tubes filled mainly with Xenon. Transition radiation will be emitted at the boundary of the radiator material, which is made of polypropylene and polyethylene fibers. This radiation corresponds to X-ray photons with an energy of $5 - 30$ keV and is strictly proportional to the relativistic factor $\gamma = E/m$ of the incident particle. Therefore, for a given momentum, it will be much higher for electrons than for pions and muons. This difference in response is a crucial input to the discrimination between electrons and pions. The TRT is capable of performing measurements in the R-ϕ coordinates with a precision of 170 μm per straw-tube. This is worse than the one from the SCT and the pixel, but this lack of precision is compensated by the higher number of hits in this sub-detector, as there are 73 parallel planes of straw-tubes in the barrel and 80 planes for each endcap.

4.2.4 The ATLAS Calorimeters

Calorimeters are particle detectors that measure energies of particles. They were initially invented for the study of cosmic-ray phenomena and later developed for use in collider experiments. Calorimeters are blocks of material in which the incident particles are fully absorbed, and their energy converted into a measurable signal. Particles entering the calorimeter initiate a particle shower of secondary particles that depend on the type and energy of the incident particles, as will be shown below.

Typically, calorimeters are segmented transversely with respect to the origin of the particles to provide information about the direction of the particles in addition to the energy deposited. They can also be longitudinally segmented to provide information about the identity of the particle based on the shape of the shower as it develops as well as standalone direction information. Calorimeters can be classified as:

- *sampling* calorimeters, consisting of alternating layers of an absorber which is a dense material used to degrade the energy of the incident particle and convert it to other particles in an active medium that provides the detectable signal through the interactions of the extra particles with the active medium.
- *homogeneous* calorimeters that are built entirely of one type of material that performs both tasks.

In terms of function, calorimeters can be classified into: electromagnetic calorimeters, used to measure mainly electrons and photons through their electromagnetic interactions, and hadronic calorimeters used to measure mainly hadrons through their strong and electromagnetic interactions. The ATLAS experiment includes the following types of calorimeters, sketched in Fig. 4.7:

- An electromagnetic sampling calorimeter [23] covering up to $|\eta| = 3.2$. The electromagnetic calorimeter has a cylindrical shape composed of a barrel section for $|\eta| < 1.475$ (EMB) and two endcaps for $1.375 < |\eta| < 3.2$ (EMEC).

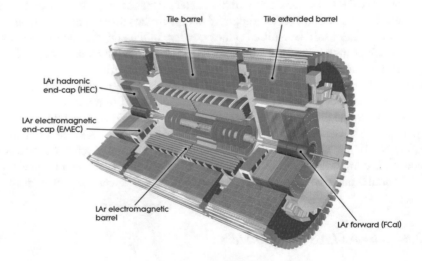

Fig. 4.7 Scheme of the ATLAS liquid argon electromagnetic and tile calorimeter [10]

- A hadronic sampling calorimeter (HCal) composed of a tile calorimeter [24], a Hadronic end-cap (HEC) covering $|\eta| < 3.9$ and forward calorimeters (FCal) extending over the pseudorapidity range $3.1 < |\eta| < 4.9$.

4.2.4.1 The Electromagnetic Calorimeter

The ATLAS electromagnetic calorimeter is the main photon detector in ATLAS, and hence, it is a key part of the measurements presented in this thesis. In this section, the details of the different components of the ATLAS electromagnetic calorimeter are described, after an overview of the mechanism of electromagnetic shower development.

Electromagnetic shower development To understand the electromagnetic shower development, we review the interactions between electromagnetic particles and matter. This is a very diverse branch of physics which depends on the energies of the particles involved in the process. We are only interested in the interactions of high-energy photons and electrons. The average energy lost by electrons in lead (used as an example as it is the same material used as an absorber in the ATLAS electromagnetic calorimeter) and the photon interaction cross section are shown in Fig. 4.8 as a function of their energies. For energies larger than $\sim$10 MeV, the primary source of electron energy loss is bremsstrahlung. Photons, on the other hand, interact mainly by producing electron-positron pairs. For energies above 1 GeV, both these processes become roughly energy independent. At low energies, electrons lose their energy mainly through collisions with the atoms and molecules of the material thus giving

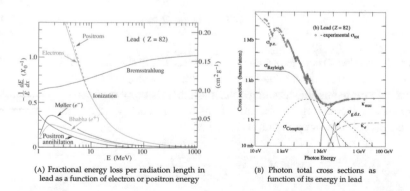

(A) Fractional energy loss per radiation length in lead as a function of electron or positron energy

(B) Photon total cross sections as function of its energy in lead

Fig. 4.8 **A** Fractional energy loss per radiation length in lead as a function of electron or positron energy showing the different contributions. **B** The total cross section of photons as function of photon energy in lead. The different contributions are: $\sigma_{\mathrm{p.e.}}$ is the atomic photoelectric effect (ionization and excitation), $\sigma_{\mathrm{Rayleigh}}$ is the Rayleigh scattering where the atom is neither ionized nor excited, $\sigma_{\mathrm{Compton}}$ is incoherent Compton scattering off an electron, κ_{nuc} is pair production (nuclear field), κ_{e} is pair production (electron field), and $\sigma_{\mathrm{g.d.r}}$ the photonuclear interaction (Giant Dipole Resonance) [25, §33.4]

rise to ionization and thermal excitation, whereas photons lose their energy through Compton scattering and the photoelectric effect. Although the Compton scattering process has a non-negligible cross section up to energies of a few GeVs, in this energy range the primary mechanism for photon energy loss is the conversion of photons to e^+e^- pairs, which can only happen in a medium.

Therefore, electrons and photons of sufficiently high energy (≥ 1 GeV) incident on a block of material produce secondary photons by bremsstrahlung, or secondary electrons and positrons by pair production. These secondary particles, in turn, produce other particles by the same mechanisms, thus giving rise to a cascade (*shower*) of particles with progressively degraded energies. This shower is then used to detect photons and electrons and to measure their energies in the EM calorimeters. Most of the physics of these showers may be described using two scales: a distance scale that is the radiation length X_0 describing the distance at which the incident particle loses $1/e = 0.37$ of its energy, and an energy scale E_c which is the so-called critical energy at which the energy lost by an electron due to bremsstrahlung is equal to the energy lost by ionization. This critical energy also defines the energy scale at which the shower development will stop as for energies below E_c the electron will mainly ionize the medium and will not generate more electromagnetic particles. Using these parameters, the description of the shower development of electrons and photons in matter can be achieved via simple empirical formulas [25]. The measurement of the particle energy is then based on the fact that the energy released in the detector material by the charged particles of the shower, mainly through ionization and excitation, is proportional to the energy of the incident particle.

Energy resolution of an electromagnetic calorimeter The energy resolution of an electromagnetic calorimeter can be described by:

$$\frac{\sigma_E}{E} = \frac{a}{\sqrt{E}} \oplus \frac{b}{E} \oplus c, \tag{4.5}$$

where the symbol $\oplus$ denotes a sum in quadrature. The first term $a/\sqrt{E}$ is the so-called *stochastic* term from the fluctuation of the shower development resulting from purely statistical arguments. In sampling calorimeters, this term is affected by a *sampling* contribution accounting for the ratio between the energy deposited in active and passive layers. The sampling ratio is defined as

$$f_{\text{samp}} = \frac{E_{\text{active}}}{E_{\text{active}} + E_{\text{passive}}}, \tag{4.6}$$

where E_{active} and E_{passive} are the energy deposits in the active and passive layers. Controlling this ratio (during the design phase of calorimeter) can thus improve the resolution as the resolution $\sigma/E \propto \sqrt{1/f_{\text{samp}}}$. In a sampling calorimeter the energy resolution from this term is typically in the range $5 - 20\%/\sqrt{E(\text{GeV})}$.

The second term b/E is known as the *noise* term which arises from the electronic noise of the signal readout chain and pileup noise. The noise term is also affected

Fig. 4.9 Scheme of the ATLAS liquid argon electromagnetic calorimeter [10]

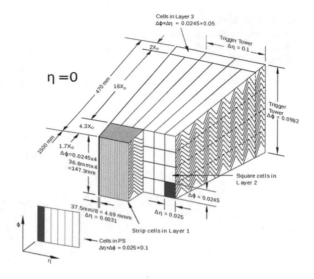

by the sampling fraction f_{samp} as increasing it will result in larger active regions and hence lead to higher signal-to-noise ratio, reducing the noise term resolution. The contribution of this term is in general negligible in the high energy ranges. The last term c is known as the *constant* term, and it arises from instrumental defects that cause variations of the calorimeter response (independent of the particle energy) giving rise to nonuniformities. Therefore, the typical contribution of this term is characteristic of each calorimeter.

4.2.4.2 The ATLAS EM Calorimeter

The main part of the ATLAS EM calorimeter, shown in Fig. 4.9, is a Lead–liquid Argon (LAr) sampling detector with accordion-shaped electrodes and lead absorbers [23]. The showers are mainly generated in the lead layer, which is the denser material, and the LAr layers are primarily used as an active material in which the ionization electrons will drift toward an electrode where the produced signals are measured. The overall target resolution of the ATLAS electromagnetic calorimeter is

$$\frac{\sigma}{E} = \frac{10\%}{\sqrt{E}} \oplus 0.7\%, \tag{4.7}$$

where E is measured in GeV. The noise term is measured from dedicated calibration runs and is typically $0.25/E(\text{GeV})$ [26]. Given the energy dependence of the different terms in the resolution, the most important at high energies is the constant term. The contribution of the stochastic term varies as a function of η, reaching a maximum value of 17%.

Fig. 4.10 Accordion
structure of the barrel part of
the LAr electromagnetic
calorimeter [27]

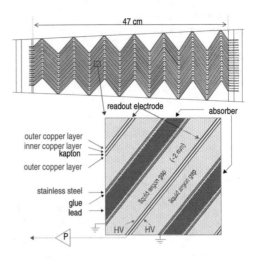

The accordion The ATLAS EM calorimeter has an accordion-shaped geometry,
sketched in Fig. 4.10, consisting of interleaved layers of the LAr gaps, electrodes
and lead absorbers. This design has the advantage that it avoids having readout
systems on the side of the electrodes, which will result in cracks at specific regions
where the readout takes place. Instead, the accordion design allows the readout to be
performed either in the front or the back of the calorimeter providing full coverage
in ϕ. The lead absorbers are reinforced with two stainless-steel sheets that provide
mechanical strength. The lead plates in the barrel have thickness of 1.53 mm for
$|\eta| < 0.8$ and 1.13 mm for $|\eta| > 0.8$. This change in thickness limits the decrease
of the sampling fraction as $|\eta|$ increases [10].

The barrel region of the electromagnetic calorimeter is divided into two half-
barrels centered around the z-axis covering the regions $0 < \eta < 1.475$ and $-1.475 <
\eta < 0$. A single half-barrel is composed of 1024 accordion-shaped absorbers,
with readout electrodes interleaved between the absorbers with varying thickness
expressed in terms of radiation length X_0 that aim to contain the total energy of
the electromagnetic shower. Similarly, the endcap region, extending in the region
$1.375 < |\eta| < 3.2$ consists of 768 lead absorbers interleaved with readout modules.
An ATLAS EM calorimeter module (sketched in Fig. 4.9) in the central region
($|\eta| < 2.5$) has three layers in depth : front, middle and back as viewed from the
interaction point. This longitudinal segmentation allows the precise measurement of
the EM shower longitudinal development. The layers have different properties as
follow:

- Front layer (L1), also known as the *strips* layer, has a thickness of about 4.4 X_0
  at $\eta = 0$. It has very fine segmentation along η, with varying width depending on
  the η position in the barrel. The width $\Delta\eta$ varies between $\approx$0.025/8 for $|\eta| < 1.4$
  and 0.025 for $1.4 < |\eta| < 1.475$. In the endcap the width varies between 0.025/8
  for $|\eta| < 2.4$ and up to 0.025 for $2.4 < |\eta| < 2.5$ as detailed in Ref. [10]. The

fine segmentation in most of the barrel and the end-caps provides a means to discriminate between prompt photons and photons from $\pi^0 \to \gamma\gamma$ decays that can have a similar signature when the π^0 is boosted (small opening angle between the decay photons).

- Middle layer (L2) where the bulk of the energy deposit takes place with thickness up to 22 X_0. The cells in the middle layer are of size $\Delta\eta \times \Delta\phi = 0.025 \times 0.025$ in the barrel an $\Delta\eta \times \Delta\phi = 0.1 \times 0.1$ in the endcaps.
- Back layer (L3) is a thin layer of about 2 X_0 that is used to measure the energy leakage to the hadronic calorimeter. The cells in the back layer are of size $\Delta\eta \times \Delta\phi = 0.05 \times 0.05$ in the barrel and the endcaps.

The presampler In addition to these layers, the electromagnetic calorimeter is complemented with an additional module known as the *presampler* (PS) [28]. The presampler is a thin layer (11 mm in the barrel and 5 mm in the endcap) of LAr layers with no lead absorbers, covering up to $|\eta| < 1.8$. The presampler has the purpose of measuring the energy lost upstream of the calorimeter due to interaction with material and hence providing corrections to the energies measured in the accordion. The presampler layer is made of 64 identical azimuthal sectors (32 per half-barrel). Each sector is 3.1 m long and 0.28 m wide thus covering the half-barrel length [10]. The presampler in the barrel is composed of eight modules of different sizes with increasing length to obtain constant η−granularity of $\Delta\eta = 0.2$ for each module. Each PS module is divided into eight cells in η and two cells in ϕ, leading to a total of 16 cells per module with granularity $\Delta\eta \times \Delta\phi = 0.025 \times 0.1$. In the endcaps, 32 identical modules cover the region $1.5 < |\eta| < 1.8$. The cell granularity in the PS endcaps is $\Delta\eta \times \Delta\phi = 0.025 \times 0.1$, similar to the barrel. The presampler modules are made of an interleaved cathode and anode electrodes glued between glass-fiber composite plates. A negative high voltage potential of 2 kV is applied to the outer layers of the anodes, and the signal is read out through capacitive coupling to the central layer at ground potential as shown in Fig. 4.11.

Fig. 4.11 A scheme of presampler modules in the endcap [23]

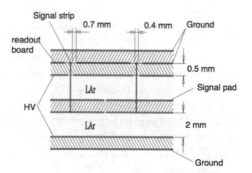

4.2.4.3 The Hadronic Calorimeter

In addition to the EM calorimeter, a hadronic calorimeter is used to measure jet energies. Jets are sprays of particles that originate as free (colored) partons leave a hard scattering event due to quark confinement as detailed in Sect. 1.2.2. The measurement of the jet energy requires a measurement of the energy of the hadrons inside the jet, and this requires specific calorimeters as the main interaction of hadrons with the detector is the strong interaction, and the nuclear interaction length of the LAr EM calorimeter is too small to absorb and measure with sufficient precision the hadron energy.

Hadronic shower development Similar to electromagnetic showers, hadronic showers are started by streams of hadrons resulting from hadronic interactions with the detector. The hadronic interaction is characterized by the *interaction length*, λ, giving the mean free path between interactions required to reduce the numbers of relativistic charged particles by the factor $1/e = 0.37$. The interaction length varies with the type of material used as $\lambda \approx 35A^{1/3}$ g cm^{-2} where A is the mass number [26]. The produced secondary hadrons carry a fraction of the initial hadron momentum (GeV scale). A significant part of the primary energy is then consumed through nuclear processes such as excitation, nucleon evaporation, and spallation, resulting in particles with characteristic nuclear energies at the MeV scale.

The ATLAS hadronic calorimeter Different hadronic calorimeters have been included in ATLAS: a barrel tile calorimeter up to $|\eta| = 1$ and an extended barrel with the same technology up to $|\eta| = 1.7$, a LAr-copper calorimeter for the endcap, and a forward calorimeter. In the range $|\eta| < 1.6$, the ATLAS hadronic calorimeter is an iron-scintillating tiles calorimeter. For rapidity larger than 1.6, the hadronic calorimeter is a LAr calorimeter, mainly because of the intrinsic radiation hardness of this technology.

The Tile Calorimeter [24] consists of a sampling calorimeter as the EM calorimeter; however, it is made out of steel absorbers to create the particle showers, and of plastic scintillator tiles as the active medium to measure them as shown in Fig. 4.12. The energy measured in this calorimeter is the ultraviolet scintillation light that is produced when a charged particle crosses the active medium. For each tile, this scintillation light is collected by a wavelength-shifting optical fiber and is converted into visible light in this fiber. The output of the fiber is connected to a photo-multiplier where the signal is measured. The tile calorimeter is segmented into three layers in depth, and the total depth of this system is about 7λ. This ensures a negligible amount of hadronic leakage to the muon spectrometer, especially given that it should be added to the depth of the LAr calorimeter and the services, giving a total of more than 10 λ in front of the muon spectrometer everywhere in the detector.

Similar to the EM calorimeter, there exists a hadronic endcap calorimeter (HEC) [10] which is also a sampling calorimeter based on a LAr technology, except that its absorbers are made of copper instead of lead. Each HEC is built as two separate wheels, the first of which is composed of a series of 25 mm thick flat copper layer and 8.5 mm wide LAr gaps. In the second wheel, the main change is the thickness

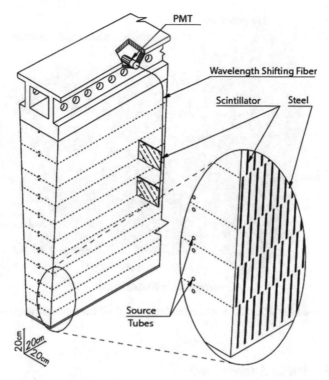

Fig. 4.12 Scheme of the ATLAS tile calorimeter [24]

of the copper layers, which is 50 mm. The HEC is approximatively 10 interaction lengths deep.

4.2.4.4 Forward Calorimeters

In addition to the hadronic and electromagnetic calorimeters, a forward calorimeter extends the acceptance of the calorimeter up to $|\eta| = 4.9$. The forward calorimeter (FCal) [10] is a sampling calorimeter based on LAr as an active medium. The detector is segmented into three layers in depth. The first layer uses copper absorber layers and targets the measurement of forward electromagnetic particles. The two other layers that are further downstream use tungsten absorbers. The size of the LAr gap varies between 0.27 mm in the first sampling to 0.51 mm in the third. The thickness of the absorbers in the FCal is optimized to achieve high absorption, approximatively 10 interaction lengths deep.

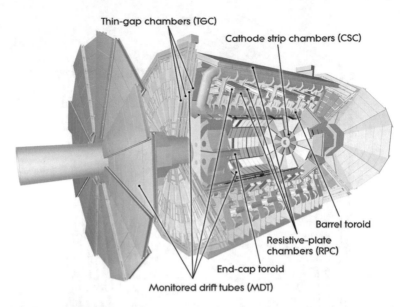

Fig. 4.13 Scheme of the ATLAS muon spectrometer with the different detector technologies indicated [10]

4.2.5　The Muon Spectrometer

The outermost part of the ATLAS detector is the muon spectrometer (MS), which is composed of several gaseous chambers, whose technology varies as a function of their usage and of their position as shown in Fig. 4.13.

Up to $|\eta| = 2$, the momentum measurement is done by monitored drift tubes (MDT). One MDT consists of straw tubes in which a gas is ionized by the incident muon, and the ionization electrons are then collected at a wire in the center of the tube. At every ϕ there are at least 3 layers of chambers of this kind. For $2 < |\eta| < 2.7$ the innermost layer changes and uses cathode strip chambers (CSC), which allow for a higher segmentation of the chamber. The CSC are multi-wire proportional chambers, with the cathode segmented into strips, and the direction of the strip is perpendicular to one of the wires. This allows for two independent measurements of the muon: one for the ionization electrons that are collected at the wire, the other one from the induced signal collected at the strips. This also gives the two coordinates of the muon, which in the MDTs comes from the trigger chambers. The trigger chambers extends up to $|\eta| = 2.4$. For $|\eta| < 1.05$ the trigger is done by Resistive Plate Chambers (RPCs), in which two parallel plates are separated by a thin layer of gas that the crossing muon will ionize. This ionization signal will drift toward one of the two metallic plates at which it will be measured. Beyond $|\eta| = 1.05$ the trigger uses thin gap chambers (TGCs), in which a layer of anode wires at high voltage lies between two parallel plates that are at ground.

4.2.6 Forward Detectors

In addition to the main ATLAS subdetectors detailed before, three small detectors are built in the very forward region ($|\eta| > 5$). The forward detectors include:

- *LUCID*: Luminosity measurement using Cherenkov Integrating Detector [10]. The LUCID detector is composed of two modules located at ± 17 m from the IP providing coverage in the region $5.5 < |\eta| < 5.9$. Each LUCID module is composed of 1.5 m long aluminum tubes surrounding the beam pipes and filled with C_4F_{10} gas giving a Cherenkov threshold of 10 MeV for electrons and 2.8 GeV for pions. LUCID has a fast timing response system (few ns) providing counting measurements of the individual bunches estimating the luminosity through the inelastic cross sections measurements. This cross section is determined by so-called Van Der Meer scans [29].
- *ZDC*: Zero degree calorimeter [10]. The ZDC detector is located at ± 140 m from the IP providing coverage for $|\eta| > 8.3$ neutral particles. The ZDC detector includes four modules (one EM and three hadronic) with each module made of tungsten plates.
- *ALFA*: Absolute Luminosity For ATLAS [10]. The ALFA detector is located at ± 240 m from the IP. It is made of two Roman pot stations measuring the elastic scattering cross section at small angles.
- *AFP*: ATLAS Forward Proton (AFP) [30]. The AFP detector was installed in 2017 to measure diffractive protons leaving under very small angles $\approx 100\,\mu$rad. In these processes, one or both protons remain intact. Such processes are associated with elastic and diffractive scattering.

4.2.7 The Trigger and Data Acquisition System

The LHC has a design instantaneous luminosity of 10^{34} cm^{-2}s^{-1} which amounts to a frequency of collisions of approximately 1.7 GHz given the 40 MHz bunch crossing rate with 25 interactions per bunch crossing. Therefore, it is technically impossible to store the huge amounts of data, and hence, a trigger system is used. The trigger system in ATLAS aims at reducing the 40 MHz rate to 1 kHz [31], which is much more manageable. The trigger system in ATLAS is composed of the following subsystems:

- A hardware Level-1 trigger built from fast electronics. The level-1 trigger selects high transverse momentum physics objects from trigger information coming from the different detectors such as muons (from the muon spectrometer), electrons and photons (from the calorimeter), jets and missing transverse energy. The level-1 trigger defines a region of interest (RoI) which are the regions in the detector (η, ϕ) where its selection process has identified interesting features. The information in the ROIs is then passed to higher-level triggers, as shown in Fig. 4.14.
- Software high-level trigger (HLT) seeded by the RoI information provided by the Level-1 trigger. The HLT selections use the full granularity and precision, all the

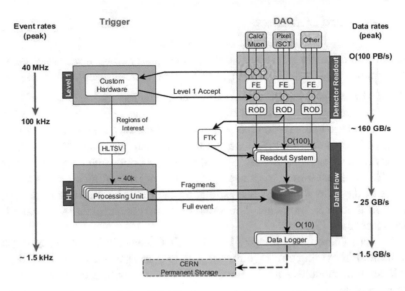

Fig. 4.14 Block diagram of the ATLAS trigger and data acquisition system (DAQ) during Run-2 [32]

available detector data within the RoI. The HLT triggers reduce the event rate to approximately 1.0 kHz.

The output of the trigger system is then passed to the data acquisition system in ATLAS via the scheme shown in Fig. 4.14. Events that are accepted by the hardware Level-1 trigger are transferred to the HLT via an HLT farm supervisor node (implemented in Run-2) that includes assembly of Regions of Interest. Event data from the detector front-end electronics systems are simultaneously sent to the Readout System (ROS) via optical links from the Readout Drivers (RODs) in response to a Level-1 trigger accept signal. These data are then buffered in the ROS and made available for sampling by algorithms running in the HLT. Once the HLT accepts an event, it is sent to permanent storage via the Data Logger [32].

4.3   Reconstruction of Physics Objects

In this section, we will review the reconstruction of the different physics objects used in this thesis. In addition, the identification and isolation algorithms for photons will be summarized as they represent the main final state particles for $pp \rightarrow H \rightarrow \gamma\gamma$ cross section measurement presented in Chap. 7.

Fig. 4.15 Shapes of the LAr
calorimeter current pulse in
the detector and of the signal
output from the shaper chip.
The dots indicate an ideal
position of samples
separated by 25 ns [33]

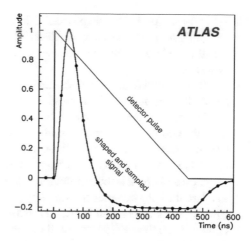

4.3.1 Electron and Photon Reconstruction

As detailed in Sect. 4.2.4.1, electrons and photons develop EM showers as they
interact with the lead absorbers in the LAr calorimeter. These EM showers ionize
the LAr in the gaps between the absorbers, resulting in electrons that drift via an
applied HV (nominally at 2 kV). The drifting electrons from the ionization processes
induce an electrical signal on the copper electrodes that is proportional to the energy
deposited. The signal is then read-out by the Front End Boards (FEB). The resulting
signal has a triangular shape with an amplitude that is proportional to the energy
deposited by the incident particles. The signal is then pre-amplified and shaped via
a $CR - (RC)^2$ multi-gain bipolar filter, as shown in Fig. 4.15. The resulting signal
shape is amplified with three linear gains: low (LG), medium (MG), and high (HG).
These gains are optimized to accommodate a broad dynamic range, and the signal
shape is chosen such that it reduces the total noise due to electronics and inelastic
pp collisions coming from previous bunch crossings (i.e. out-of-time pileup).

The resulting signals are then sampled with a 40 MHz clock frequency and stored
temporarily on a switched capacitor array until the Level-1 trigger decision is taken.
Once the Level-1 trigger decision is received, the sample corresponding to the maximum amplitude of the physical pulse stored in MG is first digitized by a 12-bit
analog-to-digital converter (ADC). Based on this sample, a hardware gain selection
is used to choose the most suitable gain. The samples with the chosen gain are digitized and transferred to the read-out drivers via optical fibers [34]. The resulting
signal amplitude is then converted to measured cell energy in MeV via:

$$E_{\text{cell}} = F_{\mu A \to \text{MeV}} \times F_{\text{DAC} \to \mu A} \times \frac{1}{\frac{M_{\text{phys}}}{M_{\text{cali}}}} \times G \times \sum_{j=1}^{N_{\text{samples}}} a_j (s_j - p), \qquad (4.8)$$

where

- s_j are the samples of the shaped ionization signal digitized in the selected electronic gain, measured in ADC counts in time slices spaced by 25 ns ($N_{samples} = 4$).
- p is the read-out electronic pedestal, measured for each gain in dedicated calibration runs.
- a_j are the optimal filtering coefficients (OFC) derived from the predicted shape of the ionization pulse and the noise autocorrelation, accounting for both the electronic and the pileup components [35].
- G is the cell gain, computed by injecting a known calibration signal and reconstructing the corresponding cell response in dedicated calibration runs.
- $\frac{M\text{phys}}{M\text{cali}}$ is a correction factor to the gain G where Mphys is the ionization pulse response and Mcali is the calibration pulse corresponding to the same input current, to adapt it to physics-induced signals.
- $F_{DAC\to\mu A}$ is a conversion factor converting the digital-to-analog converter (DAC) counts set on the calibration board to a current in μA.
- $F_{\mu A\to}$ MeVconverts the ionization current to the total deposited energy at the EM scale and is determined from test-beam studies [36].

The measured cell energies are clustered, providing a seed for different clustering algorithm as detailed below.

Sliding-Window Algorithm

The *sliding-window* clustering algorithm [37] can be summarized in the following sequence:

1. The calorimeter is divided into a grid of $N_\eta \times N_\phi = 200 \times 256$ elements known as *towers* of size $S_{tower} = \Delta\eta \times \Delta\phi = 0.025 \times 0.025$. Each tower is built by summing all the cell energies of the longitudinal layers.
2. A scan is performed using a fixed-size window of 3×5 towers. A seed is then selected with the window energy is above 2.5 GeV.
3. The cluster is then built summing the energies of all cells within 3×7 (5×5) of width $\Delta\eta \times \Delta\phi = 0.025 \times 0.025$ in the barrel (endcap) around a barycentre that is layer-dependent.

Using this algorithm, the efficiency to reconstruct EM-cluster candidates in the electromagnetic calorimeter associated with produced electrons is given by the number of reconstructed EM calorimeter clusters $N_{cluster}$ divided by the number of produced electrons N_{all}. This efficiency is evaluated entirely from simulation as the reconstructed cluster is associated with a "true" electron produced at generator-level. The reconstruction efficiency of this clustering algorithm varies as a function of $|\eta|$ and E_T, ranging from 65% at $E_T = 4.5$ GeV, to 96% at $E_T = 7$ GeV, to more than 99% above $E_T = 15$ GeV, as can be seen in Fig. 4.16a (red triangles). The number of reconstructed electron candidates relative to the number of EM-cluster candidates $N_{cluster}$ gives the reconstruction efficiency which is above 98% for $E_T > 15$ GeVas shown in Fig. 4.16b.

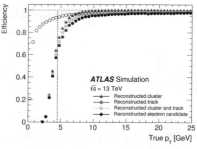
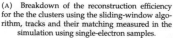
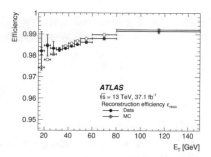

(A) Breakdown of the reconstruction efficiency for the the clusters using the sliding-window algorithm, tracks and their matching measured in the simulation using single-electron samples.

(B) Reconstruction efficiency relating the number of reconstructed electrons to the number of EM-clusters as estimated from $Z \rightarrow ee$ for data and simulation.

Fig. 4.16 A The total reconstruction efficiency for simulated electrons in a single-electron sample is shown as a function of the true (generator) transverse momentum p_T for each step of the electron-candidate formation: seed-cluster reconstruction (red triangles), track reconstruction using the Global Track Fitter (blue open circles), both of these steps together but instead using GSF tracking (yellow squares), and the final reconstructed electron candidate, which includes the track-to-cluster matching (black closed circles). As the cluster reconstruction requires uncalibrated cluster seeds with $E_T > 2.5$ GeV, the total reconstruction efficiency is less than 60% below 4.5 GeV (dashed line). **B** The reconstruction efficiency relative to reconstructed clusters, $\varepsilon_{\mathrm{reco}}$, as a function of electron transverse energy E_T for $Z \rightarrow ee$ events, comparing data (closed circles) with simulation (open circles). The inner uncertainties are statistical while the total uncertainties include both the statistical and systematic components [38]

Fig. 4.17 Diagram of an example supercluster showing a seed electron cluster and a satellite photon cluster [39]

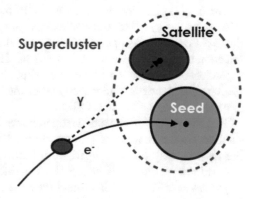

Dynamical Topological Cell Clustering Algorithm

Since 2017, a new clustering algorithm based on topological clusters was implemented in ATLAS [39]. Historically, the fixed-sized window was used since the calibration of variable-sized clusters was not feasible as it requires multi-variate techniques. The main advantage of using topological clusters is their ability to recover low-energy deposits from bremsstrahlung photons and associate them to the electron cluster, forming what is known as *supercluster*. A scheme of this procedure is shown in Fig. 4.17.

Unlike the fixed-size window algorithm, the topological clustering algorithm relies on cell energy significance defined as:

$$\varsigma_{cell}^{EM} = \left| \frac{E_{cell}^{EM}}{\sigma_{noise,cell}^{EM}} \right|, \tag{4.9}$$

where $\left| E_{cell}^{EM} \right|$ is the absolute cell energy at the EM energy scale and $\sigma_{noise,cell}^{EM}$ is the expected cell noise, including both electronic and pileup noise. The clustering algorithm follows:

1. A *proto-cluster* seeded from calorimeter cells with $\varsigma_{cell}^{EM} \geq 4$ is formed and used as initial seed.
2. All immediate neighboring cells with $\varsigma_{cell}^{EM} \geq 2$ around the proto-cluster are added.
3. All cells that are immediate neighbors of those added in steps 1 and 2 are added, regardless of the ς_{cell}^{EM} value.

The resulting clusters are commonly known as "4-2-0" topo-clusters. This algorithm includes also cells from the hadronic calorimeter. Therefore, a selection on the EM fraction is performed as follows:

$$f_{EM} = \frac{E_{L1} + E_{L2} + E_{L3}}{E_{Cluster}}, \tag{4.10}$$

where $E_{L1,L2,L3}$ are the energy deposits in the first, second and third layers. Only clusters with $f_{EM} > 0.5$ and EM energy greater than 400 MeV are considered. The threshold on f_{EM} was optimized using simulated samples in order to achieve large rejection of pileup as shown in Fig. 4.18a. The reconstruction efficiency for using superclusters is shown in Fig. 4.18b using simulated single electron samples. The resulting efficiency is higher than that of the sliding-window algorithm for low E_T as expected (detailed in Fig. 4.20).

From the EM topoclusters, a supercluster is built from a seed topo-cluster after satellite cluster candidates around the seed candidate are resolved. These satellite clusters may have emerged from bremsstrahlung radiation or other material interactions. A cluster is accepted as a satellite if it falls within a window of $\Delta\eta \times \Delta\phi = 0.075 \times 0.125$ around the seed cluster barycentre. An identified satellite cluster is then vetoed from future usage.

Track to Cluster Matching

The reconstruction of electrons and photons relies as well on the tracking information. Depending on the number of reconstructed tracks and on the matching of those tracks to the EM clusters, one can identify electrons and unconverted and converted photons. The matching procedure for an electron candidate is sketched in Fig. 4.19.

The reconstruction of tracks is based on the standard track pattern reconstruction algorithm [41] that is first performed everywhere in the inner detector. The track

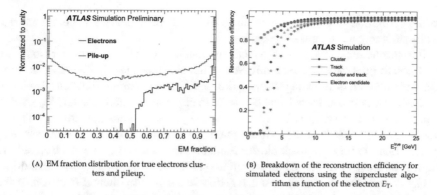

(A) EM fraction distribution for true electrons clusters and pileup.

(B) Breakdown of the reconstruction efficiency for simulated electrons using the supercluster algorithm as function of the electron E_T.

Fig. 4.18 **A** Distributions of f_{EM} for simulated true electron (detailed in Sect. 5.2) clusters (black) and pile-up (red) with $\langle \mu \rangle \leq 30$. **B** The cluster, track, cluster and track, and electron reconstruction efficiencies as a function of the generated electron E_T [40]

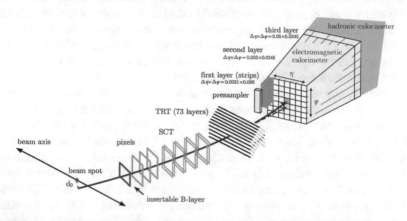

Fig. 4.19 A schematic illustration of the path of an electron through the detector (passive material not shown for clarity). The red trajectory shows the hypothetical path of an electron, which first traverses the tracking system (pixel detectors, then silicon-strip detectors and lastly the TRT) and then enters the electromagnetic calorimeter. The dashed red trajectory indicates the path of a photon produced by the interaction of the electron with the material in the tracking system [38]

reconstruction relies on a silicon track seed defined as a set of silicon detector hits used to start a track. The track is then identified after accounting for energy loss due to material interaction using a Kalman filter [42], allowing for up to 30% energy loss at each material intersection. Track candidates are then fitted with the global χ^2 fitter [43], allowing for additional energy loss in cases that the standard track fit fails. This track fitter was upgraded in 2012 to a Gaussian Sum Filter (GSF) [44] that re-fits all the loosely matched tracks. The re-fitted tracks are then used to compute the final matching with the seed cluster and to compute the electron four-momentum. The GSF algorithm is of most importance for low-p_T electrons where the contribution to the resolution of the four-momentum of the track parameter is dominant, and bremsstrahlung effects are stronger.

The reconstructed tracks are then loosely matched to an EM calorimeter cluster by extrapolating the track to the second layer of the calorimeter using either the measured track momentum or rescaling the magnitude of the momentum to match the cluster energy. A track is considered matched if, with either momentum magnitude, $|\Delta\eta| <$ 0.05 and $-0.10 < \text{sign(track)} \cdot (\phi_{\text{track}} - \phi_{\text{clus}}) < 0.05$, where sign(track) refers to the sign of the charge of the particle making the track. The momentum rescaling is performed to improve track-cluster matching in cases where the electron suffers significant energy loss due to bremsstrahlung radiation in the tracker. In case multiple tracks are matched to the same cluster, they are sorted by their ΔR match to the cluster in the second layer of the calorimeter, and the best one is chosen. Using this information, one can classify the physics objects as follows:

- **Electrons** for fixed-size clusters within $|\eta| < 2.47$, matched to a well-reconstructed ID track originating from a vertex found in the beam interaction region. Similarly, for superclusters, an electron supercluster seed is required to have a minimum E_T of 1 GeV and must be matched to a track with at least four hits in the silicon tracking detector. The "best-matched" tracks for satellite clusters are required to be the best-matched track for the seed cluster.
- **Converted photons** for clusters matched to conversion vertices with tracks loosely matched to the clusters. Both tracks with silicon hits (denoted Si tracks) and tracks reconstructed only in the TRT (denoted TRT tracks) are used for the conversion reconstruction. Two-track conversion vertices are reconstructed from two opposite-charged tracks with an invariant mass of zero ($m_\gamma = 0$). Single-track vertices are tracks without hits in the innermost sensitive layers. Conversion vertices made of Si tracks are denoted "Si conversions", and those made of TRT tracks "TRT conversions". For photon superclusters with conversion vertices made up only of tracks containing silicon hits, a cluster is added as a satellite if either its best-matched (electron) track is one track of the conversion vertex associated to the seed cluster, or if it has the same matched conversion vertex as the seed cluster. The determination of the satellite clusters, in this case, relies on tracking information to discriminate distant radiative photons or conversion electrons from pileup noise or other unrelated clusters.
- **Unconverted photons** for fixed-size clusters without matched tracks. For superclusters, an additional requirement that the cluster must have E_T greater than 1.5 GeV to qualify as supercluster seed is applied.

The performance of the supercluster algorithm is compared with that of the fixed-size clusters by comparing their the width (resolution) of the energy response, using the *effective interquartile range* (IQE), defined as:

$$\text{IQE} = \frac{Q_3 - Q_1}{1.349}, \tag{4.11}$$

where Q_1 and Q_3 are the first and third quartiles of the distribution of $E_{\text{calib}}/E_{\text{true}}$, where E_{calib} is the calibrated energy response as will be detailed in Chap. 5 and E_{true} is the generator-level particle energy. The normalization factor is chosen such that the IQE of a Gaussian distribution would equal its standard deviation. The IQE of the energy response distributions as a function of the particle energy is shown in Fig. 4.20 for electrons, unconverted and converted photons. The IQE is chosen to characterize the energy resolution in single particles since it factors the asymmetric behavior in the tails of the energy response more properly with respect to a Gaussian fit around the peak of the response [39].

4.3.1.1 Electron and Photon Identification

Before the electron and photon candidates can be used in the different analyses, further quality criteria, referred to as *identification*, are defined to select a pure sample of prompt electrons and photons. Prompt photons and electrons are those coming from the hard scattering vertex and not the result of hadronic activity. The identification criteria require that the longitudinal and transverse shower profiles of the candidates are consistent with those expected for EM showers induced by such particles. For this purpose, various discriminant variables based on the different shower shape parameters, shown in Table 4.1, are used.

Photons Different calorimeter discriminating variables are used, exploiting the information in the different layers of the calorimeter, using a cut-based approach, sketched in Fig. 4.21. For example, the variables using the EM strip layer play a particularly important role in rejecting π^0 decays into two highly collimated photons. Three working points are chosen for the photon identification: *loose*, *medium* and *tight*. The loose working point is typically used for the single and diphoton triggers, and it uses the R_{had}, R_{had_1}, R_η, and w_{η_2} shower shape variables. The medium working point includes all the selection used for the loose working point in addition to a loose cut on E_{ratio} and is used as well for triggering (mainly in high pileup conditions). The tight selection is the primary photon identification selection used in offline analyses. It exploits the full granularity of the calorimeter, including the fine segmentation of the first sampling layer. The tight identification selection is optimized using a multivariate algorithm, and is performed separately for converted and unconverted photons (loose and medium identification are the same for converted and unconverted). The main difference in the shower shapes of converted photons and unconverted photons is due to the opening angle of the e^+e^- conversion pair, which is amplified by the magnetic field, and from the additional interaction of the conversion pair with the material upstream of the calorimeters.

The efficiency of the photon identification is measured in data and in simulated samples using three different processes: photons from radiative Z-boson decays, $Z \to ll\gamma$, a matrix method based on inclusive-photon production, and using electrons from $Z \to ee$ decays with their shower shapes modified to resemble photons [40]. A combination of the three methods is used to provide an estimate of the photon identification efficiency in data and simulation, and the difference is used as a cor-

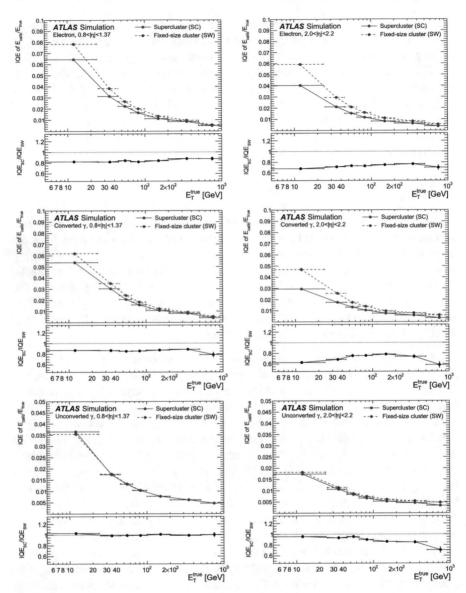

Fig. 4.20 Calibrated energy response resolution, expressed in terms of IQE, for simulated electrons (top), converted photons (middle), and unconverted photons (bottom) with no pileup. The plots on the left are for the central calorimeter ($0.8 < |\eta| < 1.37$), while the plots on the right are for the endcaps. The response for fixed-size clusters based on the sliding window method is shown in dashed red, while the supercluster-based one is shown in full blue [40]

Table 4.1 Discriminating variables used for electron and photon identification. The usage column indicates if the variables are used for the identification of electrons, photons, or both. For the variables in the EM strip layer (shown in Fig. 4.7), if the cluster has more than one strip in the ϕ direction for a given η, then the two most central strips with that η are merged, and all references below to strips refer to these potentially merged strips [45, 46]

Category	Description	Name	Usage
Hadronic leakage	Ratio of E_T in the first sampling of the hadronic calorimeter to the E_T of the EM cluster (used over the ranges $|\eta| < 0.8$ and $|\eta| > 1.37$)	R_{had_1}	e/γ
	Ratio of the E_T in the hadronic calorimeter to E_T of the EM cluster (used over the range $0.8 < |\eta| < 1.37$)	R_{had}	e/γ
EM third layer	Ratio of the energy in the third layer to the total energy in the EM calorimeter	f_3	e
EM middle layer	Ratio between the sum of the energies of the cells contained in a 3×7 $\eta \times \phi$ rectangle (measured in cell units) and the sum of the cell energies in a 7×7 rectangle, both centered around the most energetic cell	R_η	e/γ
	Lateral shower width, $\sqrt{(\Sigma E_i \eta_i^2)/(\Sigma E_i) - ((\Sigma E_i \eta_i)/(\Sigma E_i))^2}$, where E_i is the energy and η_i is the pseudorapidity of cell i and the sum is calculated within a window of 3×5 cells	$w_{\eta 2}$	e/γ
	Ratio between the sum of the energies of the cells contained in a 3×3 $\eta \times \phi$ rectangle (measured in cell units) and the sum of the cell energies in a 3×7 rectangle, both centered around the most energetic cell	R_ϕ	e/γ
EM strip layer	Lateral shower width, $\sqrt{(\Sigma E_i (i - i_{max})^2)/(\Sigma E_i)}$, where i runs over all strips in a window of 3 strips around the highest-energy strip, with index i_{max}	$w_{s\,3}$	γ
	Total lateral shower width, $\sqrt{(\Sigma E_i (i - i_{max})^2)/(\Sigma E_i)}$, where i runs over all strips in a window of $\Delta\eta \approx 0.0625$ and i_{max} is the index of the highest-energy strip	$w_{s\,tot}$	e/γ
	Fraction of energy outside core of three central strips but within seven strips	f_{side}	γ
	Difference between the energy of the strip associated with the the second maximum in the strip layer and the energy reconstructed in the strip with the minimal value found between the first and second maxima	ΔE_s	γ

(continued)

Table 4.1 (continued)

Category	Description	Name	Usage
	Ratio of the energy difference between the maximum energy deposit and the energy deposit in the second maximum in the cluster to the sum of these energies	E_{ratio}	e/γ
	Ratio of the energy measured in the first sampling of the electromagnetic calorimeter to the total energy of the EM cluster	f_1	e/γ
Track conditions	Number of hits in the innermost pixel layer	$n_{\text{innermost}}$	e
	Number of hits in the pixel detector	n_{Pixel}	e
	Total number of hits in the pixel and SCT detectors	n_{Si}	e
	Transverse impact parameter relative to the beam-line	d_0	e
	Significance of transverse impact parameter defined as the ratio of d_0 to its uncertainty	$|d_0/\sigma(d_0)|$	e
	Momentum lost by the track between the perigee and the last measurement point divided by the momentum at perigee	$\Delta p/p$	e
	Likelihood probability based on transition radiation in the TRT	eProbabilityHT	e
Track–cluster matching	$\Delta\eta$ between the cluster position in the first layer of the EM calorimeter and the extrapolated track	$\Delta\eta_1$	e
	$\Delta\phi$ between the cluster position in the second layer of the EM calorimeter and the momentum-rescaled track, extrapolated from the perigee, times the charge q	$\Delta\phi_{\text{res}}$	e
	Ratio of the cluster energy to the track momentum	E/p	e

rection factor for the simulation [40]. The correction factor is derived in bins of $|\eta|$ and E_T. The measured photon identification efficiency in data and simulation is shown in Fig. 4.22 using 81 fb$(^{-1})$ of data collected in 2015–2017 for photons with $0 < |\eta| < 0.6$ as a function of the photon E_T.

Electrons The identification of prompt electrons relies on a likelihood discriminant constructed from quantities measured in the inner detector and the calorimeter. A detailed description is given in Ref. [48]. The quantities used in the electron identification are chosen based on their ability to discriminate prompt isolated electrons from energy deposits from hadronic jets, from converted photons, and from genuine

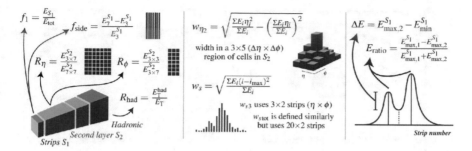

Fig. 4.21 Schematic representation of the photon identification discriminating variables [47]

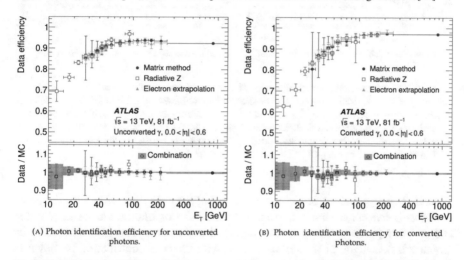

(A) Photon identification efficiency for unconverted photons.

(B) Photon identification efficiency for converted photons.

Fig. 4.22 The photon identification efficiency, and the ratio of data to MC efficiencies, for **A** unconverted photons and **B** converted photons with a `FixedCutLoose` isolation requirement applied as preselection, as a function of E_T for $0 < |\eta| < 0.6$. The combined correction factor, obtained using a weighted average of scale factors from the individual measurements, is also presented; the band represents the total uncertainty on the correction factor [40]

electrons produced in the decays of heavy-flavor hadrons. Similar to photon identification, three working points are provided: *loose*, *medium* and *tight*. The efficiency of the identification is measured in data and simulation using $Z \rightarrow ee$ and $J/\psi \rightarrow ee$ decays in bins of $|\eta|$ and E_T [40] and shown in Fig. 4.23.

4.3.1.2 Electron and Photon Isolation

To further suppress the background from hadronic decay, *isolation* criteria based on the transverse energy deposits around the electron or photon candidate or the transverse momenta of the tracks around the electron (or photon) candidate is required. The calorimeter isolation variable ($E_{T,raw}^{isol}$) [49] is the sum of the transverse energy

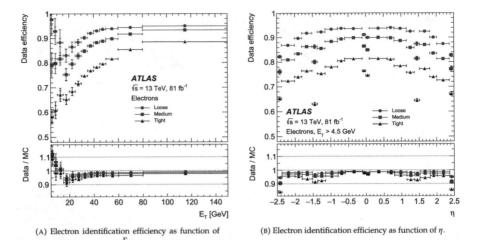

(A) Electron identification efficiency as function of E_T.

(B) Electron identification efficiency as function of η.

Fig. 4.23 The electron identification efficiency in data as a function of **A** E_T and **B** η for the loose, medium, and tight operating points. The efficiencies are obtained by applying data-to-simulation efficiency ratios measured in $J/\psi \to ee$ and $Z \to ee$ events to $Z \to ee$ simulation. The inner uncertainties are statistical, and the total uncertainties are the statistical and systematic uncertainties in the data-to-simulation efficiency ratio added in quadrature. For both plots, the bottom panel shows the data-to-simulation ratios [40]

of positive-energy topological clusters whose barycenter falls within a cone, known as the *isolation cone*, centered around the electron or photon cluster barycenter. The raw EM particle energy is also included in this cone and hence needs to be subtracted. A scheme of the isolation cone with the core contribution is shown in Fig. 4.24a. The subtraction is done by removing EM calorimeter cells contained in a $\Delta\eta \times \Delta\phi = 5 \times 7$ (in layer-2 cells) rectangular cluster around the barycenter of the EM particle cluster. If the shower is wider, energy leakage in the isolation cone, outside of the core 5×7 window, can occur and hence a Monte-Carlo based leakage correction (as a function of E_T and $|\eta|$) is performed. In addition, the contribution from pileup and underlying-event is estimated and subtracted [50]. The final, corrected, calorimeter isolation variable is thus:

$$E_T^{\mathrm{coneXX}} = E_{T,\mathrm{raw}}^{\mathrm{isolXX}} - E_{T,\mathrm{core}} - E_{T,\mathrm{leakage}}(E_T, \eta, \Delta R) - E_{T,\mathrm{pileup}}(\eta, \Delta R) \quad (4.12)$$

where XX refers to the size of the employed cone, $\Delta R = \mathrm{XX}/100$. A cone size $\Delta R = 0.2$ is used for the electron working points whereas cone sizes $\Delta R = 0.2$ and 0.4 are used for photon working points. An example of the distribution of the isolation variables for tight photons using inclusive samples in data and simulation is shown in Fig. 4.24b for prompt photons signal and a background template built by inverting the identification criteria.

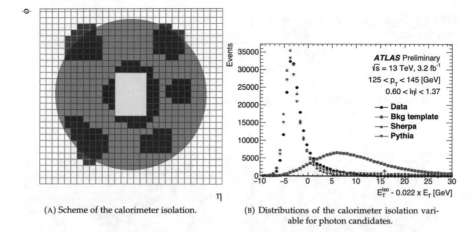

(A) Scheme of the calorimeter isolation.

(B) Distributions of the calorimeter isolation variable for photon candidates.

Fig. 4.24 **A** Scheme of the calorimeter isolation method: the grid represents the second-layer calorimeter cells in the η and ϕ directions. The candidate electron is located in the centre of the purple circle that represents the isolation cone. All topological clusters, represented in red, for which the barycentres fall within the isolation cone are included in the computation of the isolation variable. The 5×7 cells (which cover an area of $\Delta\eta \times \Delta\phi = 0.125 \times 0.175$) represented by the yellow rectangle correspond to the subtracted cells in the core subtraction method [38]. **B** Distributions of the calorimeter isolation variable for photon candidates fulfilling the tight identification criteria using inclusive prompt photon samples for data and simulation (Sherpa and Pythia). The background contribution to the 2015 data, shown as "Bkg template", has been subtracted. It has been determined using a control sample with a subset of the identification requirements inverted and normalized to the data [51]

The track isolation variable (p_T^{coneXX}) is the sum of the transverse momentum of selected tracks within a cone centered around the electron track (or the photon cluster direction) excluding tracks associated with the electron or the converted photon. The tracks considered are required to have $p_T > 1$ GeVand $|\eta| < 2.5$ and to originate from the primary vertex to avoid tracks from pileup events. The cone size varies as a function of the transverse momentum of the electron, as objects become closer to the electron for boosted topologies. Therefore, the isolation cone is reduced for larger transverse momentum, with a maximum of 0.2 via the relation:

$$\Delta R = \min \left(\frac{10\,\text{GeV}}{p_T}, 0.2 \right). \tag{4.13}$$

Photon isolation Similar to the photon identification, different working points of the isolation selection providing a different balance between prompt photon efficiency and non-prompt photon rejection have been defined as shown in Table 4.2.

The photon isolation efficiency for each of these working points is measured in data and simulation as a function of photon $|\eta|$ and E_T using photons from $Z \to ll\gamma$ ($10 < E_T < 100\,\text{GeV}$) and inclusive photon ($25\,\text{GeV} < E_T < 1.5\,\text{TeV}$) and is shown in Fig. 4.25. The difference between data and simulation for the measured efficiencies is then used to derive correction factors applied to simulation.

Table 4.2 Definition of the photon isolation working points [40]

Working point	Calorimeter isolation	Track isolation
FixedCutLoose	$E_T^{cone20} < 0.065 \times E_T$	$p_T^{cone20}/E_T < 0.05$
FixedCutTight	$E_T^{cone40} < 0.022 \times E_T + 2.45$ GeV	$p_T^{cone20}/E_T < 0.05$
FixedCutTightCaloOnly	$E_T^{cone40} < 0.022 \times E_T + 2.45$ GeV	–

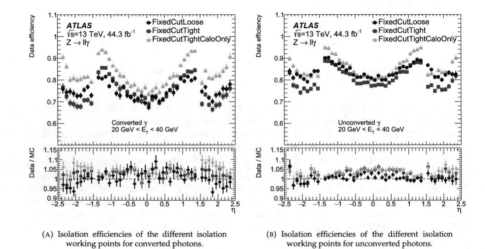

(A) Isolation efficiencies of the different isolation working points for converted photons.

(B) Isolation efficiencies of the different isolation working points for unconverted photons.

Fig. 4.25 Efficiency of the isolation working points defined in Table 4.2, using $Z \to \ell\ell\gamma$ events, for **A** converted and **B** unconverted photons as a function of photon η. The lower panel shows the ratio between the efficiency measured in data and in MC simulations. The total uncertainty is shown, including the statistical and systematic components [40]

Electron isolation Similarly, for electrons, isolation criteria are implemented using the track and calorimeter isolation variables, as shown in Table 4.3. The efficiency of the isolation is measured using electrons from $Z \to ee$ and $J/\psi \to ee$ decays, as shown in Fig. 4.26.

4.3.2 Jet Reconstruction

As detailed in Sect. 1.2.2, QCD forbids the existence of free quarks and gluons which forces them to "hadronize", producing a stream of particles known as *jets*. There are several algorithms for the reconstruction of jets [52]. ATLAS uses the anti-k_t algorithm [53] with topo-clusters built from calorimeter cells as input. ATLAS uses radius parameters $R = 0.4$ and $R = 1.0$ depending on the physics intent: small radius is used for quark and gluons jets, whereas the larger cones are used for energetic

Table 4.3 Definition of the electron isolation working points and isolation efficiency ε. In the `Gradient` working point definition, the unit of p_T is GeV. All working points use a cone size of $\Delta R = 0.2$ for calorimeter isolation and $R_{max} = 0.2$ for track isolation [40]

Working point	Calorimeter isolation	Track isolation
`Gradient`	$\varepsilon = 0.1143 \times p_T + 92.14\%$ (with E_T^{cone20})	$\varepsilon = 0.1143 \times p_T + 92.14\%$ (with $p_T^{varcone20}$)
`FCHighPtCaloOnly`	$E_T^{cone20} <$ $\max(0.015 \times p_T, 3.5 \text{ GeV})$	–
`FCLoose`	$E_T^{cone20}/p_T < 0.20$	$p_T^{varcone20}/p_T < 0.15$
`FCTight`	$E_T^{cone20}/p_T < 0.06$	$p_T^{varcone20}/p_T < 0.06$

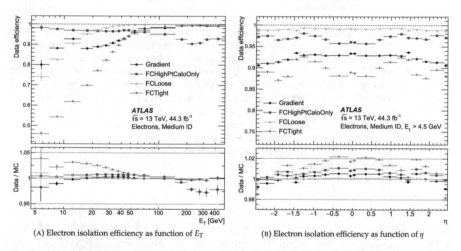

(A) Electron isolation efficiency as function of E_T

(B) Electron isolation efficiency as function of η

Fig. 4.26 Efficiency of the different isolation working points for electrons from inclusive $Z \to ee$ events as a function of **A** the electron E_T, and **B** electron η. The electrons are required to fulfill the Medium selection from the l-based electron identification. The lower panel shows the ratio between the efficiency measured in data and MC simulations. The total uncertainty is shown, including the statistical and systematic components [40]

$W \to q\bar{q}$, $Z \to q\bar{q}$, $t \to Wb \to q\bar{q}b$ decays [54]. Before reconstructed jets can be used for physics analyses, an intricate calibration chain is applied [55] to correct the jet energies for pileup effects, non-compensating calorimeter response, data and simulation differences. Besides, a residual correction is applied to the jet p_T in data using jets produced in balance with additional objects (such as photons or Z bosons) known as the Jet Energy Scale (JES) corrections [55]. A summary of the JES uncertainties using 80 fb$(^{-1})$ of data is shown in Fig. 4.27a. The resulting jet energy resolution (JER) is shown in Fig. 4.27b along with the different uncertainties.

In order to reduce the number of reconstructed jets produced by the additional pileup interactions, a selection based on the *Jet Vertex Tagger (JVT)* is used [58]. The JVT, computed for jets with $|\eta| < 2.5$ and $p_T < 120$ GeV, estimates the probability of a jet to originate from pileup or a hard scattering, based on the number of tracks

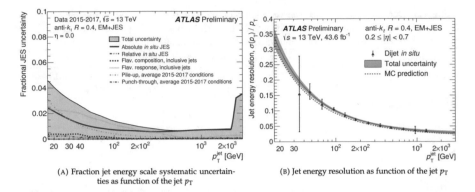

(A) Fraction jet energy scale systematic uncertainties as function of the jet p_T

(B) Jet energy resolution as function of the jet p_T

Fig. 4.27 **A** Fractional jet energy scale systematic uncertainty components as a function of jet p_T for $R = 0.4$ anti-kt jets at $\eta = 0.0$, reconstructed from electromagnetic-scale topo-clusters. The total uncertainty (all components summed in quadrature) is shown as a filled region topped by a solid black line [56]. **B** The relative jet energy resolution $\sigma(p_T)/p_T$ as a function of p_T for anti-kt jets with a radius parameter of $R = 0.4$ and inputs of EM-scale topoclusters calibrated with the EM+JES scheme followed by a residual in situ calibration and using the 2017 dataset. The error bars on points indicate the total uncertainties on the derivation of the relative resolution in dijet events, adding in quadrature statistical and systematic components. The result of the combination of the in situ techniques is shown as the dark line. The band indicates the total uncertainty resulting from the combination of in situ techniques, and includes the statistical and systematic components added in quadrature, although the statistical component is found to be negligible [57]

associated to the primary vertex. The distribution of the JVT variable for hard-scattering and pileup jets is shown in Fig. 4.28a. Selected jets are required to have a JVT value larger than certain thresholds defining three working points: loose, medium, and tight. The default working point is the medium one. The efficiencies of the JVT selection for hard-scattering jets are determined from Z+jet events. The JVT selection efficiency for data and the simulation is shown in Fig. 4.28b [58]. The ratios of efficiency between data and the simulation are used to derive correction factors as a function of the jet p_T and η that are applied to the jets in the simulation.

4.3.3   Muon Reconstruction

Muons are reconstructed using information from the inner detector (ID) and the muon spectrometers (MS). Muons are reconstructed independently in the ID and MS. The information from the individual sub-detectors is then combined to form the muon tracks. In the ID, muons are reconstructed using the same reconstruction algorithms as any charged particle [43]. In the MS, muon reconstruction starts with a search for hit patterns inside each muon chamber to form segments that are then connected using different algorithms [59]. Therefore, there are four classes of reconstructed muons [60]:

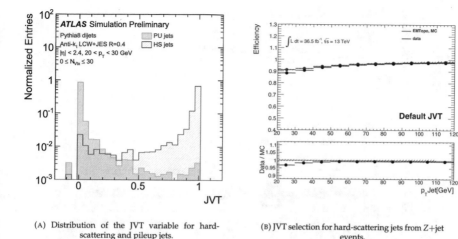

(A) Distribution of the JVT variable for hard-scattering and pileup jets.

(B) JVT selection for hard-scattering jets from Z+jet events.

Fig. 4.28 **A** Distribution of JVT for pileup and hard-scattering jets with $20 < p_T < 30$ GeV. **B** Selection efficiency of hard-scattering jets in data and simulation for the JVT requirement applied to central jets. The efficiencies are measured from Z+jet events, with $Z \to \mu\mu$. The difference between data and simulation is used to derive correction factors that are applied to jets in the simulation [58]

- Stand-alone muons (SA) where the muon trajectory is reconstructed only in MS with extrapolation of the track parameters to the interaction point taking into account energy loss in the calorimeter.
- Combined muons (CB) where the track is formed from the successful combination of the SA and Inner Detector track. CB muons have the highest purity among the different muon classes.
- Segment-Tagged (ST) muons where the ID track is identified as a muon if it is matched to at least one segment in the precision chambers.
- Calorimeter Tagged (CT) muons where calorimeter deposition is used for tagging. It is mainly used for $\eta \sim 0$. It has the lowest purity of all muon classes.

Similar to electrons and photons, muon candidates are required to pass and identification and isolation criteria that are optimized using $Z \to \mu\mu$ and $J/\psi \to \mu\mu$ decays [59]. The reconstruction efficiency as a function of the muon p_T is shown in Fig. 4.29.

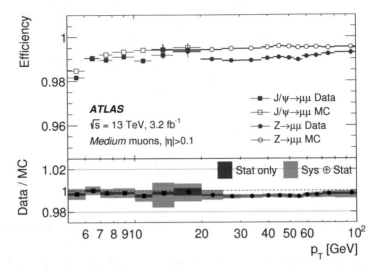

Fig. 4.29 Reconstruction efficiency for the Medium muon selection as a function of the p_T of the muon, in the region $0.1 < |\eta| < 2.5$ as obtained with $Z \rightarrow \mu\mu$ and $J/\psi \rightarrow \mu\mu$. The error bars on the efficiencies indicate the statistical uncertainty. The panel at the bottom shows the ratio of the measured to predicted efficiencies, that are used to derive correction factors for the simulation, with statistical and systematic uncertainties [59]

References

1. Evans L, Bryant P (2008) LHC machine. J. Instrum. 3.08:S08001–S08001. https://doi. org/10.1088/1748-0221/3/08/s08001.   https://doi.org/10.1088%2F1748-0221%2F3%2F08 %2Fs08001

2. Bryant PJ A brief history and review of accelerators. http://www.afhalifax.ca/magazine/wp-content/sciences/lhc/HistoireDesAccelerateurs/histoire1.pdf

3. Smith CL (2014) Genesis of the large hadron collider. Philos Trans R Soc A: Math Phys Eng Sci 373.2032:20140037–20140037. https://doi.org/10.1098/rsta.2014.0037. https://doi. org/10.1098/rsta.2014.0037

4. Arnaudon L et al (2006) Linac4 technical design report. Tech. rep. CERN-AB-2006-084. CARE-Note-2006-022-HIPPI. revised version submitted on 2006-12-14 09:00:40. Geneva: CERN. https://cds.cern.ch/record/1004186

5. Arnison G et al (1983) Experimental observation of isolated large transverse energy electrons with associated missing energy at ps = 540GeV. Phys Lett B122:103–116 [611(1983)]. https:// doi.org/10.1016/0370-2693(83)91177-2

6. Arnison G et al (1983) Experimental observation of lepton pairs of invariant mass around 95-GeV/c2 at the CERN SPS collider. Phys Lett B126:398–410 [7.55(1983)]. https://doi.org/10. 1016/0370-2693(83)90188-0

7. Marcastel F (2013) CERN's accelerator complex. La chaîEEne des accélérateurs du CERN. In: General photo. https://cds.cern.ch/record/1621583

8. Boussard D, Linnecar TPR (1999) The LHC superconducting RF system. Tech. rep. LHCProject-Report-316. CERN-LHC-Project-Report-316. Geneva: CERN. https://cds.cern. ch/record/410377

9. Parma V, Rossi L (2009) Performance of the LHC magnet system. CERN-ATS-2009-023 , 6. https://cds.cern.ch/record/1204578

10. The ATLAS Collaboration (2008) The ATLAS experiment at the CERN large hadron collider. J Instrum 3.08:S08003–S08003. https://doi.org/10.1088/1748-0221/3/08/s08003. https://doi.org/10.1088%2F1748-0221%2F3%2F08%2Fs08003
11. The CMS Collaboration (2008) The CMS experiment at the CERN LHC. J Instrum 3.08:S08004–S08004. https://doi.org/10.1088/1748-0221/3/08/s08004. https://doi.org/1088%2F1748-0221%2F3%2F08%2Fs08004
12. Aamodt K et al (2008) The ALICE experiment at the CERN LHC. JINST 3:S08002. https://doi.org/10.1088/1748-0221/3/08/S08002
13. Augusto Alves A Jr et al (2008) The LHCb detector at the LHC. JINST 3:S08005. https://doi.org/10.1088/1748-0221/3/08/S08005
14. Bailey R, Collier P (2003) Standard filling schemes for various lhc operation modes. Tech. rep. LHC-PROJECT-NOTE-323. Geneva: CERN. https://cds.cern.ch/record/691782
15. Taylor BG (2002) Timing distribution at the LHC. https://doi.org/10.5170/CERN-2002-003.63. https://cds.cern.ch/record/592719
16. The ATLAS Collaboration (2019) ATLAS online luminosity public plots. Tech. rep. Geneva: CERN. https://twiki.cern.ch/twiki/bin/view/AtlasPublic/LuminosityPublicResultsRun2
17. Aad G et al (2012) Observation of a new particle in the search for the standard model Higgs boson with the ATLAS detector at the LHC. Phys Lett B716:1–29. https://doi.org/10.1016/j.physletb.2012.08.020. arXiv: 1207.7214 [hep-ex]
18. Yamamoto A et al (2000) Progress in ATLAS central solenoid magnet. IEEE Trans Appl Supercond 10.1:353–356. ISSN: 1051-8223. https://doi.org/10.1109/77.828246
19. Capeans M et al (2010) ATLAS insertable B-layer technical design report. Tech. rep. CERN-LHCC-2010-013. ATLAS-TDR-19. https://cds.cern.ch/record/1291633
20. ATLAS Collaboration (2017) Technical design report for the ATLAS inner tracker pixel detector. Tech. rep. CERN-LHCC-2017-021. ATLAS-TDR-030. Geneva: CERN. https://cds.cern.ch/record/2285585
21. ATLAS Collaboration (2017) Technical design report for the ATLAS inner tracker strip detector. Tech. rep. CERN-LHCC-2017-005. ATLAS-TDR-025. Geneva: CERN. https://cds.cern.ch/record/2257755
22. Particle identification performance of the ATLAS transition radiation tracker. Tech. rep. ATLAS-CONF-2011-128. Geneva: CERN (2011). https://cds.cern.ch/record/1383793
23. ATLAS liquid-argon calorimeter: Technical design report. Technical design report ATLAS. Geneva: CERN (1996). https://cds.cern.ch/record/331061
24. ATLAS tile calorimeter: technical design report. Technical design report ATLAS. Geneva: CERN (1996). https://cds.cern.ch/record/331062
25. Tanabashi M et al (2018) Review of particle physics. Phys Rev D 98:030001. https://doi.org/10.1103/PhysRevD.98.030001. https://link.aps.org/doi/10.1103/PhysRevD.98.030001
26. Wolfgang Fabjan C, Gianotti F (2003) Calorimetry for particle physics. Rev Mod Phys 75. CERN-EP-2003-075, 1243–1286. 96 p. https://doi.org/10.1103/RevModPhys.75.1243. https://cds.cern.ch/record/692252
27. Gabaldon C (2010) Drift time measurement in the ATLAS liquid argon electromagnetic calorimeter using cosmic muons. Tech. rep. ATL-LARG-PROC-2010-013. Geneva: CERN. https://cds.cern.ch/record/1301524
28. Andrieux ML et al (2002) Construction and test of the first two sectors of the ATLAS barrel liquid argon presampler. Nucl Instrum Methods Phys Res Sect A: Accel Spectrom Detect Assoc Equip 479.2:316–333. ISSN: 0168-9002. https://doi.org/10.1016/S0168-9002(01)00943-3. http://www.sciencedirect.com/science/article/pii/S0168900201009433
29. van der Meer S (1968) Calibration of the effective beam height in the ISR. Tech. rep. CERN-ISR-PO-68-31. ISR-PO-68-31. Geneva: CERN. https://cds.cern.ch/record/296752
30. Adamczyk L et al (2015) Technical design report for the ATLAS forward proton detector. Tech. rep. CERNLHCC-2015-009. ATLAS-TDR-024. https://cds.cern.ch/record/2017378
31. Bernius C (2017) The ATLAS trigger algorithms upgrade and performance in run-2. In: Proceedings, meeting of the APS division of particles and fields (DPF 2017): Fermilab, Batavia, Illinois, USA, July 31–August 4, 2017. arXiv: 1709.09427 [hep-ex]

32. Panduro Vazquez W, ATLAS Collaboration (2017) The ATLAS data acquisition system in LHC run 2. Tech. rep. ATL-DAQ-PROC-2017-007. 3. Geneva: CERN. https://doi.org/10.1088/1742-6596/898/3/032017. https://cds.cern.ch/record/2244345

33. Abreu H et al (2010) Performance of the electronic readout of the ATLAS liquid argon calorimeters. JINST 5:P09003. https://doi.org/10.1088/1748-0221/5/09/P09003. https://cds.cern.ch/record/1303004

34. Aad G et al (2010) Readiness of the ATLAS liquid argon calorimeter for LHC collisions. Eur Phys J C70:723–753. https://doi.org/10.1140/epjc/s10052-010-1354-y. arXiv: 0912.2642

35. Cleland WE, Stern EG (1994) Signal processing considerations for liquid ionization calorimeters in a high rate environment. Nucl Inst Methods A338:467–497

36. Aharrouche M et al (2010) Measurement of the response of the ATLAS liquid argon barrel calorimeter to electrons at the 2004 combined test-beam. Nucl Inst Methods A614:400–432. https://doi.org/10.1016/j.nima.2009.12.055

37. Lampl W et al (2008) Calorimeter clustering algorithms: description and performance. Tech. rep. ATLLARG-PUB-2008-002. ATL-COM-LARG-2008-003. Geneva: CERN. https://cds.cern.ch/record/1099735

38. Aaboud M et al (2019) Electron reconstruction and identification in the ATLAS experiment using the 2015 and 2016 LHC proton-proton collision data at ps = 13 TeV. Eur Phys J C79.8:639. https://doi.org/10.1140/epjc/s10052-019-7140-6. arXiv: 1902.04655 [physics.ins-det]

39. Electron and photon reconstruction and performance in ATLAS using a dynamical, topological cell clusteringbased approach. Tech. rep. ATL-PHYS-PUB-2017-022. Geneva: CERN (2017). https://cds.cern.ch/record/2298955

40. Aad G et al (2019) Electron and photon performance measurements with the ATLAS detector using the 2015-2017 LHC proton-proton collision data. arXiv: 1908.00005 [hep-ex]

41. Cornelissen TG et al (2007) Concepts, design and implementation of the ATLAS New Tracking (NEWT). ATL-SOFT-PUB-2007-007. Geneva. https://cds.cern.ch/record/1020106

42. FrüFChwirth R (1987) Application of Kalman filtering to track and vertex fitting. Nucl Instrum Meth A 262:444–450. https://doi.org/10.1016/0168-9002(87)90887-4

43. Cornelissen TG et al (2008) The global c2 track fitter in ATLAS. J Phys Conf Ser 119: 032013. https://doi.org/10.1088/1742-6596/119/3/032013

44. Improved electron reconstruction in ATLAS using the Gaussian Sum Filter-based model for bremsstrahlung. Tech. rep. ATLAS-CONF-2012-047. Geneva: CERN (2012). https://cds.cern.ch/record/1449796

45. Proklova N et al (2018) Measurements of photon efficiencies in pp collision data collected in 2015, 2016 and 2017 at ps = 13 TeV with the ATLAS detector. Tech. rep. ATL-COM-PHYS-2018-1604. Geneva: CERN. https://cds.cern.ch/record/2647979

46. Anastopoulos C et al (2019) Electron identification and efficiency measurements in 2017 data. Tech. rep. ATL-COM-PHYS-2018-1727. Geneva: CERN. https://cds.cern.ch/record/2652163

47. Saxon J (2014) Discovery of the Higgs boson, measurements of its production, and a search for Higgs boson pair production. Presented 13 06 2014. https://cds.cern.ch/record/1746004

48. Aaboud M et al (2019) Electron and photon energy calibration with the ATLAS detector using 2015-2016 LHC proton-proton collision data. JINST 14.03:P03017. https://doi.org/10.1088/1748-0221/14/03/P03017. arXiv: 1812.03848 [hep-ex]

49. Aaboud M et al (2019) Measurement of the photon identification efficiencies with the ATLAS detector using LHC Run 2 data collected in 2015 and 2016. Eur Phys J C79.3:205. https://doi.org/10.1140/epjc/s10052-019-6650-6. arXiv: 1810.05087 [hep-ex]

50. Cacciari M, Salam GP (2008) Pileup subtraction using jet areas. Phys Lett B 659:119–126. https://doi.org/10.1016/j.physletb.2007.09.077. arXiv: 0707.1378 [hep-ph]

51. Aaboud M et al (2016) Search for resonances in diphoton events at ps=13 TeV with the ATLAS detector. JHEP 09:001. https://doi.org/10.1007/JHEP09(2016)001. arXiv: 1606.03833 [hep-ex]

52. Atkin R (2015) Review of jet reconstruction algorithms. J Phys Conf Ser 645.1:012008. https://doi.org/10.1088/1742-6596/645/1/012008

53. Cacciari M, Salam GP, Soyez G (2008) The Anti-k(t) jet clustering algorithm. JHEP 0804:063. https://doi.org/10.1088/1126-6708/2008/04/063. arXiv: 0802.1189 [hep-ph]

54. Schramm S (2017) ATLAS jet reconstruction, energy scale calibration, and tagging of Lorentz-boosted objects. https://cds.cern.ch/record/2284807

55. Jet Calibration and Systematic Uncertainties for Jets Reconstructed in the ATLAS Detector at ps = 13 TeV. Tech. rep. ATL-PHYS-PUB-2015-015. Geneva: CERN (2015). https://cds.cern.ch/record/2037613

56. The ATLAS Collaboration. Jet energy scale and uncertainties in 2015-2017 data and simulation. Tech. rep. Geneva: CERN (2018). https://atlas.web.cern.ch/Atlas/GROUPS/PHYSICS/PLOTS/JETM-2018-006/

57. The ATLAS Collaboration. Jet energy resolution in 2017 data and simulation. Tech. rep. Geneva: CERN (2018). https://atlas.web.cern.ch/Atlas/GROUPS/PHYSICS/PLOTS/JETM-2018-005/

58. Tagging and suppression of pileup jets with the ATLAS detector. In: ATLAS-CONF-2014-018 (2014). https://cds.cern.ch/record/1700870

59. Aad G et al (2016) Muon reconstruction performance of the ATLAS detector in proton-proton collision data at ps =13 TeV. Eur Phys J C76.5:292. https://doi.org/10.1140/epjc/s10052-016-4120-y. arXiv: 1603.05598 [hep-ex]

60. VranješA N (2013) Muon reconstruction in ATLAS and CMS. https://indico.in2p3.fr/event/6838/contributions/39638/attachments/32073/39614/LHCFranceMuonRecoNVranjes.pdf

Chapter 5
Energy Calibration of the ATLAS Electromagnetic Calorimeter with 2015 and 2016 pp Collisions

5.1 Overview of the Calibration Procedure

The calibration of the reconstruction and identification performance of a particle detector is an essential procedure in any particle physics measurement. The calibration usually exploits control samples of particles of a well-defined type, selected in data using kinematic requirements that do not bias the performance that has to be calibrated. In this chapter, the procedure used to calibrate the energy of electrons and photons reconstructed by the ATLAS electromagnetic calorimeter is reviewed. Photons constitute the main final state particles for the processes that are measured in this thesis. Particular emphasis will be given to the calibration of the energy response of the presampler layer of the electromagnetic calorimeter, which was the work that I performed during the first year of my Ph.D. in order to qualify as an ATLAS author.

Given the complexity of the ATLAS electromagnetic calorimeter, the calibration of the energy measurement of electrons and photons is an intricate process. The result of the calibration is a precise energy measurement of electrons and photons, which is a crucial requirement for several analyses, including studies of the Higgs boson in the two-photon and four-lepton decay channels as well as precise studies of the properties of the W and Z bosons. The energy calibration of the ATLAS electromagnetic calorimeter is a sequence of several steps, sketched in Fig. 5.1, aims at minimizing the bias, and achieving the best resolution on the measured energy of electrons and photons. The calibration procedure is as follows:

- The first step of the calibration procedure is a multi-variate (MVA) regression algorithm based on a Monte-Carlo simulation of the ATLAS detector. The MVA algorithm aims at estimating the true energy of electrons and photons from the calorimeter cluster properties measured by the detector. The optimization of the MVA algorithm is performed separately for electrons, converted and unconverted photons and is detailed in Sect. 5.2. An essential requirement for this simulation-based calibration is that the simulations describe well the electromagnetic calorimeter

© The Editor(s) (if applicable) and The Author(s), under exclusive license to Springer Nature Switzerland AG 2020
A. Tarek Abouelfadl Mohamed, *Measurement of Higgs Boson Production Cross Sections in the Diphoton Channel*, Springer Theses, https://doi.org/10.1007/978-3-030-59516-6_5

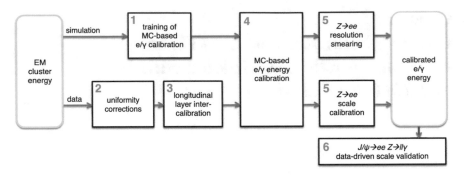

Fig. 5.1 Schematic overview of the procedure used to calibrate the energy response of electrons and photons in ATLAS [1]

geometry and the material budget upstream of the calorimeter. The details of the estimation procedure of the material budget are given in Sect. 5.4.3.

- The second step of the calibration chain is the correction of the energy response for small details that were not included in the simulation. They include, for instance, specific regions of the detector that were affected by data-taking issues such as operation at non-nominal high voltage, geometric effects such as the inter-module widening (IMW), or biases associated with the calorimeter electronics calibration. These corrections are detailed in Sect. 5.3.
- The third step is the longitudinal layer energy inter-calibration, detailed in Sect. 5.4. The ATLAS electromagnetic calorimeter is segmented longitudinally into a few layers, and the electron and photon energies are determined from the energy deposited in each of these layers. A mismatch of the relative energy response of different layers between data and the simulation can lead to biases in the calibrated energy. This step equalizes the energy scales of the different longitudinal layers between data and the simulation.
- The subsequent step is the extraction of the global energy correction factors applied to electrons and photons (steps 4 and 5). The overall electron response in data is calibrated so that it agrees with the expectation from simulation, using a large sample of electrons from Z boson decays. The details of this procedure are shown in Sect. 5.5.
- The final step of the calibration chain is the validation of the extrapolation of the calibration to low-p_T electrons using $J/\psi \to ee$ decays, and to photons using $Z \to ll\gamma$ decays, as described in Sect. 5.6.

A summary of the calibration uncertainties and of their effect on the Higgs boson mass measurement is given in Sect. 5.7.

5.2   Multivariate Monte Carlo Based Calibration

The energy of an electromagnetic particle absorbed by the ATLAS electromagnetic calorimeter is not fully converted into an electric signal. The accordion geometry of interleaved layers of active (LAr) and passive (lead) layers means that a sizable part of the energy of incident particles will be absorbed in the lead layers and not converted to measurable ionization energy in the LAr layers. In order to recover this energy, various methods were developed.

At the beginning of Run-1 the default method used in ATLAS was the "calibration-hits" method [2], in which an empirical function was used to parameterize the corrections to the energy and position of a cluster (as the material budget changes along ϕ). An upgrade of this method, based on a multivariate (MVA) regression technique, was developed around the end of Run-1 and replaced the calibration-hits method as the default energy calibration procedure of electromagnetic particles. The MVA regression tool reconstructs the energy of incident particles from a set of input variables using boosted decision trees (BDT) with gradient boosting training [3]. The MVA calibration provides a tool to derive a new set of corrections that can be updated relatively easily in cases of changes in the detector simulation or the geometry description. The MVA also makes it easy to extend the set of input variables, providing a systematic way for improving the calibration. Another advantage of the MVA algorithm is that it takes into account the correlations – even non-linear ones – among the input variables, therefore improving the energy resolution when these correlations are significant.

The MVA regression algorithm is trained using simulated event samples of single particles interacting with the ATLAS detector. The simulation of the detector response is performed using GEANT 4 [4]. Particles are generated with transverse energy up to 3 TeV, with a larger fraction of particles in the energy range between 7 and 200 GeV as shown in Fig. 5.2. The MVA training aims at estimating the generated energy of the particle, E_{gen}, from the total raw cluster energy measured in the accordion calorimeter, E_{acc}. The ratio E_{gen}/E_{acc} is the target variable of the MVA algorithm, from which the calibrated particle energy E_{calib} can be easily computed through multiplication by the measured value of E_{acc}.

The following variables are used as an input to the MVA algorithm for electrons and photons:

- **Total raw cluster energy of the accordion**: $E_{acc} = E_1^{raw} + E_2^{raw} + E_3^{raw}$, defined as the sum of the uncalibrated energies of the three accordion layers (strips, middle and back).
- **Ratio of the energy in the presampler to the energy in the accordion**: E_0^{raw}/E_{acc}, used only for clusters within the geometric range of the presampler, $|\eta| < 1.8$.
- **Ratio of the energy in the first accordion layer to the energy in the second one**: E_1^{raw}/E_2^{raw}. This variable provides insight on the longitudinal shower depth, to which it is largely correlated.

Fig. 5.2 Input E_T
distributions of electrons at
truth-level used to generate
the samples for the MVA
calibration. In black is the
distribution used for Run-1
calibration, labeled "MC12",
whereas the red curve is the
one used for the Run-2
calibration detailed in this
thesis, labeled "MC15" [5]

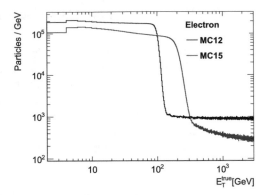

- **Pseudorapidity in the ATLAS frame**: η_{cluster}. This variable takes into account the misalignment of the detector, in order to correct for the variation of the material upstream of the accordion.
- **Cell index**: $\eta_{\text{calo}}/\Delta\eta$, η_{calo} is the pseudorapidity of the cluster in the calorimeter frame and $\Delta\eta = 0.025$ is the size along η of one cell in the middle layer. The cell index is a variable sensitive to non-uniformities of the calorimeter.
- **η with respect to the cell edge**: η_{calo} modulus $\Delta\eta = 0.025$. This variable allows correcting for the variation of the lateral energy leakage due to the finite cluster size. Such variation is larger for particles that hit the cell close to the edges.
- **ϕ with respect to the lead absorbers**: ϕ_{calo} modulus $2\pi/1024$ ($2\pi/768$) in the barrel (end-cap) section of the calorimeter, corresponding to the periodicity of lead sheets in each region. This variable allows correcting for the slight variations of the sampling fraction depending on the particle trajectory.

For converted photon candidates, i.e. with a reconstructed radius of conversion $0 < R < 800$ mm, the following variables are also used:

- **Radius of conversion** R: used only for $p_T^{\text{conv}} > 3$ GeV, where p_T^{conv} is the sum of the transverse momenta of the conversion tracks.
- **Ratio of the conversion transverse momentum** (p_T^{conv}) **to the transverse energy** E_T **reconstructed in the accordion calorimeter**: $E_T^{\text{acc}}/p_T^{\text{conv}}$, where $E_T^{\text{acc}} = E_{\text{acc}}/\cosh(\eta_{\text{cluster}})$
- **Fraction of p_T^{conv} carried by the highest-p_T track**.

The input variables for the MVA algorithm are binned to adjust the MVA response in regions of the phase space with different behaviors. The MVA algorithm yields an improvement in the energy resolution for electrons and photons in comparison with the calibration-hits method used in Run-1, translating into an improved invariant mass resolution of diphoton and dielectron resonances, as shown in Fig. 5.3.

In addition to the previous variables, specific variables are used for the transition region ($1.4 < |\eta| < 1.6$) between the barrel and the end-cap sections, known as the 'crack' region (the exact region definition depends on the type of the study). This region has a large amount of material that will be traversed by the particles

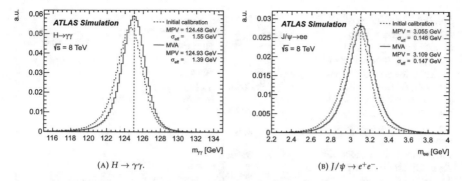

(A) $H \rightarrow \gamma\gamma$.　　　　　　　　　　　(B) $J/\psi \rightarrow e^+e^-$.

Fig. 5.3 Comparison between the MVA calibration and the calibration-hits method. **A** The diphoton invariant mass distribution $m_{\gamma\gamma}$ is shown for Standard Model Higgs boson with mass of 125 GeV decaying to two photons. **B** The dielectron invariant mass is shown for J/ψ decaying to a pair of electrons. The dashed black line indicates the mass of the simulated particle [1]

before reaching the first active layer of the calorimeter (from 5 to almost 10 radiation lengths), yielding a degradation in the energy resolution. The energy calibration of electrons and photons traversing the crack uses as an additional input the energy measured by the E_4 scintillators [6] installed in this transition region. This leads to an improvement in the resolution of about 20%, as shown in Fig. 5.4. This region, however, is omitted from all analyses in this work.

The energy resolution in the simulation, after the MVA calibration, is shown in Fig. 5.5. The resolution is defined as the interquartile range of $E_{\text{calib}}/E_{\text{gen}}$, i.e. the interval excluding the first and last quartiles of the $E_{\text{calib}}/E_{\text{gen}}$ distribution in each bin, converted to the equivalent standard deviation of a Gaussian distribution through division by 1.35.

For unconverted photons, the energy resolution in the simulated samples used for the training of the MVA regression, which do not include any simulated pileup events, closely follows the expected sampling term of the calorimeter ($\approx 10\%/\sqrt{E/\text{GeV}}$ in the barrel section and $\approx 15\%/\sqrt{E/\text{GeV}}$ in the end-cap section). For electrons and converted photons, the degraded energy resolution at low energies reflects the presence of significant tails induced by interactions with the material upstream of the calorimeter. This degradation is largest in the regions with the largest amount of material upstream of the calorimeter, i.e. for $1.2 < |\eta| < 1.8$.

5.3 Uniformity Corrections

In order to achieve an optimal energy resolution in data, the uniformity of the calorimeter energy reconstruction across pseudorapidity and azimuthal angle and as a function of time and pileup conditions has to be ensured. Sources of non-uniformity generally arise from non-nominal calorimeter data-taking conditions or physical

Fig. 5.4 Distribution of the calibrated energy, E_{calib}, divided by the generated energy, E_{gen}, for electrons in the transition region $1.4 < |\eta| < 1.6$ and with energy $50 < E_T^{gen} < 100$ GeV. The solid (dashed) histogram shows the results based on the energy calibration with (without) the E_4 scintillator information. The curves represent Gaussian fits used to estimate the observed resolutions, σ [7]

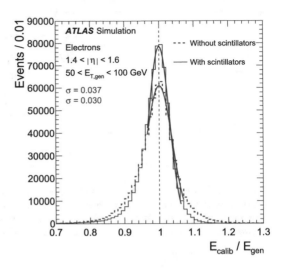

effects that are not implemented in the simulation. Examples of known sources of non-uniformity include: high-voltage inhomogeneities, inter-module widening, and relative miscalibration of the amplifier gains used in the readout of the signal. These checks also constitute a prerequisite for the passive material determination and energy scale measurement presented in Sects. 5.4 and 5.5. Below is a summary of the main sources of non-uniformity.

Intermodule widening effect This effect describes the widening of the gaps between the 16 barrel modules due to gravity. The effect is a consequence of the gaps between the barrel modules, as these gaps are slightly larger than the other LAr gaps.

These inter-module gaps are located in ϕ at $(2n - 15)\pi/16$ with n ranging from 0 to 15. The energy leaking in these gaps will yield an underestimation of the energy in data with respect to the simulation. In particular, these gaps will show up as dips in the ϕ profile of the E/p distribution of electrons from $Z \rightarrow ee$ decays, where E is their energy measured by the electromagnetic calorimeter and p their momentum reconstructed by the inner detector. Since this effect is not modeled in the simulation, a dedicated correction is applied. An example of such corrections is shown in Fig. 5.6.

HV effect In a small number of sectors of the calorimeter (of size $\Delta\eta \times \Delta\phi = 0.2 \times 0.2$), the applied high voltage during data taking was reduced with respect to the nominal values due to short circuits in specific LAr gaps. The actual value of the high voltage is used to derive a correction at the cell level. Residual corrections, however, might be needed when large currents are drawn as they cause dips in the E/p (η, ϕ) profiles. The values of the corrections are typically between 1% and 7%, and affect about 2% of the calorimeter acceptance. These corrections were found to be similar to those derived during Run-1 [1], with the exception of a few cases where voltage settings were different between Run-1 and Run-2, as shown in Fig. 5.7.

Gain effects As detailed in Sect. 4.3.1, the output signal of the electromagnetic readout system can be reconstructed using three different amplifier gain settings:

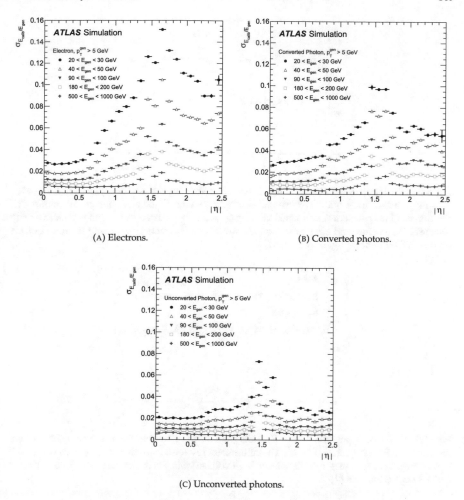

(A) Electrons.

(B) Converted photons.

(C) Unconverted photons.

Fig. 5.5 Energy resolution, $\sigma_{E_{\mathrm{calib}}/E_{\mathrm{gen}}}$, estimated from the interquartile range of $E_{\mathrm{calib}}/E_{\mathrm{gen}}$ as a function of $|\eta|$ for **A** electrons, **B** converted photons and **C** unconverted photons, for different energy ranges. The generated transverse momentum is required to be above 5 GeV [7]

high, medium, and low. This allows a wide range of measured energies. The linearity of the readout electronics was measured for each of the three gains and found to be better than 0.1% [8]. However, the relative calibration of the different readout gains is less well known, and might induce a dependence of the energy response on the energy of the particle. For this reason, a measurement of the relative calibration was performed using data taken with special detector conditions in 2015 (corresponding to an integrated pp luminosity of 12 pb^{-1}) and 2017 (60 pb^{-1}). In these special-condition runs, the high-to-medium gain switch thresholds were lowered by a factor of 5. Under these circumstances, the signal from the highest-energy layer-2 cell of electrons from $Z \rightarrow ee$ decays will be read out using medium gain, whereas in

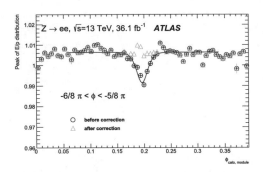

Fig. 5.6 Fitted peak of the E/p distribution for electrons from Z boson decays as a function of ϕ, for a range between the middle of two neighbouring modules in the electromagnetic calorimeter. The impact of the gap between two modules at $\phi_{\text{calo,module}} = \frac{\pi}{16}$ can be seen with a reduced energy response. Data before and after correction are shown. The correction is derived from a fit of the uncorrected data shown by the red line [7]

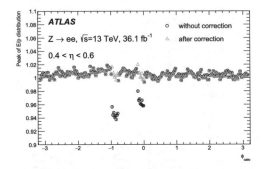

Fig. 5.7 Fitted peak position of the E/p distribution for electrons from Z boson decays as a function of ϕ for the η range 0.4–0.6. The calorimeter response is lower in two regions of size 0.2 in ϕ, corresponding to sectors with reduced high voltage values in the calorimeter. A correction is derived from these data [7]

normal runs it would be read out with the high gain. The comparison between the dielectron invariant mass distributions in these runs and in the normal runs permits the intercalibration of the medium and high gains.

Figure 5.8 shows the measured energy scale difference between the special and the nominal datasets, α_G, as a function of $|\eta|$. α_G is parameterized in a way such that $\alpha_G = 0$ if the high and medium gains were perfectly intercalibrated ($\frac{\Delta E}{E} \propto \alpha_G$). The results show a small but significant difference observed in the region $0.8 < |\eta| < 1.37$. The observed difference is assigned as a systematic uncertainty. This uncertainty is assumed to be a correction factor between the calibration of the two gains, independent of the cell energy. This translates to an uncertainty in the total energy of about 0.05% to 0.1%, depending on η, for photons of $E_T = 60$ GeV, and up to 0.2% to 1% for electrons and photons with transverse energies above a few hundreds of GeV.

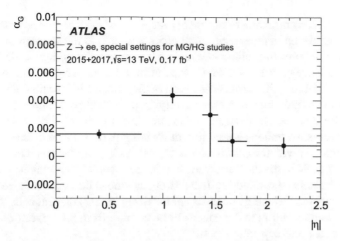

Fig. 5.8 Difference of energy scales, α_G, extracted from $Z \to ee$ events, as a function of $|\eta|$ between data recorded with the standard thresholds for the transition between high and medium gains in the readout of the layer-two cells and data with lowered thresholds. Only statistical uncertainties are shown [7]

5.4 Longitudinal Layer Inter-calibration

The precise measurements that use electrons or photons, such as the Higgs or electroweak boson property measurements, require that the linearity and absolute scale of the calorimeter measurements are understood with a high level of precision. Due to the longitudinal segmentation of the EM calorimeter, a crucial ingredient is the relative calibration of the energy response of its different longitudinal layers. The layer intercalibration is performed before the extraction of any global data-to-simulation scale factor. Intercalibrating the energy responses of the various layers will also ensure the correct extrapolation to the full p_T range as required by the various analyses using electrons and photons.

In this section, the energy intercalibration of the longitudinal layers of the EM calorimeter is discussed in detail. The relative calibration of the first (strips) and second (middle) layers of the accordion calorimeter is described in Sect. 5.4.1, while the calibration of the presampler is illustrated in Sect. 5.4.2. Given the small fraction of energy deposited in the third layer of the calorimeter, no dedicated inter-calibration of its energy response is performed.

5.4.1 The $E_{1/2}$ Inter-calibration Using Muons

The relative calibration of the energy response of the first and second layers of the accordion calorimeter, which contain the bulk of the energy deposited by an electromagnetic particle, is a necessary step before the extraction of global energy calibration scale factors. The correction of the energy response of the first two accordion layers is also a crucial step in other calibration processes such as the estimation of passive material upstream of the calorimeter and of the presampler energy scale. For these reasons, the energy scale of the first two accordion layers needs to be extracted in a manner that is immune to the mis-modeling of the detector in the simulation, in order to not bias the remaining measurements relying on it. Control samples of muons are thus used for this measurement since they are almost unaffected by upstream material. However, since muons are minimum ionizing particles (the critical energy for muons to interact with the calorimeter is $\mathcal{O}(100)$ GeV [1]), their typical signal yield is about 150 MeV in the most energetic cell in the second layer, which is only about five times the noise threshold [9]. In addition, as muons do not induce EM showers in the calorimeter, the energy deposits are localized in few adjacent cells. Moreover, a significant contribution to the noise will be coming from pileup, especially in the first layer of the end-cap calorimeter.

For the inter-calibration of the first two accordion layers, a sample of muons from $Z \to \mu\mu$ decays with at least one "Combined" muon with $p_T > 27$ GeV and $|\eta| < 2.4$ was used. Only three (two) cells were used to measure the muon energy in the first (second) accordion layer to minimize the effect of noise. In the first layer, the energy deposited by the muon is estimated from the sum of the energies measured in three neighboring cells along η, centered around the cell crossed by the extrapolated muon trajectory. In the second layer, the energy deposited by the muon is estimated by summing the energies in the cells crossed by the extrapolated muon trajectory and in the neighboring cell in ϕ with higher energy. The muon energy distribution in each layer, as shown in Fig. 5.9 for the case of the first accordion layer, is modeled with the convolution of a Landau distribution describing the energy deposit, and of a template accounting for the electronic and pileup noise. The noise has significant contribution from pileup that results in positive energy deposits, with an average of zero; hence the most probable value of the noise template is negative (as shown in Fig. 5.9). As expected, the final resolution of the muon energy measurement is significantly affected by the noise, due to the low signal-to-noise ratio resulting from the small energy deposit.

Two alternative methods are used for the estimation of the energy deposit of muons:

- The most probable value (MPV) from a fit using a Landau function convoluted with a template for the noise. The noise templates are built using "zero-bias events" triggered on random LHC bunch crossings, thus taking into account also the effect of pileup, and are determined in bins of both η and average number of interactions per bunch crossing $\langle \mu \rangle$.
- Truncated mean, estimated from a truncated range of the distribution containing 90% of the energy. The truncated mean is used to minimize sensitivity to the tails.

The $E_{1/2}$ layer intercalibration constant $\alpha_{1/2}$ is then defined as

$$\alpha_{1/2} = (\langle E_1 \rangle^{\text{data}} / \langle E_1 \rangle^{\text{MC}})/(\langle E_2 \rangle^{\text{data}} / \langle E_2 \rangle^{\text{MC}}),$$

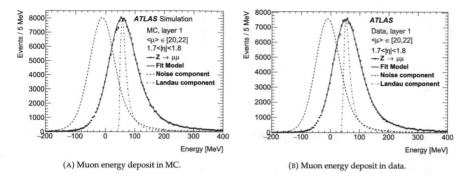

(A) Muon energy deposit in MC.   (B) Muon energy deposit in data.

Fig. 5.9 Muon energy deposit distributions for the $1.7 < |\eta| < 1.8$ region in data and the simulation for the first layer. The energy distribution is fitted with a convolution of the noise distribution and a Landau function. The result of the fit is shown together with the individual components. The distributions are shown for an average number of interactions per bunch crossing $\langle \mu \rangle$ in the range from 20 to 22 [7]

where $\langle E_1 \rangle$ ($\langle E_2 \rangle$) is the estimated energy deposited in the first (second) layer. The final value of $\alpha_{1/2}$ is given by the average of the two methods used to estimate the muon energy deposit with their difference used as a systematic uncertainty. The relative calibration $\alpha_{1/2}$ is estimated in bins of the absolute value of pseudorapidity $|\eta|$, as the intercalibration constant values are found to be consistent between the positive and negative η. The correction $\alpha_{1/2}$ is then applied as a function of $|\eta|$ on data such that $E_2^{\text{corr}} = E_2 \times \alpha_{1/2}$.

As mentioned before, a significant fraction of the noise originates from pileup. To reduce the dependence on pileup mis-modeling, the measurements are performed as a function of the average number of interactions per bunch crossing $\langle \mu \rangle$ and then extrapolated to the case of no pileup ($\langle \mu \rangle = 0$) to measure the intrinsic energy scale of each calorimeter layer for a pure signal. The extrapolation procedure itself was validated by performing an identical procedure in the simulation and comparing the extrapolated result to the prediction of a simulation without pileup. Any difference between the extrapolated result to $\langle \mu \rangle = 0$ and the prediction of the simulation without pileup is taken as a systematic uncertainty. The difference was found to range from 0.2% to 0.5% depending on $|\eta|$. An example of the extrapolation procedure is shown in Fig. 5.10.

The final result of $\alpha_{1/2}$ for both the fitting and truncated-mean methods, including the total statistical and systematic uncertainties, is shown in Fig. 5.11. In addition to the systematic uncertainties mentioned before, the following systematic uncertainties were found to affect the layer inter-calibration $\alpha_{1/2}$:

- Energy leakage resulting from the modeling of the energy loss outside the cells used for the measurement. Muons with trajectories close to the boundaries in ϕ (η) between the first (second) layer cells can have a significant fraction of their energy deposit outside the used cells. The uncertainty from the modeling of these effects is computed by repeating the analysis using only muons crossing the center of the

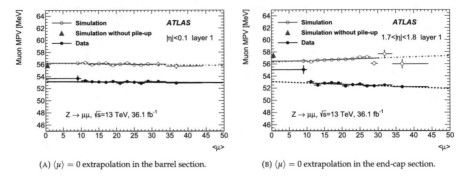

(A) $\langle \mu \rangle = 0$ extrapolation in the barrel section.

(B) $\langle \mu \rangle = 0$ extrapolation in the end-cap section.

Fig. 5.10 Distribution of the fitted MPV of the muon energy deposit in two $|\eta|$ intervals ($< E_1 >$) in **A** the barrel region and **B** the end-cap region as a function of the average number of pileup interactions per bunch crossing $\langle \mu \rangle$. The plot shows the MPV obtained from data (full circles) and simulation (empty circles). The plots include also the MPV using simulations without pileup (triangles) [7]

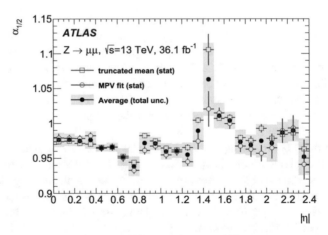

Fig. 5.11 Ratio $\alpha_{1/2} = (\langle E_1 \rangle^{\text{data}} / \langle E_1 \rangle^{\text{MC}})/(\langle E_2 \rangle^{\text{data}} / \langle E_2 \rangle^{\text{MC}})$ as a function of $|\eta|$, measured with the fitting (open circles) or the truncated-mean (open square) methods. The final average measurement with its total uncertainty is also shown (full circles) [7]

first (second) layer cells within 0.04 (0.008) in the ϕ (η) direction. The uncertainty in $\alpha_{1/2}$ varies from 0.5% to 1%.

- Geometry effects resulting from the choice of the cell in ϕ in the second layer. The uncertainty is computed by estimating $\alpha_{1/2}$ using the neighboring cell closest to the extrapolated muon trajectory instead of the one with the highest energy. This yields an uncertainty of around 0.2%.
- Truncation range for the truncated-mean method. The resulting uncertainty is 0.5%.

5.4.2 The Presampler Layer Calibration

The presampler (PS) energy scale, α_{PS}, is defined as the ratio of the PS energy (E_0) in data and simulation and is estimated using electrons from $Z \to ee$ decays.

To interpret the ratio of the PS energy between data and simulation as an energy scale effect, various corrections are applied beforehand to the simulation to remove other sources of mis-modeling of the energy deposits in the PS. In particular, the energy deposited in the presampler depends on the upstream material. The ratio $E_{1/2}$ of the energy deposits in the front and middle layers of the accordion calorimeter is thus used to obtain an improved estimate E_0^{corr} which accounts for material mis-modeling from the nominal value of E_0 predicted by the simulation (E_0^{nom}). The idea is that the presence of more upstream material in data in comparison to the simulation causes an earlier shower development, resulting in larger energy deposits in the first layer and therefore larger $E_{1/2}$ in data with respect to the simulation. To use the $E_{1/2}$ distribution to correct for mis-modeling of material upstream of the PS in the simulation, two factors have to be taken into account. The first one is the correlation between $E_{1/2}$ and the PS energy at a given η, due to material upstream of the PS. The second factor is a correction of $E_{1/2}$ to account for mis-modeling in the simulation of material that affects $E_{1/2}$ but not E_0, i.e. between the PS and the accordion.

The correction for the passive material effects is performed exploiting the difference in response and shower development, sketched in Fig. 5.12, of two different particle types:

- Electrons that are sensitive to all detector material crossed along their trajectory, from the interaction point up to the first layer of the calorimeter (L1).
- Photons that did not convert to e^+e^- pairs before reaching the PS, i.e. reconstructed unconverted photon candidates with small energy deposit in the PS, that are insensitive to the material upstream of the presampler, making such photons specifically sensitive to passive material between the PS and L1.

In the following, the procedure used for the PS calibration is summarised in Sect. 5.4.2.1. The data and simulation samples used for the measurement and the event selection are given in Sects. 5.4.2.2 and 5.4.2.3. The corrections related to passive material up to the first layer of the accordion calorimeter and the material between the presampler and this layer are described in Sects. 5.4.2.4 and 5.4.2.5. A closure study of the procedure on simulated event samples is performed in Sect. 5.4.2.6. Finally, the results of the PS energy scale calibration in data are shown in Sect. 5.4.2.7, and checks of its stability in η and ϕ and as a function of the high voltage applied are performed in Sect. 5.4.2.8.

5.4.2.1 Formalism

The presampler energy scale correction is defined as:

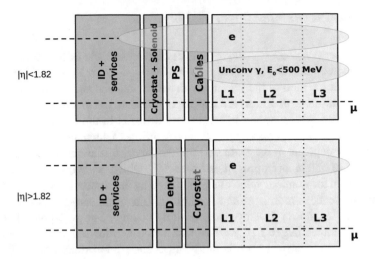

Fig. 5.12 Sketch of the electromagnetic shower development for the different particles used in the calibration/material estimation. The top figure corresponds to the pseudorapidity interval $|\eta| < 1.82$ within the PS acceptance, whereas the bottom figure is for $|\eta| > 1.82$. The pp interaction point in which the particles are produced is assumed to be on the left of the figures [1]

$$\alpha_{\mathrm{PS}} = \frac{E_0^{\mathrm{data}}(\eta)}{E_0^{\mathrm{corr}}(\eta)}, \tag{5.1}$$

where $E_0^{\mathrm{corr}}(\eta)$ is the PS energy in the simulation after applying the various material corrections, and $E_0^{\mathrm{data}}(\eta)$ is the PS energy deposit in data. The total material correction to be applied to the PS energy in the simulation is computed from the following linear parametrisation:

$$\frac{E_0^{\mathrm{corr}}(\eta)}{E_0^{\mathrm{raw}}(\eta)} = 1 + A(\eta)\left(\frac{E_{1/2}^{\mathrm{data}}(\eta)}{E_{1/2}^{\mathrm{nom}}(\eta)b_{1/2}(\eta)} - 1\right), \tag{5.2}$$

where :

- $A(\eta)$ is the correlation factor between E_0 and $E_{1/2}$ under variation of the material upstream of the presampler, and is estimated with electrons from $Z \to ee$ decays from simulation samples with different amounts of additional material upstream of the PS.
- $b_{1/2}(\eta)$ is a correction applied to the double ratio $E_{1/2}$ in data over MC, to correct for imperfect modeling of passive material between the PS and accordion, and is estimated from control samples of unconverted photons with low PS activity.

This parameterisation is derived from a systematic study of simulated samples in which the effect on E_0 and $E_{1/2}$ of passive material added upstream of the accordion is examined [10]. A priori, the correction factors A and $b_{1/2}$ are functions of η. The total material correction is then computed and E_0^{corr} is extracted to compute the PS energy scale, α_{PS}. The PS energy scale is determined as a function of η in bins matching the size of the presampler modules leading to $\Delta\eta = 0.2$ in the barrel modules and $\Delta\eta = 0.25$ for the end-cap. The transition region $1.4 < |\eta| < 1.55$ is excluded from the PS energy scale study.

5.4.2.2 Data and Simulation Samples

The determination of the PS energy scale is based on a comparison of the PS energy of electrons from Z boson decays in data and simulation. Data were collected in 2015 and 2016, using a dielectron trigger with E_T thresholds of 12 and 17 GeV, respectively, and requiring the electron candidates to pass identification criteria looser than those applied off-line. For the simulation, a large sample (17 million events) of $Z \to ee$ events generated with POWHEG interfaced with PYTHIA for the parton shower and underlying event model, reweighed with pileup profiles matching those of 2015 and 2016 data, was used. In addition to the nominal geometry simulations, Monte-Carlo samples with additional material (Table 5.1) were used for the study of the correlation between E_0 and $E_{1/2}$. These samples are re-reconstructed with the same 2015–2016 conditions, and each contains about 10 million events. In these alternative simulations, extra material was added to either:

- the whole inner detector $0 < |\eta| < 2.4$ (ID column in Table 5.1);
- the ID end-cap, in the pseudorapidity region $1.8 < |\eta| < 2.4$) (ID-EC column);
- the pixel or SCT services (Pixel S, SCT S);
- the end of the SCT or TRT end-caps, in the pseudorapidity region $1.6 < |\eta| < 2.2$ (SCT-EC, TRT-EC);
- the region between the PS and the accordion either in the barrel (PS/S1-B) or in the end-caps (PS/S1-EC);
- the barrel cryostat before the calorimeter ($0 < |\eta| < 1.6$) (Cryo 1);
- the EM calorimeter end-caps (Calo-EC).

Two additional simulations included either an improved description of the IBL geometry (IBL) or additional material in the IBL and Pixel services (PP0).

The PS energy of electron candidates in the simulation is corrected for mismodeling of the detector material, based on the correlation between E_0 and E_1/E_2 induced by material upstream of the presampler. However, to remove the effect on E_1/E_2 of a mis-modeling of the material between the PS and the first accordion layer, control samples of unconverted photon candidates from $Z \to \mu\mu\gamma$ decays or inclusive $pp \to \gamma + X$ production are studied. Data collected in 2015 and 2016 were used, recorded using either muon and dimuon triggers ($Z \to \mu\mu\gamma$), or single-photon triggers requiring the presence of at least one photon candidate with E_T larger than 140 GeV and passing *loose* photon identification criteria. Simulated $Z \to \mu\mu\gamma$ were generated with SHERPA, using leading-order matrix elements for the real emission of up to three additional partons. In addition, samples of γ+jet events from the hard subprocesses $qg \to q\gamma$ and $qq \to g\gamma$ and photon bremsstrahlung in LO QCD dijet

Table 5.1 The different distorted geometries used for the estimation of the material correction factor $A(\eta)$. The samples have scaled radiation length (X_0) with respect to the nominal MC. The table includes absolute change in number of X_0 for all configurations except for configuration A where there is +5% relative material scaling to the entire Inner Detector

Config	ID	ID-EC	Pixel S	SCT S	SCT/TRT-EC	PS/S1-B	PS/S1-EC	Cryo 1	Calo-EC
A	5%	–	–	–	–	–	–	–	–
N	–	–	–	–	–	–	0.05	–	–
C'+D'	–	–	0.1	0.1	–	–	–	–	–
E'+L'	–	–	–	–	0.075	–	–	0.05	–
F'+M+X	–	0.075	–	–	–	0.05	–	–	0.3
G'	5%	0.075	0.1	0.1	0.075	0.05	0.05	0.05	0.3
IBL	Improved IBL geometry								
PP0	50% increase in IBL + pixel services								

were generated with PYTHIA 8 using the leading-order matrix elements of these $2 \rightarrow 2$ processes. Around 3.6 million MC events were used for this study.

5.4.2.3 Event Selection

Electrons from $Z \rightarrow ee$ decays Events with $Z \rightarrow ee$ candidates are required to contain two electrons with $p_T > 27$ GeV and $|\eta| < 2.47$. Both electrons are required to pass the *medium* likelihood identification and *loose* isolation criteria detailed in Sects. 4.3.1.1 and 4.3.1.2.

Photons from radiative Z boson decays Events with $Z \rightarrow \mu\mu\gamma$ candidates are selected by applying the following selection criteria:

- two opposite charged muons with $p_T^\mu > 12$ GeV;
- an unconverted photon candidate with $p_T^\gamma > 10$ GeV, passing isolation and identification criteria. The photon is required to have low raw PS energy deposit, $E_0^{\text{raw}} < 0.5$ GeV;
- $m_{\mu\mu\gamma} \in [80 - 100]$ GeV and $m_{\mu\mu} \in [50 - 83]$ GeV, to select photons from final state radiation, thus reducing background photons from misidentified hadronic jets in Z+jet(s), $Z \rightarrow \mu\mu$ events;
- ratio of the photon energy in the first accordion layer to the total photon energy $f_1 > 0.1$, in order to reduce hadronic backgrounds.

Inclusive photon samples Events are required to contain at least a photon candidate with transverse momentum $p_T > 147$ GeV to avoid the efficiency turn-on of the photon trigger. The photon is required to pass the *tight* identification criteria. In addition, selected photons must pass a p_T-dependent requirement on the isolation variable `topoetcone40`, which is the sum of the transverse energies of the topological clusters within a cone of $\Delta R = 0.4$ around the photon candidate. The photons are required to pass the requirement `topoetcone40` $< 0.022 p_T + 2.45$ [GeV]. In addition, the photons are required to be unconverted, and a veto on their raw PS energy, $E_0^{\text{raw}} < 0.5$ GeV, is applied to ensure that the photon did not convert in the material between the ID and the PS. This veto results in a selection purity $>99\%$ for conversions with at least one Silicon hit in the ID, and $>95\%$ for conversions with TRT-only hits.

5.4.2.4 Upstream Material Correction $A(\eta)$

The correction factor $A(\eta)$ is estimated from the relative variations of $E_0(\eta)$ versus the corresponding variations in $E_{1/2}(\eta)$ in Monte-Carlo simulation samples with additional material. The η binning is optimized to capture the effects of added material in the different detector regions. For this purpose, ratios of the mean of E_0 and $E_{1/2}$ for the samples with additional material are plotted and fitted using:

$$\frac{E_0^{\text{dist}}(\eta)}{E_0^{\text{nom}}(\eta)} = 1 + A(\eta)\left(\frac{E_{1/2}^{\text{dist}}(\eta)}{E_{1/2}^{\text{nom}}(\eta) b_{1/2}^{\text{MC}}(\eta)} - 1\right). \tag{5.3}$$

The coefficient $b_{1/2}^{\text{MC}}$ is the offset of the linear correlation between E_0 and $E_{1/2}$ due to material after the presampler, that affects only $E_{1/2}$. This is different from the coefficient $b_{1/2}$ of Eq. (5.2), which depends on the actual amount of extra material between the PS and the accordion in data, and is determined using the unconverted photon control samples that are sensitive only to this region. The fit is done initially while fixing $b_{1/2}^{\text{MC}}$ to one to ensure that there will be no correction for the simulation with nominal geometry. Geometry configurations with material added between the presampler and the accordion are not used in the fit since they cause a shift in $E_{1/2}$ but not in E_0. Examples of such correlation plots for the pseudorapidity regions $0.6 < |\eta| < 0.7$ and $1.0 < |\eta| < 1.1$ are shown in Fig. 5.13. The slope of the linear fit is the correction factor $A(\eta)$.

The fit parametrisation used initially was constructed to ensure that no material corrections are applied to the nominal geometry, i.e. the fit was constrained to pass by the point (1,1). This was done by fixing $b_{1/2}^{\text{MC}}$ to 1 in the fit. However, repeating the fit while floating $b_{1/2}^{\text{MC}}$ resulted in a small deviation from 1, of the order of $<1\%$, as shown in Fig. 5.14a. This would result in a change in the fitted $A(\eta)$ of up to $\sim25\%$, as shown in Fig. 5.14b. The reason of this effect is due to the fact that the *likelihood*-based electron identification algorithm used in Run-2 uses as one of its input variables the quantity $f_1 = E_1/E_{\text{tot}}$, which is correlated with $E_{1/2}$. This results in an electron selection efficiency that depends on $E_{1/2}$, as shown in Fig. 5.15a, and hence introduces the bias that we observe. If the electron identification requirement is not applied, as shown in Fig. 5.15b, the correlation slope $A(\eta)$ is the same when fixing $b_{1/2}^{\text{MC}}$ to one or when floating it in the fit. Moreover, as shown in Fig. 5.15b, the values of the correlation slope $A(\eta)$ found without applying the *likelihood*-based

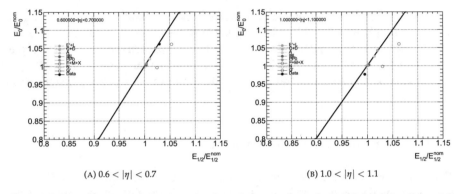

(A) $0.6 < |\eta| < 0.7$ (B) $1.0 < |\eta| < 1.1$

Fig. 5.13 Example of correlation between E_0 and $E_{1/2}$, in the regions **A** $0.6 < |\eta| < 0.7$ and **B** $1.0 < |\eta| < 1.1$. The different geometry configurations are shown, with the closed circles indicating variations only in the material upstream of the PS, that are used in the fit, whereas the open circles are configurations with material added also between PS and accordion, and thus excluded from the fit. The data point (black circle) is plotted for reference as well

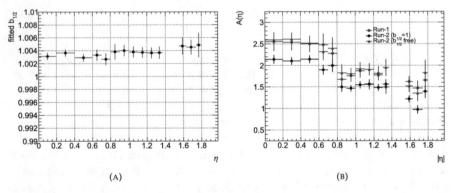

(A)   (B)

Fig. 5.14 **A** Fitted values of $b_{1/2}^{MC}$ from correlation plots. **B** Comparison between $A(\eta)$ between Run-1 (estimated using fixed $b_{1/2}^{MC}$ with a cut-based electron identification algorithm not using f_1 as input) and Run-2 (likelihood-based electron identification algorithm using f_1 as input) for free and fixed $b_{1/2}^{MC}$ where the agreement is evident for the floated $b_{1/2}^{MC}$ case

electron identification algorithm matches the fitted values of $A(\eta)$ with $b_{1/2}^{MC}$ floating. Hence, the value of $A(\eta)$ is taken as the slope of the fitted line while floating $b_{1/2}^{MC}$.

Figure 5.15 shows the final Run-2 values of $A(\eta)$ compared to those determined in Run-1. The values of $A(\eta)$ are compatible between the two LHC runs, which indicate that $A(\eta)$ is not sensitive to pileup. $A(\eta)$ is constant inside the regions $0 < |\eta| < 0.8$, $0.8 < |\eta| < 1.37$, and the end-cap region. The step observed at $|\eta| = 0.8$ is the result of the change in the thickness of the lead absorbers in the accordion. A more precise measurement is thus obtained by averaging $A(\eta)$ within each of these three regions in which it is constant. The results are summarised in Table 5.2.

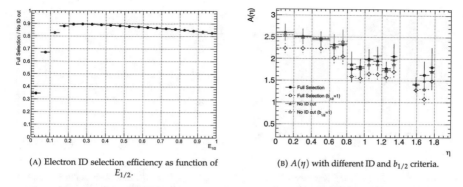

(A) Electron ID selection efficiency as function of $E_{1/2}$.

(B) $A(\eta)$ with different ID and $b_{1/2}$ criteria.

Fig. 5.15 A Electron ID selection efficiency integrated over η as function of $E_{1/2}$. **B** $A(\eta)$ shown with and without the electron ID criteria applied and for the two cases of $b_{1/2}^{MC}$ for the fits (fixed and float). The case without ID criteria (triangles) shows the same behavior for free (solid) and fixed $b_{1/2}^{MC}$ (open)

Table 5.2 Average values of $A(\eta)$ in the three η ranges indicated. The error shown comes only from the fit

η region	$A(\eta)$ Run-2	$A(\eta)$ Run-1
$0.0 < |\eta| < 0.8$	2.51 ± 0.07	2.48 ± 0.09
$0.8 < |\eta| < 1.37$	1.85 ± 0.06	1.65 ± 0.05
$1.55 < |\eta| < 1.8$	1.5 ± 0.1	1.59 ± 0.09

5.4.2.5 $E_{1/2}$ Intercalibration Correction $b_{1/2}$ Using Unconverted Photons

The second part of the material correction procedure is the residual $E_{1/2}$ intercalibration correction, $b_{1/2}$. This correction factor is necessary to account for material mis-modeling between the presampler and the accordion. The measurement of $b_{1/2}$ requires a control sample that is sensitive to sources affecting $E_{1/2}$ but not E_0. This is achieved by using unconverted photon candidates, with an upper limit on the associated raw PS energy to remove upstream material effects as shown in Fig. 5.12. Photons reconstructed as unconverted (conversion radius > 800 mm) will not be sensitive to inner detector material, and to further remove the effect of conversions happening in the material between the inner detector and the PS, such as the calorimeter cryostat and the solenoid, an upper limit is required on the raw PS energy, E_0^{raw}. For the estimation of $b_{1/2}$, low-p_T photons from $Z \rightarrow \mu\mu\gamma$ decays, and an inclusive sample of higher-p_T photons, mainly from QCD Compton scattering $qq \rightarrow q\gamma$, are used to cover a wide range of photon p_T. Once the photon candidates are selected, as described in Sect. 5.4.2.3, the correction $b_{1/2}$ is computed as the double ratio of $E_{1/2}$ between data and simulation.

EC HV problem Investigations of $E_{1/2}^{data} / E_{1/2}^{MC}$ showed an unexpected discrepancy between data and simulation in the end-cap bin compared to Run-1. The requirement

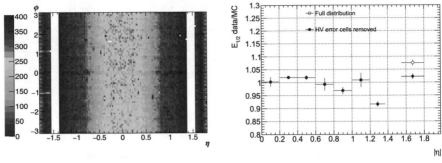

(A) η-ϕ distribution of the average raw PS energy deposits E_0^{raw} in data.

(B) $E_{1/2}$ data/MC ratio with faulty HV cells after removing faulty HV cells.

Fig. 5.16 **A** $\eta - \phi$ distribution of the PS energy for $E_0^{\mathrm{raw}} < 0.5$ GeV. **B** Data/MC ratio of $E_{1/2}$ using unconverted photons with the application of an upper limit on raw PS energies, $E_0^{\mathrm{raw}} < 0.5$ GeV, with and without the HV cells with the wrong mapping

on E_0^{nom} led to an increase in the double ratio $E_{1/2}^{\mathrm{data}} / E_{1/2}^{\mathrm{MC}}$ for both the inclusive photon and $Z \rightarrow \mu\mu\gamma$ control samples. This is not expected since there was no material added in the region between the PS and the accordion between Run-1 and Run-2. Investigating the $\eta - \phi$ distribution of raw PS energy deposits E_0^{raw} in data, shown in Fig. 5.16a, revealed the localisation of problematic cells in ϕ extending across the negative η region of the end-caps (at about $\phi = \pm 1$).

This issue was traced back to a wrong interpretation of the HV mapping in the endcap. The HV system in the PS is organized into 32 modules in ϕ with 2 cells in each module and 2 gaps per cell. One HV line powers the two gaps of one cell. However, in the simulation, it is assumed that one HV line powers one gap of each of the two cells in ϕ (similar to other parts of the calorimeter). Whenever the HV settings are changed from the nominal values, a correction factor for the drift time dependence on the HV is applied to the simulation. However, due to the mismatch in the description of HV lines, the correction was averaged over the two cells instead of having a proper correction per cell. For this reason, it was decided to exclude these faulty cells from this study. Removing these cells makes the observed $E_{1/2}^{\mathrm{data}} / E_{1/2}^{\mathrm{MC}}$ in the endcap much closer to one, as shown in Fig. 5.16b, consistently with previous studies.

Combination of the two methods and systematic uncertainties The values of the double ratio $E_{1/2}^{\mathrm{data}} / E_{1/2}^{\mathrm{MC}}$ measured with inclusive photons and photons from radiative $Z \rightarrow \mu\mu\gamma$ decays are shown in Fig. 5.17. They are in good agreement with each other and close to one within less than 5%. They are then combined using a weighted average, and the combined value is used as the final correction factor $b_{1/2}$ for the measurement of the PS energy scale in data.

The following systematic uncertainties affect the measurement of $b_{1/2}$:

- Choice of raw PS energy upper limit. Different E_0^{raw} upper limits were investigated to optimize the trade-off between selection efficiency and bias due to remain-

Fig. 5.17 Double ratio $E_{1/2}^{\text{data}}/E_{1/2}^{\text{MC}}$ for inclusive photons (in blue), photons from $Z \to \mu\mu\gamma$ decays (in red), and their combination (in black) estimated using a weighted average of the two separate measurements

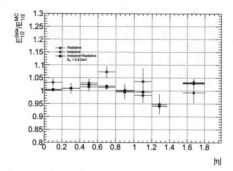

ing upstream material effects. Studies on Monte-Carlo simulated samples with added material between the PS and the accordion (configuration $F' + M + X$ in Table 5.1) and with material upstream of the calorimeter and between the PS and the accordion (configuration G' in Table 5.1) showed that a veto up to a value of 1.2 GeV on the raw PS energy can remove the effects of upstream material. The ratios of $E_{1/2}$ in the simulation with the added material to the nominal simulation obtained with either a 0.5 GeV veto or a 1.2 GeV veto were found in agreement, as shown in Fig. 5.18a. However, the double ratio $E_{1/2}^{\text{data}}/E_{1/2}^{\text{MC}}$ using the two alternative vetoes yielded a difference of ∼1% in the barrel, as shown in Fig. 5.18b. This difference is not the result of the material mis-modeling, as good closure is observed in the simulation, and is thus considered as a systematic uncertainty on $b_{1/2}$.

• In the region $1.2 < |\eta| < 1.37$ the double ratio $E_{1/2}^{\text{data}}/E_{1/2}^{\text{MC}}$ for photons was found to decrease, while no similar effect was observed for electrons. For this reason, the nominal central value of $b_{1/2}$ for this bin was taken from the previous η bin, and the full difference between the two bins was assigned as a systematic uncertainty.

5.4.2.6 Closure Test

A closure test of the procedure used to determine the PS energy scale was performed. This is done by applying the previous procedure to extract the PS energy to the simulation based on the detector configuration G' of Table 5.1. Configuration G' includes additional material upstream of the PS, and between the PS and the accordion. The same values of $A(\eta)$ estimated in Table 5.2 are used since they are estimated only from the simulation. On the other hand, $b_{1/2}$ is estimated from simulated event samples of inclusive photons in the same modified geometry (G') and extracted as the double ratio of $E_{1/2}$ between configuration G' and the nominal sample. The material correction is calculated using the same formula but using $E_{1/2}^{G'}/E_{1/2}^{\text{nom}}$ from $Z \to ee$ as shown in Fig. 5.19. The final PS energy scale is thus expected to be one if the procedure works well. The estimated PS energy scale was found to be compatible with $\alpha_{\text{PS}} = 1$, as shown in Fig. 5.20. A slight deviation of about 2% is observed in

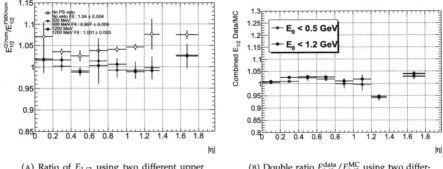

(A) Ratio of $E_{1/2}$ using two different upper limits on PS energy using simulated samples with added material.

(B) Double ratio $E_{1/2}^{\text{data}}/E_{1/2}^{\text{MC}}$ using two different upper limits on PS energy.

Fig. 5.18 **A** Ratio of $E_{1/2}^{\text{dist/nom}}$ between config-G' and F'+M+X under the two raw PS energy upper limits. The averaged (fitted) results of the 1.2 GeV and 0.5 GeV vetoes are found to be compatible. **B** Ratio of data/MC $E_{1/2}$ using different vetoes. The plot shows a difference of ∼1% between the two upper limits in the barrel region. This difference is taken as a systematic uncertainty on $b_{1/2}$ and propagated to the final PS energy scale

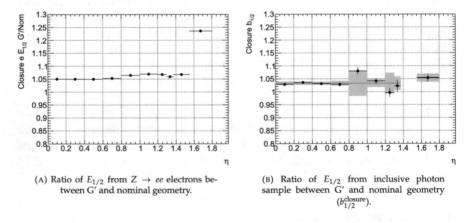

(A) Ratio of $E_{1/2}$ from $Z \to ee$ electrons between G' and nominal geometry.

(B) Ratio of $E_{1/2}$ from inclusive photon sample between G' and nominal geometry ($b_{1/2}^{\text{closure}}$).

Fig. 5.19 **A** Ratio of $E_{1/2}$ from $Z \to ee$ electrons between G' and nominal geometry. The high ratio in the endcap is from the 30% scaled X_0 of the material in front of the endcaps in geometry G'. **B** fitted value of $b_{1/2}$ estimated from $E_{1/2}$ ratio between G' and nominal MC using inclusive photons after PS veto

the barrel. This can be justified by the difference of ≈1–2% in $E_{1/2}$ between photons and electrons in the simulation with added material after the PS [10].

Another check was performed on the PS energy scale formula to check the effect on the PS energy scale of the $E_{1/2}$ intercalibration corrections derived in Sect. 5.4.1. The $E_{1/2}$ intercalibration corrections are applied to E_2 of the G' sample used for the closure test. The PS energy scale estimation formula is parameterized by correcting the $E_{1/2}$ data/MC ratio to any residual effects of material after the PS using $b_{1/2}$. Both terms $E_{1/2}$ and $b_{1/2}$ include layer intercalibration corrections, and hence the

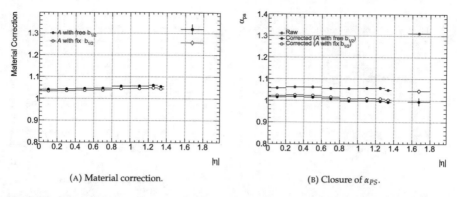

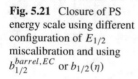

(A) Material correction.

(B) Closure of α_{PS}.

Fig. 5.20 A Total material correction estimated from $A(\eta)$, $b_{1/2}$ and $E_{1/2}$ double ratios between G' and nominal MC. **B** Final PS energy scale in geometry G' compatible with $\alpha_{PS} = 1$ within the $1 - 2\%$ uncertainty on $E_{1/2}$ due to differences between electrons and photons [10]

Fig. 5.21 Closure of PS energy scale using different configuration of $E_{1/2}$ miscalibration and using $b_{1/2}^{barrel,EC}$ or $b_{1/2}(\eta)$

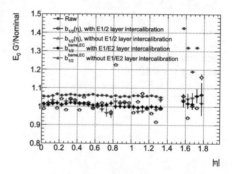

corrections cancel out, as shown in Fig. 5.21. This cancellation yields a measurement of the PS energy scale that is independent of the $E_{1/2}$ corrections. However, a requirement for this cancellation is that the material correction is performed using $b_{1/2}$ from photons with the same binning as $E_{1/2}$ from electrons. The final PS energy scale estimation uses an average $b_{1/2}$ values of in the barrel and the endcap ($b_{1/2}^{barrel}$, $b_{1/2}^{EC}$). This is done to reduce the effect of the fluctuations of finely binned $b_{1/2}$ on the final PS energy scale. Figure 5.21 shows the effect of the different cases of the $b_{1/2}$ binning on the PS energy scale.

5.4.2.7 Total Material Correction and Extraction of the PS Energy Scale

The final material correction is computed after plugging in the values of $A(\eta)$ and $b_{1/2}$ with $E_{1/2}^{data}/E_{1/2}^{MC}$ from $Z \rightarrow ee$ with binning $\Delta\eta = 0.05$ along with E_0^{data}/E_0^{MC}, as shown in Fig. 5.22. The final PS energy scale α_{PS} is then extracted. A module average is then computed in bins of width $\Delta\eta = 0.2$ in the barrel and $\Delta\eta = 0.25$

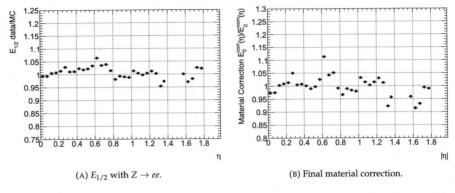

(A) $E_{1/2}$ with $Z \to ee$.

(B) Final material correction.

Fig. 5.22 **A** Data/MC ratio of $E_{1/2}$ from electrons. **B** Final material correction estimated from $A(\eta)$, $b_{1/2}$ and $E_{1/2}$ from electrons as function of η

Fig. 5.23 Effect of different $b_{1/2}$ averages on the final PS energy scale compared to Run-1 results (shaded blue uncertainty). The difference is largest in the bin $1.2 < |\eta| < 1.37$, where $b_{1/2}$ shows a sudden decrease

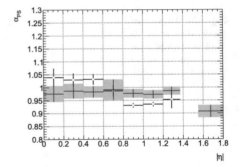

in the endcap. The RMS of the measurements in a bin is then taken as a systematic uncertainty and added in quadrature with the statistical and systematic uncertainties propagated from the $A(\eta)$ and $b_{1/2}(\eta)$ estimates, to obtain the total error for the energy scale in each module.

The PS energy scale was found to be sensitive to the choice of the $b_{1/2}$ η binning, as expected from the closure test studies. This is shown in Fig. 5.23, where a trade-off is observed between taking finer bins, which induces larger statistical fluctuations in the PS energy scale, and using a value of $b_{1/2}$ averaged over the whole barrel, which gives more stable results, but can hide some η-dependent features. It was chosen to use for the final PS energy scale measurement the value of $b_{1/2}$ computed in bins of width $\Delta\eta = 0.2$ (red points in Fig. 5.23).

The final values of the PS energy scale are shown in Fig. 5.24. They were found to agree with those measured in Run-1. The behavior of the PS energy scale in the region $\eta \in [1.3 - 1.37]$ is a residual effect of the $E_{1/2}$ intercalibration corrections and of $b_{1/2}$ that show a sudden decrease in this region, and do not entirely cancel with electrons as explained in Sect. 5.4.2.6. The final PS energy scale systematic uncertainties are summarised in Fig. 5.25.

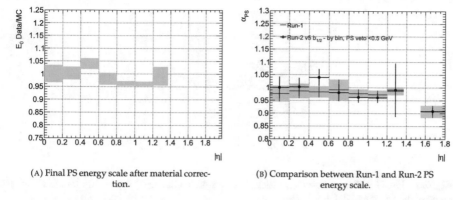

(A) Final PS energy scale after material correction.

(B) Comparison between Run-1 and Run-2 PS energy scale.

Fig. 5.24 **A** Final PS energy scale after material correction in binning of $\Delta\eta = 0.05$ and final average in η modules is shown in blue lines for 2015+2016 data, the different systematic uncertainties are shown for each module. **B** Comparison between Run-2 PS energy scale values per cells compared to Run-1 module average, the plots shows compatibility between Run-1 and Run-2. The Run-2 errors are the sum in quadrature of the different error sources

Fig. 5.25 Systematic errors for the PS modules energy scales (width $\Delta\eta = 0.2$). The total error is taken as the sum in quadrature of the different error sources

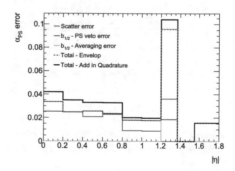

5.4.2.8 PS Energy Scale Stability

PS energy scale variation in η and ϕ The PS energy scale was estimated using E_0 and $E_{1/2}$ as function of the pseudorapidity (η) to examine the symmetry of the energy scale around $\eta = 0$. The energy scale was found to be symmetric within the uncertainty of the module average, as shown in Fig. 5.26, hence the final PS energy scale is estimated only as function of the absolute value of the pseudorapdidity.

The dependence of the PS energy scale on the azimuthal angle ϕ was checked using a ϕ-dependent material correction of E_0^{nom} of the kind:

$$\frac{E_0^{\mathrm{corr}}(\eta, \phi)}{E_0^{\mathrm{nom}}(\eta, \phi)} = 1 + A(\eta)\left(\frac{E_{1/2}^{\mathrm{data}}(\eta, \phi)}{E_{1/2}^{\mathrm{nom}}(\eta, \phi)b_{1/2}(\eta)} - 1\right). \tag{5.4}$$

The correction was done using the ϕ integrated values of $A(\eta)$ and $b_{1/2}(\eta)$. The material correction in this case properly corrects the $E_0^{\mathrm{data}}/E_0^{\mathrm{nom}}$ distributions since

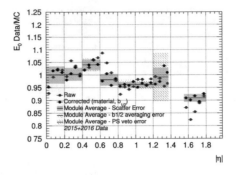

Fig. 5.26 Final PS energy scale after material correction in binning of $\Delta\eta = 0.05$ for the whole η range of the PS using 2015+2016 data

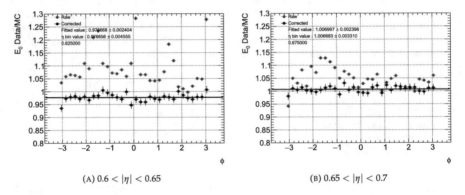

(A) $0.6 < |\eta| < 0.65$ (B) $0.65 < |\eta| < 0.7$

Fig. 5.27 Examples of the final PS energy scale after material correction in binning of $\Delta\eta = 0.05$ using 2015+2016 data. The central η value of each cell is indicated in each plot

they are flat as a function of ϕ. An example of the PS energy scale distribution along ϕ is shown in Fig. 5.27. No significant dependence on ϕ is observed.

Material mis-modeling at $\eta \simeq 0.6$ Investigating data/MC ratios for E_0 and $E_{1/2}$ as function of ϕ, an unexpected periodic structure was observed around $\eta = 0.6$ for $\phi \approx 0, \pm\pi/2$, and $\pm\pi$, as shown in Fig. 5.27. This discrepancy between data and the simulation was observed for both E_0 and $E_{1/2}$, shown for $E_{1/2}$ in the left plot of Fig. 5.28, meaning that the discrepancy is due to upstream material. The discrepancy was properly corrected using the material correction procedure of Sect. 5.4.2.7 as can be seen in Fig. 5.27. An estimation of the discrepancy in terms of change in radiation length ΔX_0, shown in Fig. 5.28, was performed using the material estimation formula

$$\Delta X_0(\eta) = \Delta E_{1/2}^{\text{data}}(\eta)\left(\partial X / \partial_{rel} E_{1/2}\right)(\eta),$$

where $\Delta E_{1/2}^{\text{data}}(\eta)$ is the relative E_1/E_2 differences between data and nominal MC simulations, and $\partial X / \partial_{rel} E_{1/2}(\eta)$ is the sensitivity of E1/E2 to differences in passive

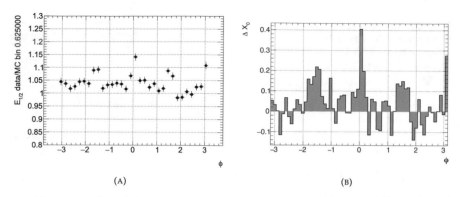

Fig. 5.28 **A** Ratio of $E_{1/2}$ between data and simulation for the bin in $0.6 < |\eta| < 0.65$. **B** Material mis-modeling in terms of ΔX_0 in the bin in $0.6 < |\eta| < 0.65$ as a function of ϕ

material from MC simulated samples with added material (detailed in Sect. 5.4.3). Further investigations showed that this mis-modeling is related to TRT services. The detailed TRT services are not included in the simulation but rather are enveloped, and the periodic structure is related to aluminum pillars used to slide the TRT barrel in case of LAr leakage. The effect of the mis-modeling is mitigated using the material correction and hence does not affect the PS energy scale.

PS high voltage change effect The HV modules in the PS are organized in 8 regions in η, and 32 regions in ϕ. Due to noise issues that first appeared in 2010, the nominal HV in some PS barrel sectors was lowered from 2000 V to 1600 V and then later to 1200 V, as discussed in Sect. 5.3. A correction factor is applied to the drift time to account for its dependence on the HV. The PS energy scale was checked to determine if further corrections are needed. In June 2016, 25 HV lines were changed from 1200 V to 1000 V, two lines from 800 V to 700 V. Most of the time both gap sides changed at the same time. Therefore, the effect of the lowered HV on the PS energy scale was checked by splitting data and simulation into two data sets corresponding to data-taking conditions before and after the HV lowering. The PS energy scale values estimated in the two subsets of the data and the simulation were found to be in agreement, as shown in Fig. 5.29. The exact value of PS energy scales for the modules with the HV changes was also checked, and the resulting PS energy scale shows that the material correction is sufficient to correct the E_0 data/MC ratio and that no residual correction is needed for the HV change, as shown in Fig. 5.30.

5.4.3 Passive Material Estimation

As shown in the previous sections, a good description of the detector geometry is an essential part of the Monte Carlo-based calibration of the energy response of the ATLAS EM calorimeter. Therefore, using the L1/L2 calibration corrections of

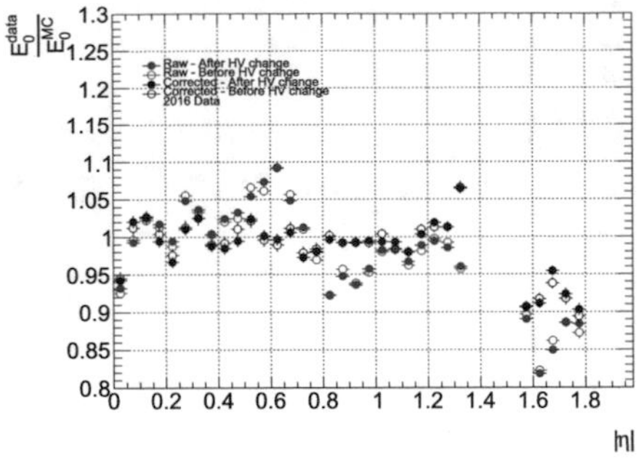

Fig. 5.29 Comparison between the PS energy scale estimated using the 2016 data before and after the HV change

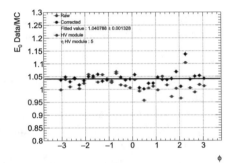

Fig. 5.30 An example of PS energy scale in HV modules. The results show that the E_0 data/MC ratio is properly corrected using only the material correction, with no need for further correction. The blue dots show the results for the exact HV modules that were affected by the change

Sect. 5.4.1, the $E_{1/2}$ distribution observed for EM showers in data can be used to quantify the amount of detector material upstream of the calorimeter as higher values of $E_{1/2}$ in data would indicate earlier shower development, and hence a local excess of material in comparison with the simulation. The usage of $E_{1/2}$ to estimate the passive material came after attempts of estimating the material using shower shape variables of high p_T electrons [11]. Using shower shape variables to estimate the material budget did not succeed because the GEANT 4 simulation poorly models the lateral shower profile (R_η, $\omega_{\eta 2}$, F_{side} described in Sect. 4.3.1.1) [12].

An estimation of the relative difference in radiation length $\Delta X/X_0$ with respect to the nominal simulation can be computed from the following formula

$$\Delta X/X_0 = \Delta E_{1/2}^{data} \left(\frac{\partial X/X_0}{\partial_{rel} E_{1/2}} \right). \tag{5.5}$$

with $\Delta E_{1/2}^{data}$ denoting the relative difference of $E_{1/2}$ between data and the simulation after calibration corrections detailed in Sect. 5.4.1. The term $\left(\frac{\partial X/X_0}{\partial_{rel} E_{1/2}} \right)$ is a material sensitivity factor relating the change in $E_{1/2}$ to the change in material in terms of X_0. This sensitivity factor is derived from Monte Carlo simulations with distorted geometries similar to those detailed in Table 5.1. The method is based on the evaluation of $E_{1/2} \left(X_0^{injected} \right)$ where $X_0^{injected}$ is the amount of added material in the distorted geometry simulations. The material sensitivity curve using electrons

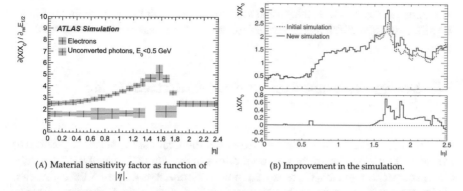

(A) Material sensitivity factor as function of
$|\eta|$.

(B) Improvement in the simulation.

Fig. 5.31 **A** Material sensitivity factor $\frac{\partial X/X_0}{\partial_{rel} E_{1/2}}$ plotted as a function of $|\eta|$, for material variations upstream of the PS using electrons, and for variations between the PS and strips for unconverted photons with $E_0 < 500$ MeV. The shaded bands represent the systematic uncertainty due to the dependence on the location of the material additions. **B** Amount of material traversed by a particle (X/X_0) as a function of $|\eta|$ for the base simulation used in Run-1 and the improved simulation used in Run-2 [1]

Fig. 5.32 The data-Monte
Carlo differences between
the material estimate
$(\Delta X/X_0)$ using Run-2 data
[13]

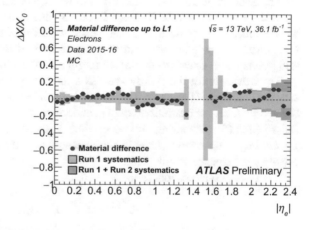

and unconverted photons with low PS activity is shown in Fig. 5.31a. A detailed estimation of the material was performed using Run-1 data and results were used also to improve the detector simulation to better describe the data, namely an improvement on the description of the SCT heater tubes as shown in Fig. 5.31b.

An estimation of the passive material was performed in Run-2 to assess the material variation of the detector during the shutdown (LS1) between Run-1 and Run-2. The main material changes in Run-2 are the insertion of IBL and the installation of a new Pixel patch panel (PP0). Therefore, dedicated distorted geometries were used with X_0 scaling of these regions. The systematic uncertainties of the material estimation arise from various sources, specifically from the uncertainties of the sensitivity curves and the L1/L2 calibration. Any residual discrepancy is then taken as final uncertainty; therefore, additional systematic uncertainties were estimated from the difference in the PP0 region simulation geometry, this is shown in Fig. 5.32.

5.5 Energy Scale and Resolution Measurement Using $Z \rightarrow ee$ Decays

After the application of the Monte Carlo-based multivariate regression algorithm to data and simulation, the uniformity corrections to the simulation, and the layer calibration corrections to data, the next step in the calibration procedure is the estimation of the difference in energy scale between data and simulation. Despite the various corrections applied, there remain discrepancies between data and the simulation. The sources of these discrepancies are not precisely known, and they are corrected using energy scale factors, α, measured in-situ. The Z-boson decays to electrons are used to estimate the energy scale factors, as the Z-boson mass is known very precisely from the LEP experiments, $m_Z = 91.1875 \pm 0.0021$ GeV [14].

The Z-boson is used as a standard "candle", given its copious production, and its clean dielectron final state $Z \rightarrow ee$. However, due to the absence of similar candles for photons, $Z \rightarrow ee$ decays are also used to estimate photon energy scale factors through an extrapolation procedure that accounts for differences in the reconstruction and possible non-linearity effects that might affect electrons and photons differently.

The extraction of the energy scale is a delicate process, as various sources can bias the outcome of the result. This is due to the interplay between the resolution and the energy scale, as resolution effects can induce shifts in the measured value of the mass peak [15]. Therefore, any uncertainty in the detector resolution will result in uncertainty in the energy scale. For this reason, both the energy scale and resolution are always extracted simultaneously.

The energy scale factors α are defined as:

$$E^{\text{data}} = E^{\text{MC}}(1 + \alpha_i), \tag{5.6}$$

where E^{data} and E^{MC} are the electron energy in data and simulation. The i index represent bins of pseudorapidity. For $Z \rightarrow ee$ decays, the dielectron invariant mass m_{ee} is computed from $m_{ee} = \sqrt{2E_1 E_2(1 - \cos\theta_{12})}$, where θ_{12} is the opening angle between the two electrons measured by the tracker and E_1, E_2 are their energies. Using Eq. (5.6), one finds that

$$m_{ee}^{\text{data}} = m_{ee}^{\text{MC}} \sqrt{(1 + \alpha_i)(1 + \alpha_j)}, \tag{5.7}$$

where i and j are the pseudorapidity bins of the two electron candidates. Expanding in α and keeping only first order terms and assuming that θ_{12} is known with a resolution significantly better than the energies E_1 and E_2, Eq. (5.7) can be approximated as:

$$m_{ee}^{\text{data}} \simeq m_{ee}^{\text{MC}}(1 + \frac{\alpha_i + \alpha_j}{2}) \equiv m_{ee}^{\text{MC}}(1 + \alpha_{ij}), \qquad (5.8)$$

with

$$\alpha_{ij} \equiv \frac{\alpha_i + \alpha_j}{2} \qquad (5.9)$$

Likewise, the discrepancy in the resolution of the calorimeter between data and the simulation is characterized by an additional constant term c_i' (detailed in Sect. 4.2.4.1). This term will *smear* the energy of electrons in the simulation with a Gaussian distribution, $\mathcal{N}(\mu = 0, \sigma = 1)$, as follows:

$$E_i^{\text{data}} = E_i^{\text{MC}}(1 + c_i \times \mathcal{N}(0, 1)) \qquad (5.10)$$

The smearing will lead to a larger width of data with respect to the simulation :

$$\left(\frac{\sigma(E)}{E} \right)_i^{\text{data}} = \left(\frac{\sigma(E)}{E} \right)_i^{\text{MC}} \oplus c_i. \qquad (5.11)$$

The correlation between the scale and resolution will then manifest itself because the resolution smearing propagates to the invariant mass shape of two electrons falling in pseudorapidity bins i and j:

$$m_{ee}^{\text{data}} = m_{ee}^{\text{MC}} \sqrt{(1 + c_i \times \mathcal{N}_i(0, 1)) (1 + c_j \times \mathcal{N}_j(0, 1))}. \qquad (5.12)$$

Consequently, the relation between the dielectron invariant mass resolution $\frac{\sigma(m)}{m}$ in data and each electron's smeared energy resolution will be given by:

$$
\begin{aligned}
\left(\frac{\sigma(m)}{m} \right)_{data}^2 &\simeq \frac{1}{4} \left((\frac{\sigma(E_1)}{E_1})_{MC}^2 + c_i^2 + (\frac{\sigma(E_2)}{E_2})_{MC}^2 + c_j^2 \right) \\
&= \left(\frac{\sigma(m)}{m} \right)_{MC}^2 + \frac{c_i^2 + c_j^2}{4} \\
&= \left(\frac{\sigma(m)}{m} \right)_{MC}^2 + \frac{c_{ij}^2}{2},
\end{aligned}
\qquad (5.13)
$$

with

$$c_{ij}^2 \equiv \frac{c_i^2 + c_j^2}{2} \qquad (5.14)$$

denoting the effective relative invariant mass resolution correction for the two electrons in pseudorapidity bins i and j.

5.5.1 Methodology

As described in the previous section, the correlation between the energy scale and the resolution impose the simultaneous extraction of both parameters (α_{ij}, c'_{ij}) in i, j pseudorapidity bins. For this purpose, two alternative methods are used, and the difference between the two methods is used as a systematic uncertainty:

- The *template fit method*, introduced in Ref. [15], is based on templates of m_{ee} from Monte Carlo simulations of Z decays obtained while shifting the mass scale and smearing the resolution in a range covering the expected uncertainty in narrow steps, resulting in a two-dimensional grid of (α_{ij}, c_{ij}). The templates are built separately for the electron pseudorapidity in pseudorapidity bins (η_i, η_j). The optimal values, uncertainties and correlations of α_{ij} and c_{ij} are then obtained by χ^2 minimization with a similar configuration of pseudorapidity bins for data. An illustration of the method is shown in Fig. 5.33. The individual electron scale factors α_i, c'_i are then obtained by an inversion procedure using Eqs. (5.5)–(5.14) as detailed in Ref. [16].
- The *lineshape method* on the other hand uses an analytic probability density function (PDF) to parameterize the invariant mass distributions in pseudorapidity (η_i, η_j). The PDF used for the parametrization of the invariant mass distributions is a sum of three Gaussian functions. The shapes of the PDFs are fixed using a fit to distribution in the bins (η_i, η_j) in the simulations. The parameters of the PDFs for the corresponding data distributions are then expressed in terms of the simulation parameter values, corrected by the energy scales α_i, α_j and the additional constant terms c_i, c_j. The parameters $\alpha_{i,j}, c_{i,j}$ are then determined from a fit to the data distributions.

5.5.2 Results and Systematic Uncertainties

Energy scale and resolution corrections measured with 2015–2016 data are shown in Fig. 5.34. The results show that the energy scale corrections range between -3% and $+2\%$ with an uncertainty between 0.02% and 1% depending on pseudorapidity. The additional constant term of the energy resolution was found to be typically less than 1% in most of the barrel region and between 1% and 2% in the end-cap region, with an uncertainty between 0.03% and 0.6%.

The sources of uncertainty are the following:

- *Method Comparison*. The difference between the results of the two methods discussed in Sect. 5.5.1, yields a systematic uncertainty of at most 0.1% for α_i and at most 0.2% for c_i. In addition, another uncertainty is due to the bias on each method estimated from pseudo-experiments varying in the range $(0.001 - 0.01)\%$ for α_i and $(0.01 - 0.03)\%$ for c_i. An additional uncertainty also results from the non-

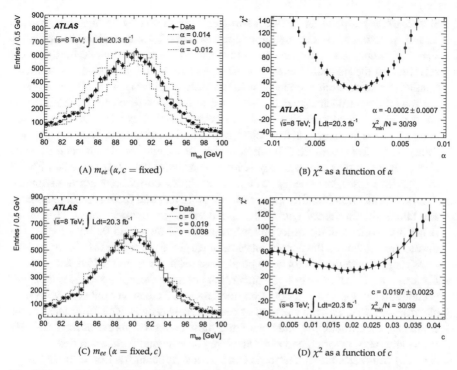

(A) m_{ee} (α, c = fixed)

(B) χ^2 as a function of α

(C) m_{ee} (α = fixed, c)

(D) χ^2 as a function of c

Fig. 5.33 An illustration of the template-fit method for the extraction of the energy scale α and resolution c from m_{ee} built from $Z \rightarrow ee$ decays in data and Mont Carlo in pseudorapidity bins $1.63 < \eta_i < 1.74$, $2.3 < \eta_j < 2.4$ [1]

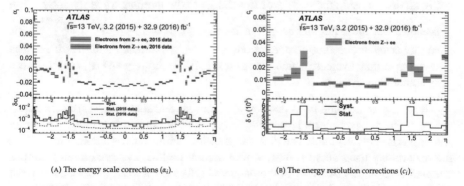

(A) The energy scale corrections (α_i).

(B) The energy resolution corrections (c_i).

Fig. 5.34 Measured **A** energy scale and **B** resolution corrections as a function of η using $Z \rightarrow ee$ events in 2015 and 2016 data. The systematic and statistical uncertainties are shown separately in the bottom panels [7]

closure of the template-fit method, estimated from samples with injected non-zero α_i and c_i, giving a non-closure uncertainty of 0.004% for α_i and 0.02% for c_i.

- *Event Selection.* These uncertainties result from biases introduced by the event selection on the measured energy scale and resolution corrections. Such uncertainties can arise from the electron identification criteria, as small correlations between the electron energy response and the quality of the electron identification are expected. The electron isolation requirement as well can be a source of such biases, since residual effects from electrons not originating from vector boson or τ decays can affect the results. The choice of the m_{ee} range can also affect the results if non-Gaussian tails of the energy resolution are not accurately modeled. For all the previous cases, the energy scale and resolution corrections are re-estimated varying the electron identification, electron isolation, and the mass range criteria, and taking the difference with the nominal values as an uncertainty. The uncertainty due to electron identification is at most 0.2% for c_i and 0.003% for α_i. For the electron isolation, the uncertainty is at most 0.5% for c_i and 0.15% for α_i. The mass range uncertainty is found to be at most 0.4% for c_i and 0.35% for α_i.

- *Background.* This is a fairly small uncertainty coming from $Z \rightarrow \tau\tau$, diboson pair production and top-quark production, leading to a dielectron final state with both electrons originating from τ-lepton or vector-boson decays. The differences in the results for α_i, c_i including and neglecting these backgrounds are at most 0.005% for α_i and 0.004% for c_i, and are considered as systematic uncertainties.

- *Effect of Bremsstrahlung.* Electrons can lose a significant fraction of their energy by bremsstrahlung before reaching the calorimeter. The effect of the modeling of bremsstrahlung on the estimation of α_i and c_i is performed by imposing requirements on the fraction of electron bremsstrahlung f_{brem}, determined from the tracking algorithm using the ratio $\frac{q}{p}$ of the charge to the momentum at the interaction point and at the outer radius of the tracker, $f_{brem} = 1 - \left(\frac{q}{p}\right)^{IP} / \left(\frac{q}{p}\right)^{out}$. The energy scale and resolution corrections are then re-estimated and the differences with the nominal values, which are at most 0.1% for both c_i and α_i, are taken as a systematic uncertainty.

The $Z \rightarrow ee$ invariant mass distributions for data and simulations after the application of the energy scale and resolution corrections are shown in Fig. 5.35. The plot shows a fair agreement between data and the simulation. The differences are within the uncertainty band being shown, which includes all the systematic uncertainties detailed before. The decrease in the data/MC ratio near a mass of 96 GeV is most likely related to imperfect modeling of the tails of the energy resolution by the simulation, which affects the extraction of the energy scale and resolution correction factors.

Fig. 5.35 A shape comparison of the invariant mass distribution of the two electrons in the selected $Z \rightarrow ee$ candidates, after the energy scale calibration and resolution corrections are applied. The uncertainty band of the bottom plot represents to the impact of the uncertainties in the energy scale and resolution correction factors [7]

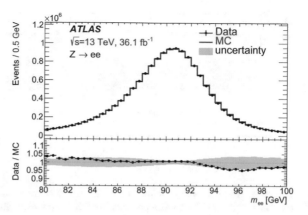

5.6 Energy Scale Validation

The global energy scale corrections extracted from $Z \rightarrow ee$ decays are assumed to be effectively correcting the electromagnetic calorimeter energy response for electrons and photons of any energy. Therefore, a verification process is needed to validate these energy scales for photons and for electrons at energies different from those of electrons from $Z \rightarrow ee$ decays.

5.6.1 Energy Scale Extrapolation to Photons

The electron and photon behavior is not identical in the electromagnetic calorimeter (as detailed in Sect. 4.3.1.1). Therefore, an electron-to-photon extrapolation is performed. The extrapolation is performed assuming that the in-situ energy scale corrections extracted from $Z \rightarrow ee$ are also valid for photons within the computed uncertainties. These energy scale corrections are then validated using a sample of photons from final state radiation in $Z \rightarrow \ell\ell\gamma$ ($\ell = e, \mu$) decays. This analysis is performed separately for unconverted, one-track, and two-track converted photons and using the electron and the muon channels separately and then combined.

The residual energy scale difference between data and simulation is quantified using $\Delta\alpha_i$. The residual correction factor $\Delta\alpha_i$ will be an additional correction to be applied for photons on top of the full calibration chain to recover the correct photon energy response. The value of $\Delta\alpha_i$ is computed by modifying the invariant-mass distribution of the three-body $\ell\ell\gamma$ system after applying all the corrections from electrons using the double ratio

$$R(\alpha_i) = \frac{\langle m(\ell\ell\gamma(\alpha_i))_{\text{data}}\rangle / \langle m(\ell\ell)_{\text{data}}\rangle}{\langle m(\ell\ell\gamma)_{\text{MC}}\rangle / \langle m(\ell\ell)_{\text{MC}}\rangle}, \qquad (5.15)$$

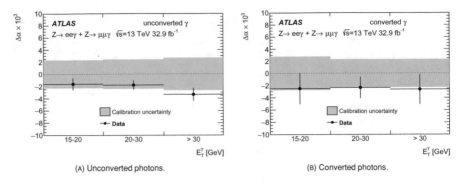

(A) Unconverted photons.   (B) Converted photons.

Fig. 5.36 Residual energy scale factor, $\Delta\alpha$, for **A** unconverted and **B** converted photons with their uncertainties. The band represents the full energy calibration uncertainty for photons from $Z \to \ell\ell\gamma$ decays [7]

where $\langle m(\ell\ell\gamma) \rangle$ and $\langle m(\ell\ell) \rangle$ are the average values of the three-body and two-body invariant mass distributions of the selected $Z \to \ell\ell\gamma$ and $Z \to \ell\ell$ candidates, respectively. The use of the double ratio $R(\alpha_i)$ suppresses the lepton energy scale uncertainties. The residual corrections were computed using 2015+2016 data. The residual corrections are shown in Fig. 5.36 as function of the photon energy, and were found to be consistent with zero within the calibration uncertainties.

Additional sources of uncertainty for the photon energy scale are the following:

- *Photon Conversion classification*. The MVA algorithm described in Sect. 5.2 is trained separately for candidates reconstructed as converted or unconverted photons. Therefore, any misclassification of the conversion category can lead to biases in the calibration. The fake rate (i.e. the fraction of unconverted photons reconstructed as converted) is typically between 1% and 4%, depending on η and pileup conditions [17]. The longitudinal shower shape of the photon candidates is used to provide statistical discrimination between genuine converted and unconverted photons and to estimate the efficiencies and fake rate in both data and simulation, with efficiency typically of 90%. The impact on the photon energy measurement is estimated from the difference between the original Monte Carlo simulation and another sample reweighted with the data-to-MC ratio of efficiencies and fake rates. This results in uncertainties in the energy scale of photons with $E_T = 60$ GeV of about 0.05%.
- *Modeling of the lateral shower shape*. This takes into account the difference between electron and photon showers related to the different interaction probabilities with the material upstream of the calorimeter. The lateral energy leakage outside of the cluster is studied for data and simulation, and the difference is taken as an uncertainty. Figure 5.37 shows the distribution of the energy leakage, defined as the difference between a larger second layer cluster size of 7×11 and the nominal size of 3×7 in units of layer-2 cells. An uncertainty is derived from the double ratio of the difference between electrons and photons in data and simulation and was found to be at maximum 0.25%.

Fig. 5.37 Distributions of the lateral leakage in data and simulation for electron and unconverted photon candidates with $E_T > 25$ GeV and $|\eta| < 0.8$. Photons from $Z \to \ell\ell\gamma$ decays are compared with electrons from $Z \to ee$ [7]

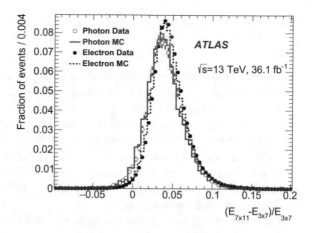

Fig. 5.38 Residual energy scale differences, $\Delta\alpha$, between data and simulation extracted from $J/\psi \to ee$ events as a function of η after the $Z \to ee$ calibration scale factors have been applied. The band shows the uncertainty of the energy calibration for the energy range of $J/\psi \to ee$ decays [7]

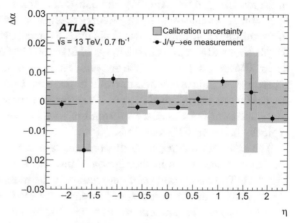

5.6.2 Energy Scale Cross-Checks with $J/\psi \to e^+e^-$

The energy scale at low electron energy ($E_T \sim 10$ GeV) is probed using electrons from $J/\psi \to e^+e^-$ decays, exploiting the fact that the mass of J/ψ resonance is well known. The corrections derived from $Z \to ee$ decays are applied to the selected electron candidates, and the residual energy scale corrections are then derived with a *line-shape* fit similar to the nominal $Z \to ee$ calibration. The difference is quantified with a residual energy difference $\Delta\alpha$, parameterized such that $\Delta\alpha = 0$ if the calibration is correct. Figure 5.38 shows the extracted values of $\Delta\alpha$. The results show a good agreement between the energy scale factors derived with low-energy electrons and the nominal calibration. The residual scale factors are consistent with zero within the systematic uncertainties of the nominal calibration.

5.7   Summary of Uncertainties and Effect on the Higgs Mass

In summary, the final energy scale derived in Sect. 5.5 has 6 main sources of systematic uncertainty (described in their respective sections). Each source is detailed with multiple variations in different regions in $|\eta|$. This yields a systematic uncertainty model with an overall 64 independent uncertainty variations [7]. In addition, a simplified model of the uncertainties is built from the addition in quadrature of the different sources (assuming they are fully correlated across η). The impact of these uncertainties on the photon energy scale is shown in Fig. 5.39. The typical uncertainty of the energy scale is between 0.2% and 0.3% for the barrel region and 0.45% to 0.8% in the end-cap region.

For the energy resolution, the systematic uncertainties arise from the uncertainties in the modeling of the sampling and constant terms in the resolution, the energy loss upstream of the calorimeter, the effect of electronics and pileup noise, and the impact of the residual non-uniformities. This leads to an uncertainty in the energy resolution for photons and electrons in the range of 30 to 60 GeV of the order of 5% to 10% and up to 50% for photons or electrons with energies of several hundreds of GeV. This represents an improvement compared to the Run-1 results reported in Ref. [1]. The impact on the energy resolution of the different sources is summarized in Fig. 5.40. The uncertainty of the energy resolution for photons from Higgs boson decays is typically between 10% and 20%.

The calibration detailed in this chapter was used to perform a measurement of the Higgs-boson mass in the diphoton and the 4-lepton channels, summarized in Fig. 5.41. The main sources of uncertainty in the mass measurement in the diphoton channel are, in decreasing order of importance: the LAr cell non-linearity, the layer calibration, and the uncertainty in non-ID material. The total relative systematic

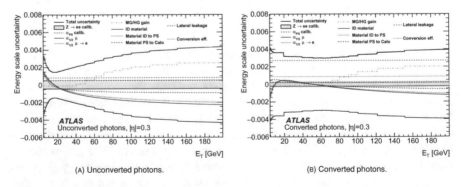

(A) Unconverted photons.   (B) Converted photons.

Fig. 5.39 Fractional energy scale calibration uncertainty for **A** unconverted photons and **B** converted photons, as a function of E_T for $|\eta| = 0.3$. The total uncertainty is shown as well as the main contributions, which are represented by the signed impact of a one-sided variation of the corresponding uncertainty [7]

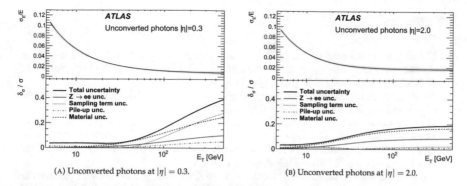

Fig. 5.40 Relative energy resolution, σ_E/E, as a function of E_T for unconverted photons at **A** $|\eta| = 0.3$ and **B** $|\eta| = 2.0$. The yellow band in the top panels shows the total uncertainty in the resolution. The breakdown of the relative uncertainty in the energy resolution, δ_σ/σ is shown in the bottom panels [7]

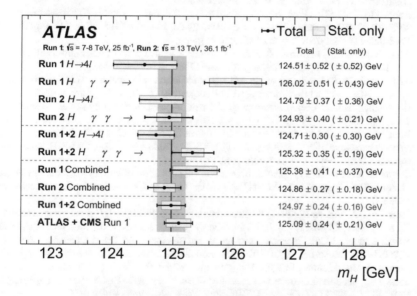

Fig. 5.41 Summary of the Higgs boson mass measurements from the individual and combined analyses using Run-2 analyses and the Run-1 combined measurement with CMS. The statistical-only (horizontal yellow-shaded bands) and total (black error bars) uncertainties are indicated. The (red) vertical line and corresponding (grey) shaded column indicate the central value and the total uncertainty of the combined ATLAS Run 1 + 2 measurement, respectively [18]

uncertainty on the estimated m_H is 0.29% [18]. On the other hand, The measurement in the four-lepton channel has a smaller systematic uncertainty as the measurement is dominated by the four-muon final state, which is affected by much smaller calibration systematic uncertainties.

References

1. The ATLAS Collaboration (2014) Electron and photon energy calibration with the ATLAS detector using LHC Run 1 data. Eur Phys J C74.10:3071. https://doi.org/10.1140/epjc/s10052-014-3071-4. arXiv:1407.5063 [hep-ex]
2. TheATLAS Collaboration (2009) Expected performance of the ATLAS experiment detector, trigger and physics. Tech. rep. arXiv: 0901.0512 [hep-ex]
3. Hoecker A et al (2007) TMVA: toolkit for multivariate data analysis. PoS ACAT, p 040. arXiv: physics/0703039
4. GEANT4 Collaboration (2003) GEANT4: a simulation toolkit. Nucl Instrum Meth A 506:250–303. https://doi.org/10.1016/S0168-9002(03)01368-8
5. Turra R, Manzoni S, Durglishvili A (2017) Monte Carlo energy calibration of electrons and photons for release 20.7. Tech. rep. ATL-COM-PHYS-2017-761. Geneva: CERN. https://cds.cern.ch/record/2268813
6. The ATLAS Collaboration (2008) The ATLAS experiment at the CERN large hadron collider. J Instrum 3.08:S08003–S08003. https://doi.org/10.1088/1748-0221/3/08/s08003. https://doi.org/10.1088%2F1748-0221%2F3%2F08%2Fs08003
7. Aaboud M et al (2019) Electron and photon energy calibration with the ATLAS detector using 2015-2016 LHC proton-proton collision data. JINST 14.03:P03017. https://doi.org/10.1088/1748-0221/14/03/P03017. arXiv:1812.03848 [hep-ex]
8. Abreu H et al (2010) Performance of the electronic readout of the ATLAS liquid argon calorimeters. JINST 5:P09003. https://doi.org/10.1088/1748-0221/5/09/P09003
9. Aad G et al (2010) Readiness of the ATLAS liquid argon calorimeter for LHC collisions. Eur Phys J C70:723–753. https://doi.org/10.1140/epjc/s10052-010-1354-y. arXiv:0912.2642
10. Boonekamp M et al (2013) Electromagnetic calorimeter layers energy scales determination. Tech. rep. ATLCOM- PHYS-2013-1423. Geneva: CERN. https://cds.cern.ch/record/1609068
11. Kuna M et al (2008) Study of material in front of the EM calorimeter with high pT electron shower shapes and tracks. Tech. rep. ATL-PHYS-INT-2008-026. ATL-COM-PHYS-2008-081. Geneva: CERN. https://cds.cern.ch/record/1107811
12. Carminati L et al (2011) Reconstruction and identification efficiency of inclusive isolated photons. Tech. rep. ATL-PHYS-INT-2011-014. Geneva: CERN. https://cds.cern.ch/record/1333390
13. Marc Hupe A, Vincter M (2017) Passive material before the ATLAS EM calorimeter in Run 2. Tech. rep. ATL-COM-PHYS-2017-759. Geneva: CERN. https://cds.cern.ch/record/2268804
14. Schael S et al (2006) Precision electroweak measurements on the Z resonance. Phys Rept 427:257–454. https://doi.org/10.1016/j.physrep.2005.12.006. arXiv:hep-ex/0509008 [hep-ex]
15. Besson N, Boonekamp M (2005) Determination of the absolute lepton scale using Z boson decays: application to the measurement of MW. Tech. rep. ATL-PHYS-PUB-2006-007. ATL-COM-PHYS-2005-072. Geneva: CERN. https://cds.cern.ch/record/910107
16. Bittrich C et al (2017) In-situ scale factors from Z ! ee events. Tech. rep. ATL-COM-PHYS-2017-757. Geneva: CERN. https://cds.cern.ch/record/2268800
17. Aaboud M et al (2018) Measurement of the photon identification efficiencies with the ATLAS detector using LHC Run 2 data collected in 2015 and 2016. In: Submitted to: Eur Phys J. arXiv:1810.05087 [hep-ex]
18. Aaboud M et al (2018) Measurement of the Higgs boson mass in the H ! $ZZ_!$ 4' and H ! gg channels with ps = 13 TeV pp collisions using the ATLAS detector. Phys Lett B784:345–366. https://doi.org/10.1016/j.physletb.2018.07.050. arXiv:1806.00242 [hep-ex]

Part III
Physics Analysis

Chapter 6
Interlude B. Unfolding

Introduction

The outcome of a measurement in particle physics is the result of an underlying physical process that we want to measure (possibly an unknown physics process) convoluted with the response of the detector. In particular, the finite resolution of the detector will smear the quantities that we want to measure. In addition, the detector has a limited acceptance and efficiency, meaning that certain events will not be captured if they produce particles that do not cross the active detector regions or due to reconstruction inefficiency. In our analysis, the measurements are based on counting events (Higgs boson signal events) in particular regions (bins) of the phase space. Therefore, the limited resolution of the detector will result in events migrating to wrong (neighboring) bins. The process of correcting for the resolution migrations and detector efficiency in order to measure quantities not affected by these effects (typically cross sections) is called *unfolding* or *deconvolution*. A sketch of the unfolding problem is shown in Fig. 6.1. Unfolding is an ill-posed problem since the most straightforward solutions can be very sensitive to data fluctuations, yielding unstable results, and thus requiring some "regularization" procedure.

The problem of unfolding is particularly challenging for particle physics, as the measurements will include statistical fluctuations and (potentially) background events. For the measurement of the fiducial integrated and differential Higgs boson production cross sections presented in this thesis, the unfolding procedure represents an essential ingredient of the analysis. Unfolding allows for an easy comparison of theory predictions with the measured cross sections. This results in long-lasting measurements that can be compared to theory models developed long after the measurement is done. In addition, unfolding allows for easier comparison and combination with the results of other experiments, as the results are deconvoluted from detector effects.

© The Editor(s) (if applicable) and The Author(s), under exclusive license
to Springer Nature Switzerland AG 2020
A. Tarek Abouelfadl Mohamed, *Measurement of Higgs Boson Production
Cross Sections in the Diphoton Channel*, Springer Theses,
https://doi.org/10.1007/978-3-030-59516-6_6

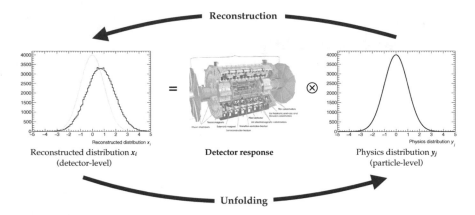

Fig. 6.1 Illustration of the unfolding procedure

Formalism

One can express the PDF of a given measurement $f_{\mathrm{meas}}(x)$ as follows:

$$f_{\mathrm{meas}}(x) = \int R(x|y) f_{\mathrm{true}}(y) \, \mathrm{d}y, \tag{6.1}$$

where $f_{\mathrm{true}}(y)$ is the PDF of the true underlying physics effects convoluted with the detector response function $R(x|y)$. In practice, we deal usually with binned observables and Eq. (6.1) becomes a matrix multiplication:

$$x_i = \sum_{j=1}^{N} R_{ij} y_j, \tag{6.2}$$

where N is the number of bins of the distribution of the true quantity y. The detector response matrix R_{ij} can be interpreted as a conditional probability:

$$R_{ij} = \mathcal{P}(\text{reconstructed in bin } i \mid \text{true value in bin } j) \tag{6.3}$$

The sum:

$$\sum_{i=1}^{M} R_{ij} = \mathcal{P}(\text{observed anywhere} \mid \text{true value in bin } j) = \epsilon_j, \tag{6.4}$$

corresponds to the efficiency for events with true value of y in bin j. In our analysis, we are only interested in the cases where the binning of true distribution y_j and the reconstructed one x_i is the same, i.e. $M = N$. The task of unfolding is to invert Eq. (6.2) to convert measured values x_i to true values y_j. For our analysis, we aim to obtain *particle-level* cross sections from the measured *detector-level* cross sections defined in Sect. 7.1.1. Several methods exist to perform such procedure, each with its own strengths and caveats. In all methods, the particle-level event yields (or cross sections) y_j are related to the detector-level quantities x_i by linear relations $y_j = U_{ij}x_i$, where U_{ij} is known as the *unfolding matrix*. In the next section, a summary of the studied unfolding methods is given.

Review of Unfolding Methods

This review will use as example histograms of Higgs boson signal events as a function of the diphoton transverse momentum $p_T^{\gamma\gamma}$, defined in Sect. 7.1.1. The $p_T^{\gamma\gamma}$ distributions are built using the Higgs boson signal simulated samples detailed in Sect. 7.2. The chosen binning for the $p_T^{\gamma\gamma}$ histograms allows us to test the limits of the different methods as migrations between bins increase with finer binning. The true (particle-level) distribution is shown in Fig. 6.2a, and the reconstructed distribution is shown in Fig. 6.2b. The difference between the two is due to the limited resolution and efficiency of the detector. One way to visualize the effect of the detector is via the response matrix, shown in Fig. 6.3. This matrix relates the reconstructed (detector-level) events and the true (particle-level) events using a 2D histogram. The off-diagonal elements of this matrix represent events that migrated to neighboring bins after the measurement by the detector.

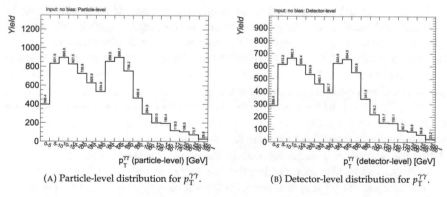

(A) Particle-level distribution for $p_T^{\gamma\gamma}$.   (B) Detector-level distribution for $p_T^{\gamma\gamma}$.

Fig. 6.2 Distribution of **A** the true (particle-level) and **B** the reconstructed (detector-level) Higgs boson event yield as a function of $p_T^{\gamma\gamma}$

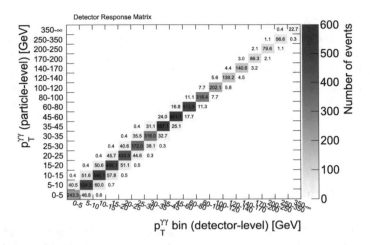

Fig. 6.3 The detector response matrix for $p_T^{\gamma\gamma}$ giving the number of events for each true (particle-level) bin yield versus the reconstructed (detector-level) yield. The off-diagonal elements of this matrix represent bin migrations

Matrix inversion

The inversion of the detector response matrix is the most straightforward approach to unfolding. The unfolding matrix U_{ij} will be set as the inverse of the detector response matrix Eq. (6.3), $U_{ij} = R_{ij}^{-1}$. The inversion of the response matrix can be done directly, or by maximizing the likelihood $(R_{ij}x)^T V_{ij}^{-1}(R_{ij}x)$, where V_{ij} is the covariance matrix of the measurement.

Despite its simplicity, this method has a major drawback. The resulting unfolded distributions have extremely large variances and strong negative correlations between neighboring bins. This is illustrated in Fig. 6.4a, where the relative statistical error of the unfolded distribution is compared with the relative statistical error of the initial measured distribution. This is more evident for bins with large migrations (the low $p_T^{\gamma\gamma}$ bins). The correlation matrix of the unfolded distribution is shown in Fig. 6.4b.

This effect is due to the non-zero off-diagonal elements in the detector response matrix. These elements can result from the finite resolution of the detector especially when the bin size is small compared to the resolution or when looking at observables with poor resolution. This can be seen in the detector response matrix in Fig. 6.3. Most of the simulated events populate the diagonal elements in the detector response matrix, whereas the off-diagonal elements are affected by large statistical uncertainties. These non-zero off-diagonal elements will then appear in the denominator in some of the elements of the inverted matrix, amplifying the statistical error of the unfolded distribution (including the diagonal elements of the response matrix).

On the other hand, the advantage of this inversion approach is that the resulting values for y, despite being affected by significant variances, are in fact unbiased from a statistical point of view. Of course, the method can be biased if the response matrix

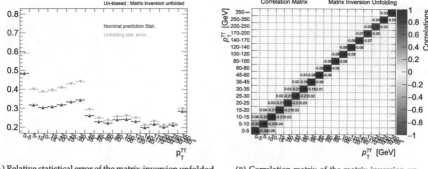

(A) Relative statistical error of the matrix-inversion unfolded distribution.

(B) Correlation matrix of the matrix-inversion unfolded distribution.

Fig. 6.4 Results of the unfolding using matrix inversion for the $p_T^{\gamma\gamma}$ distribution. **A** A comparison of the relative statistical error of the unfolded distribution (in green) with respect to the statistical error of the measured detector-level distribution (in red). The statistical error of the unfolded distribution is largely increased due to the off-diagonal elements in the detector response matrix as detailed in the text. **B** The correlation matrix of the unfolded distribution using matrix inversion. The matrix shows negative correlations between neighboring bins (most notably for the bins with large migrations)

does not reflect the actual detector response. Also, the maximum likelihood solution for the inversion has the smallest possible variance value for any unbiased estimator [1], i.e. the large variance we observe is the minimum bound for the variance for an unbiased estimator. Therefore, for the other unfolding methods detailed in this section, the strategy is that one would accept small bias (that will be added as a systematic uncertainty to the unfolded distribution) in exchange for reduction in the variance, i.e. trading statistical for systematic uncertainties.

Bin-by-Bin Unfolding

The bin-by-bin unfolding method is a simple method based on rescaling the detector-level yields with multiplicative correction factors derived from Monte Carlo simulations. Using the same notation as Eq. (6.2), the estimator for y in a given bin i is constructed as

$$y_i = C_i x_i \qquad (6.5)$$

where the correction factor C_i is

$$C_i = \frac{y_i^{MC}}{x_i^{MC}}, \qquad (6.6)$$

Fig. 6.5 The relative
statistical error of the
unfolded distribution using
the bin-by-bin unfolding
method (in green) compared
the relative statistical error
on the measured
detector-level distribution

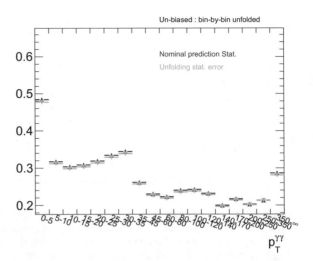

where y_i^{MC} and x_i^{MC} are the expected particle-level (true) and detector-level (reconstructed) yields from the simulation. In other words, the bin-by-bin unfolding approximates the detector response matrix to $R_{ij} \approx \delta_{ij} \frac{1}{C_j}$, and the unfolding matrix $U_{ij} = \delta_{ij} C_j$.

In order to quantify migrations in the detector response matrix we introduce the following two quantities that are sensitive to migrations due to resolution effects:

- The migration purity P_i is defined as $P_i = n_i^{ptcl+det}/n_i^{det}$, where $n_i^{ptcl+det}$ is the yield of events belonging to bin i of both the detector-level and particle-level distribution, and n_i^{det} is the yield of events in the detector-level bin i. The purity is sensitive to fake events which are, incorrectly, reconstructed in a given at the detector-level, with large purity denoting smaller migrations.
- The reconstruction efficiency ϵ_i is defined as $\epsilon_i = n_i^{ptcl+det}/n_i^{ptcl}$, with n_i^{ptcl} the number of events with y in the particle-level bin i. The efficiency is sensitive to events that are, incorrectly, reconstructed out of a given bin. Therefore, larger efficiency implies better object reconstruction.

The bin-by-bin unfolding correction C_i can then by expressed in terms of purity and efficiency:

$$C_i = \frac{P_i}{\epsilon_i} \tag{6.7}$$

The main advantage of the bin-by-bin unfolding is that it has a much smaller variance than the inversion of the migration matrix, as shown in Fig. 6.5. This is a result of the covariance matrix of the unfolded yields $U_{ij} = \text{cov}[y_i, y_j] = C_i C_j \text{cov}[x_i, x_j] = C_i C_j \text{var}[x_j]\delta_{ij} = C_i^2 \text{var}[x_i]$, with generally $C_i \sim \mathcal{O}(1)$.

On the other hand, bin-by-bin unfolding might result in a bias that is not negligible relative to its variance. The bias from the bin-by-bin unfolding is estimated from

$$E[\hat{y}_i] = C_i x_i \equiv \left(\frac{y_i^{MC}}{x_i^{MC}} - \frac{y_i}{x_i} \right) x_i + y_i, \tag{6.8}$$

therefore the bias is:

$$b_i = \left(\frac{y_i^{MC}}{x_i^{MC}} - \frac{y_i}{x_i} \right) x_i \tag{6.9}$$

The bias is zero if the simulation predicts correctly the ratio y_i/x_i, which can not be inferred prior to a measurement. In order to estimate the bias of the bin-by-bin unfolding, pseudo-data samples are generated following specific *bias scenarios* covering the potential discrepancy that might be observed between data and the nominal simulation.

Regularized Methods

As seen in the case of the matrix-inversion unfolding method, the variance for an unbiased unfolding method is large. To counterbalance this effect, one might accept a small bias in exchange for reducing the variance through a so-called *regularization* procedure. Below, we will briefly review the regularization techniques that were investigated.

Bayesian Iterative Unfolding

This method is motivated by Bayesian statistics. Bayes' theorem is used to estimate the unfolding matrix $U_{ij} = P(i^{ptcl}|j^{det})$ from the detector response matrix from the simulation $R_{ij} = P(i^{det}|j^{ptcl})$:

$$P(i^{ptcl}|j^{det}) = \frac{P(j^{det}|i^{ptcl})P_0(i^{ptcl})}{\sum_{k^{ptcl}} P(j^{det}|k^{ptcl})P_0(k^{ptcl})} \tag{6.10}$$

where $P_0(i^{ptcl})$ is the prior. We can then obtain the truth event yield distribution as

$$y_i = \frac{1}{\epsilon_i} \sum_{j^{det}} x_i P(i^{ptcl}|j^{det}) \tag{6.11}$$

The drawback of the method is that the results will depend on the chosen prior $P_0(i^{ptcl})$. To overcome this limitation, an iterative procedure is used, minimizing the model dependence. After each iteration (k) the prior probability distribution $P_0(i^{ptcl})$ is replaced with the obtained unfolded event yield distribution $y_{(k)}$, and

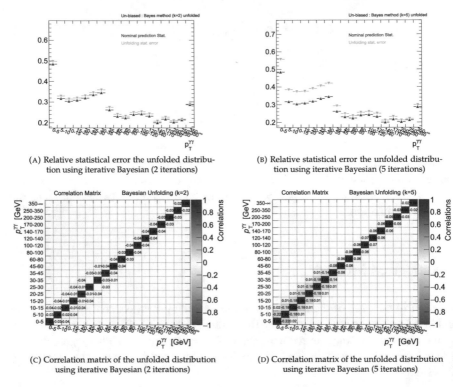

(A) Relative statistical error the unfolded distribution using iterative Bayesian (2 iterations)

(B) Relative statistical error the unfolded distribution using iterative Bayesian (5 iterations)

(C) Correlation matrix of the unfolded distribution using iterative Bayesian (2 iterations)

(D) Correlation matrix of the unfolded distribution using iterative Bayesian (5 iterations)

Fig. 6.6 Results of unfolding using iterative Bayesian. The figures on the left (right) use 2 (5) iterations. The top plots show the relative statistical error of the unfolded distribution compared to the statistical error of measured detector-level distribution. The bottom plots show the correlation matrix of the unfolded distributions. The results show that the behavior of iterative Bayesian unfolding approaches that of matrix inversion (large variances and negative correlations) with more iterations

iterated to obtain $y_{(k+1)}$. The unfolded results can be thought of as middle ground between matrix inversion and bin-by-bin, as we approach more significant variances and negative correlations (similar to matrix-inversion) with more iterations. This is illustrated in Fig. 6.6, where the correlations matrices and the relative statistical errors of the unfolded distributions are shown using 2 and 5 iterations.

Singular Value Decomposition (SVD)

The singular value decomposition (SVD) unfolding provides an alternative way to reduce the large variances resulting from the matrix inversion [2]. The method is based on SVD factorization of the response matrix $R = ASB^T$, where A and B are orthogonal matrices, and S is a diagonal matrix with real positive entries $s_i =$

S_{ii} known as the singular values. The SVD factorization can be considered as a generalization to any matrix of the eigendecomposition of positive definite matrices. The singular values contain very valuable information about the properties of the matrix: for example, the small singular values are associated with the enhancement of the statistical fluctuations. The statistical fluctuations can then be dampened by replacing s_i^2 by $s_i^2/(s_i^2 + \tau)$, where τ is a parameter determining the strength of the regularization. If τ is too large, this will cause *over regularization*, meaning that the unfolded distribution will be biased towards the shape from simulation. A too-small τ, on the other hand, will lead to *under regularized* results, yielding large oscillations. Therefore, τ is tuned to be $\tau = s_k^2$, where k is the effective rank of the system.

The detector response matrix is re-scaled with the uncertainties of the measured spectrum Δx_i, $\tilde{x}_i = x_i/\Delta x_i$ and $\tilde{R}_{ij} = \hat{R}_{ij}/\Delta x_i$, or analogously with the full covariance matrix in case of non-zero correlations before the SVD decomposition of the detector response matrix. Using our $p_T^{\gamma\gamma}$ distribution, the SVD unfolding seems to show over-regularization, resulting in positive correlations between neighboring bins and reduced statistical error for the unfolded distribution (with respect to the statistical error of the measurement). This is shown in Fig. 6.7 for different tuning parameter values $k = 5$ and 18. The over-regularization is reduced by increasing the k (approaching more matrix-inversion). However, even the largest value of $k = n_{\text{bins}} = 18$ slightly reduces the error and results in positive correlations between neighboring bins.

Choice of Unfolding Method

The choice of the unfolding method is made by applying the different unfolding methods on pseudo-data distributions. The pseudo-datasets are required to match the expected level of variations between our simulated samples and data. These datasets are known as *bias scenarios* and are generated for each of the differential variables we investigate. The details of the variables and the selected binning are shown in Sect. 7.3.4. The studies are performed using the different variables in order to check the performance of the different unfolding methods under different scenarios of resolution and statistical precision as follows:

$$\begin{pmatrix} \text{Poor Resolution} \\ \text{Good Resolution} \end{pmatrix} \times \begin{pmatrix} \text{Good statistical precision} \\ \text{Poor statistical precision} \end{pmatrix}. \tag{6.12}$$

Bias Scenarios

In order to estimate the bias of the different unfolding methods, we generate pseudo-data samples following a selection of bias scenarios, in which the central values of the SM expectation are shifted by some amount. The bias scenarios include realistic scenarios (i.e. introducing distortions similar to fluctuations we expect in data). In

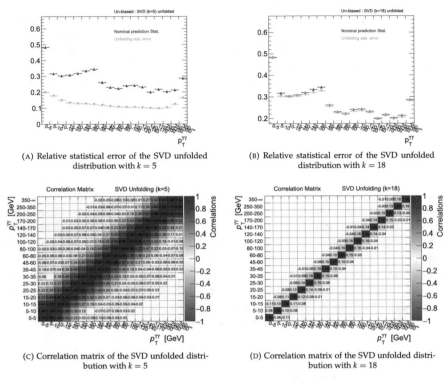

(A) Relative statistical error of the SVD unfolded distribution with $k = 5$

(B) Relative statistical error of the SVD unfolded distribution with $k = 18$

(C) Correlation matrix of the SVD unfolded distribution with $k = 5$

(D) Correlation matrix of the SVD unfolded distribution with $k = 18$

Fig. 6.7 Results of the unfolding using SVD with different tuning parameter $k = 5, 18$ for the $p_T^{\gamma\gamma}$ distribution. The top plots show the relative statistical error of the unfolded (in green) compared to the statistical error of the measurement (in red). The SVD results in over-regularization and reduces the error of the measurement. The bottom plots show the correlation matrix of the unfolded distributions. The matrices show positive correlations between neighboring bins. The correlations are reduced by increasing the tuning parameters k

addition, we also check extreme bias scenarios with large distortions used as stress tests for the unfolding methods. The bias scenarios are produced by simultaneously reweighting the SM expectations of the Higgs boson $p_T^{\gamma\gamma}$ and $|y_{\gamma\gamma}|$ distributions to match the observed distributions from a previous measurement. These distributions ($p_T^{\gamma\gamma}$ and $|y_{\gamma\gamma}|$) are chosen for the reweighting since they are uncorrelated to a good approximation. The biases estimated using these scenarios will not be used for the actual cross section measurement. The recipe to compute the unfolding bias uncertainty for the final cross section is detailed in Sect. 7.5.3.1. The following bias scenarios are investigated:

- *both scenario* includes a simultaneous re-weighting of the Higgs boson $p_T^{\gamma\gamma}$ and $|y_{\gamma\gamma}|$ spectra using the $H \rightarrow \gamma\gamma$ 80 fb^{-1} dataset [3]. This bias scenario is thought to be a reasonable representation of the biases that might be expected in real data.
- *both_ZZ scenarios* includes simultaneous re-weighting of the Higgs boson $p_T^{\gamma\gamma}$ and $|y_{\gamma\gamma}|$ spectra using the $H \rightarrow ZZ^*$ 36 fb^{-1} dataset [4].

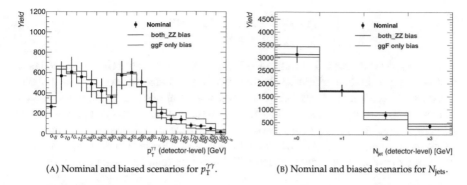

(A) Nominal and biased scenarios for $p_T^{\gamma\gamma}$.   (B) Nominal and biased scenarios for N_{jets}.

Fig. 6.8 The SM prediction compared to the various bias scenarios for **A** $p_T^{\gamma\gamma}$ and **B** N_{jets}. The error bars on the nominal predictions are estimated by scaling the signal yield fit errors to 140 fb^{-1}, as detailed in the text

- *ggF only* scenario includes a modification of the Higgs boson signal composition that includes only gluon-fusion.

For the choice of the methods we concentrated on the bias scenarios with large variations (i.e. *both_ZZ* and *ggF only*). The effect of the bias scenario on the $p_T^{\gamma\gamma}$ and N_{jets} distributions is shown in Fig. 6.8.

Bias and Expected Error

In order to assess the effect of the unfolding method on the uncertainty of the unfolded distributions, we assign a realistic error for the yield of each bin. In the actual data unfolding, this error will be the result of the fit procedure detailed in Sect. 7.6. To obtain an estimate for the statistical uncertainty for a given measurement, we scale the statistical uncertainty from the previous measurement using 80 fb^{-1} of 13 TeV data to the luminosity of the current measurement (140 fb^{-1}) as follows:

$$\delta_i^{140\,\text{fb}^{-1}} = \delta_i^{80\,\text{fb}^{-1}} \times \sqrt{140\,\text{fb}^{-1} / 80\,\text{fb}^{-1}} \qquad (6.13)$$

The estimated statistical error is shown in Fig. 6.9.

Ensembles of pseudo data are generated in order to estimate the error coverage of each unfolding method (i.e. ensuring that the error of the unfolded distribution is not underestimated). These pseudo-datasets are generated for the nominal SM expectation as well as for the different bias scenarios. We use them to compute the bias in each bin b_i from the following formula:

$$b_i = \frac{\langle N_{\text{toys}} \rangle - N_{\text{particle-level}}}{N_{\text{particle-level}}}, \tag{6.14}$$

where $\langle N_{\text{toys}} \rangle$ is the mean of the unfolded signal yield for the different pseudo-datasets and $N_{\text{particle-level}}$ is the particle-level (true) signal yield that we expect. The pseudo-datasets are generated by smearing the detector-level spectrum using a Gaussian with a mean equal to the detector-level nominal (or bias-scenario) yield and standard deviation equal to the expected error. Each of these pseudo-datasets is unfolded using the nominal response matrix, and the unfolded distribution is compared to the underlying particle-level (truth) spectra. This procedure is performed 1000 times.

Figures of Merit

Various figures of merit can be used to compare the performance of the different methods, as well as of the same method with different parameter settings (number of iterations for iterative methods or tuning of the regularization for SVD). The chosen figures of merit stem from the following considerations:

- The (statistical) uncertainty of the unfolded distribution should neither be significantly reduced nor strongly increased by the unfolding with respect to the statistical uncertainty of the measured distribution, i.e. the ratio $\frac{\delta_{i,\text{stat.}}^{\text{unfold}}/\mu_i^{\text{true}}}{\delta_{i,\text{stat.}}^{\text{pred}}/\mu_i^{\text{pred}}}$ should be close to one.
- The systematic bias introduced by the unfolding should be small compared to the statistical and other experimental uncertainties, $b_i < \delta_{i,\text{stat.}}$. The bias will be accounted for as an additional uncertainty on the unfolded distribution.

Therefore, the two figures of merit that we chose to quantify the previous arguments for the distribution as a whole are:

- The sum of absolute biases $\Sigma_i |b_i|$.
- The sum of absolute biases divided by the effective total statistical error taking into account correlations, $\frac{\Sigma_i |b_i|}{\sqrt{\Sigma_{i,j} Cov_{i,j}(\text{stat.})}}$. This provides the metric for checking that the total bias due to the unfolding is smaller than the statistical uncertainty for the overall distribution.

Summary of Results and Choice of Unfolding Method

The choice of unfolding method is based on comparing the different figures of merit for the different methods. We focused on the bin-by-bin, iterative Bayesian, and matrix inversion unfolding methods.

The relative statistical uncertainty of the unfolded distribution using the different methods is shown in Fig. 6.9, where the statistical uncertainty from the input measure-

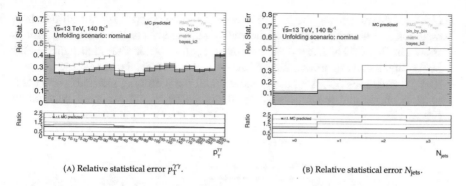

(A) Relative statistical error $p_T^{\gamma\gamma}$.   (B) Relative statistical error N_{jets}.

Fig. 6.9 The relative statistical error of the unfolded distribution for the different unfolding methods (shown with different colors) for **A** $p_T^{\gamma\gamma}$ and **B** N_{jets} distributions. The relative statistical uncertainty of the unfolded distribution is compared with the relative statistical error of the measurement in shaded purple

ment is also shown for reference. For all variables, the relative statistical uncertainty is mostly unchanged by the bin-by-bin unfolding, whereas the matrix inversion yields larger statistical uncertainty. This effect is more evident for distributions with large migrations, such as the N_{jets} distribution where the resulting statistical uncertainty can be up to double that of the measurement. The iterative Bayesian method with two iterations provides a middle ground, preserving the statistical uncertainty of the measurement at the cost of increasing the bias.

The coverage of the statistical error of the unfolded distributions was checked for the different methods. This check is performed to make sure that the statistical uncertainty of the unfolded distribution does not underestimate the true statistical uncertainty. The coverage test was performed by checking the RMS of the pulls of the pseudo-datasets. An RMS compatible with unity ensures that the statistical uncertainty of the unfolded distribution achieves coverage. The check was performed for different bias scenarios. The RMS of the pulls for the different bias scenarios using the different unfolding methods was found to be compatible with one within the statistical uncertainty from the number of pseudo-datasets. An example is shown in Fig. 6.10 for the N_{jets} distribution using bin-by-bin unfolding for different bias scenarios. To summarize, from the statistical error point of view, the bin-by-bin unfolding method seems to be the best option, provided that it does not yield large biases.

The biases of the different methods for the different variables are summarized below:

- $p_T^{\gamma\gamma}$: no significant biases were found using the different bias scenarios. In general, the introduced biases are much smaller than the statistical uncertainty. This is shown in Fig. 6.11a, where the biases from the different methods are compared to the relative statistical error.

- $N_{\text{jets}}^{\geq 30\text{Gev}}$: the bias using the *both_ZZ* is found to be comparable for all the three methods and is small relative to the statistical uncertainty for every bin. This is shown in Fig. 6.11b. The extreme scenario (*ggF only*) on the other hand, yields larger biases that are around 20% of the expected statistical uncertainty.
- **The jet variables** $p_T^{j_1}$, $\Delta\phi_{jj,\text{signed}}$, and m_{jj}: all the unfolding methods yield comparable biases that are relatively small compared to the expected uncertainty (Fig. 6.12).

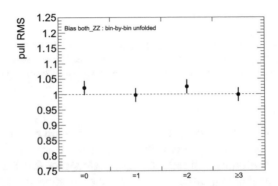

Fig. 6.10 The RMS of the bias pulls for bin-by-bin unfolding estimated from pseudo-datasets for the *both_ZZ* bias scenario. The RMS of the pulls is consistent with 1 within the statistical error resulting from the number of generated pseudo-datasets. This is an evidence for the coverage of the unfolded statistical error using bin-by-bin unfolding for this analysis

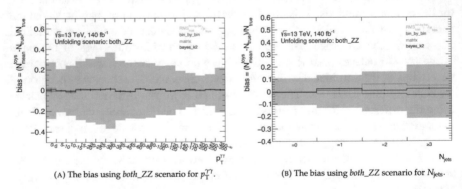

(A) The bias using *both_ZZ* scenario for $p_T^{\gamma\gamma}$.   (B) The bias using *both_ZZ* scenario for N_{jets}.

Fig. 6.11 The bias of the different unfolding methods for **A** $p_T^{\gamma\gamma}$ and **B** N_{jets} using the *both_ZZ* bias scenario. The biases are compared with the relative statistical error of the input measurement shown in shaded purple. In general, the biases were found to be very small with respect to the statistical error

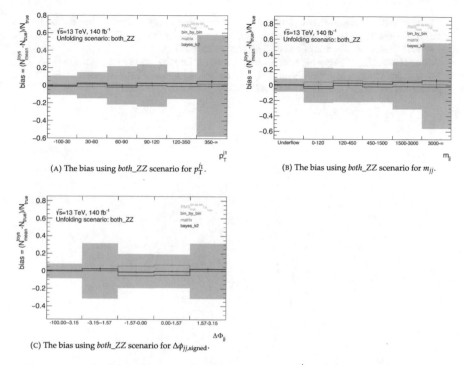

(A) The bias using *both_ZZ* scenario for p_T^{j1}.

(B) The bias using *both_ZZ* scenario for m_{jj}.

(C) The bias using *both_ZZ* scenario for $\Delta\phi_{jj,\text{signed}}$.

Fig. 6.12 The bias of the different unfolding methods for **A** p_T^{j1}, **B** m_{jj} and **C** $\Delta\phi_{jj,\text{signed}}$ using the *both_ZZ* bias scenario. The biases are compared with the relative statistical error of the input measurement shown in shaded purple. In general, the biases were found to be very small with respect to the statistical error

From these results, given its small bias relative to the statistical uncertainty of the measurement, and comparing with matrix inversion, the bin-by-bin unfolding method is chosen for the unfolding of the $H \rightarrow \gamma\gamma$ cross sections studied in this thesis. The actual unfolding bias that will enter as uncertainty on the measurement will be estimated from the observed differences between data and simulation. These differences will be used to reweight the simulation to match the data and then measure the bias with respect to the nominal simulation. This procedure is detailed in Sect. 7.5.3.1.

References

1. Cowan G (1998) Statistical data analysis. Oxford science publications, Clarendon Press. ISBN 9780198501558. https://books.google.ch/books?id=ff8ZyW0nlJAC
2. Hocker A, Kartvelishvili V (1996) SVD approach to data unfolding. Nucl Instrum Meth A372:469–481. https://doi.org/10.1016/0168-9002(95)01478-0. arXiv: hep-ph/9509307 [hep-ph]

3. Measurements of Higgs boson properties in the diphoton decay channel using 80 fb^{-1} of pp collision data at $\sqrt{s} = 13$ TeV with the ATLAS detector. Tech rep ATLAS-CONF-2018-028. Geneva: CERN (2018). https://cds.cern.ch/record/2628771
4. Aaboud M et al (2017) Measurement of inclusive and differential cross sections in the $H \rightarrow ZZ^*$ $\rightarrow 4l$ decay channel in pp collisions at $\sqrt{s} = 13$ TeV with the ATLAS detector. JHEP 10:132. https://doi.org/10.1007/JHEP10(2017)132. arXiv: 1708.02810 [hep-ex]

Chapter 7
Measurement of the Higgs Boson Fiducial Inclusive and Differential Cross Sections in the $H \rightarrow \gamma\gamma$ Channel

7.1 Introduction

Following the discovery of the Higgs boson in July 2012 by ATLAS [1] and CMS [2], several studies were performed using Run-1 and Run-2 data in order to characterize the newly discovered particle. The first of these studies measured ratios of the total production rate over the SM expectation (i.e. signal strengths) [3]. Other studies followed, measuring the different Higgs boson production and decay rates and coupling strengths modifications with respect to the SM expectation within the so-called κ-framework [4]. These measurements were performed based on the assumption that the Higgs boson production and decay kinematics are the same as those expected for the SM Higgs boson.

This approach was extended during Run-2 with the development of the simplified template cross section (STXS) framework [5]. The STXS are cross sections for the different Higgs boson production modes in well defined kinematic regions (bins) of the phase space. These kinematic bins are defined using the SM as a template and rely only on properties related to the Higgs boson production, such as the Higgs boson p_T or the number of accompanying particle-level jets. The STXS framework has the advantage that it further disentangles the theoretical interpretation (i.e. SM or BSM) from the measurement, hence reducing the theory dependence of the results on the SM prediction. In addition, the STXS bins can be split in finer granularity, providing a systematic way to increase the sensitivity and reduce any residual theory uncertainty. Furthermore, the bins are defined independently of the Higgs boson decay mode; therefore, measurements with different decay channels can be combined.

A different approach based on the measurement of the fiducial inclusive and differential cross sections is used in this chapter. This approach consists in choosing the quantities to measure such that they are largely model-independent, as will be detailed in the following sections. These quantities are inclusive, and differential Higgs boson production cross sections in a fiducial region that is as close as possible

A. Tarek Abouelfadl Mohamed, *Measurement of Higgs Boson Production Cross Sections in the Diphoton Channel*, Springer Theses, https://doi.org/10.1007/978-3-030-59516-6_7

to the detector-level selection. The measurements, which are also documented in a public note [6], are performed using the full LHC Run-2 pp collision data collected with the ATLAS detector.

7.1.1   Motivation and Strategy

The measurement of the fiducial inclusive and differential Higgs boson cross section provides an alternative framework to measure the properties of the Higgs boson. In this framework, the Higgs boson cross sections are measured in regions of the phase space called "fiducial volumes" defined by selections that match as closely as possible the experimental selection. Performing the measurement in well defined "fiducial volumes" has several advantages:

1. It does not require extrapolation to the full phase space. This removes all the additional theoretical uncertainties associated with such extrapolations. Compared to the STXS approach, the fiducial region of these cross sections also includes requirements on the final state particles from the Higgs boson decay, further reducing the extrapolation and corresponding theoretical uncertainties.
2. The model-dependence of the signal efficiencies within the fiducial volumes can be reduced to be smaller than the overall experimental uncertainty, thus minimizing the impact of theoretical variations on the signal efficiency [7].
3. Fiducial differential cross section measurements are particularly powerful for testing BSM scenarios that would affect the Higgs boson kinematic distributions, which can not be probed only by a simple scaling of couplings as in the κ-framework. A detailed example of such procedures to probe BSM scenarios using effective field theory models using the results of the measurements presented in this chapter is shown in Chap. 8.
4. Model-independent measurements can be compared to different theoretical models and to other models that may be developed after the measurement without the need to re-analyze the data.

The measurements presented in this manuscript rely on data collected during the LHC Run-2 and simulation samples, described in Sect. 7.2. The strategy used to measure the fiducial inclusive and differential cross sections, illustrated in Fig. 7.1, is as follows:

- A selection is applied to the events and the different physics objects that constitute the fiducial volume. The detector-level and corresponding particle-level selections are detailed in Sect. 7.3.
- After the selection, the main discriminant variable between the Higgs signal and the background is the diphoton invariant mass, $m_{\gamma\gamma}$. The distributions of this quantity for signal and background are studied in Sect. 7.4, where analytical models describing these shapes are described.

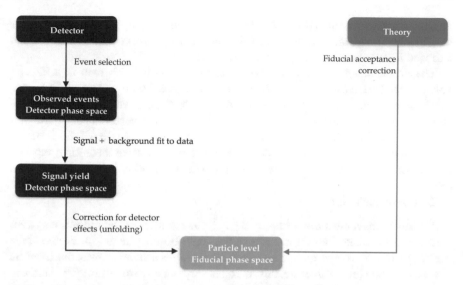

Fig. 7.1 Schematic representation of the analysis strategy. Stages of processing of quantities measured by the detector are shown in dark red, and stages involving quantities predicted by theory are shown in blue. Measured quantities at the particle level fiducial volume are compared to theory. This stage is shown in green

- The Higgs boson signal yield in the selected sample is determined by fitting the diphoton invariant mass distribution in data using the previously found signal and background parameterizations. The details of the fit are given in Sect. 7.6.
- The extracted signal in each bin of the differential distributions is corrected for the detector resolution and migration effects among bins (unfolded). Different unfolding methods have been studied and the one providing the best compromise between the added variance and the added bias to the results has been chosen. The details of the different unfolding methods are shown in *Interlude B*.
- The uncertainties associated with the signal extraction and unfolding steps of the measurement are summarised in Sect. 7.5.
- The particle-level cross sections are obtained from the unfolded yields and compared with different theoretical predictions in Sect. 7.7.

A similar strategy was used in the previous Higgs boson cross section measurements using $\sqrt{s} = 8$ TeV and $\sqrt{s} = 13$ TeV [8, 9]. The cross section measurement uses a "blind-analysis" approach, meaning that the optimization and validation of the analysis chain is performed before looking at the data in the signal region, defined as the events that pass all the selection criteria and whose reconstructed diphoton invariant mass lie in a narrow window around the mass of the Higgs boson. The analysis optimization and validation is performed using simulation samples and data in control regions, either consisting of events in which one or both photon candidates fail some of the selection criteria (control regions enriched in "reducible" background in which at least one photon candidate arises from the misidentification of a hadronic jet), or of events passing all the selection criteria but with diphoton invariant mass in a "sideband" region away from the value of the Higgs boson mass (control region enriched in "irreducible" background from non-resonant diphoton events).

The measurement of the Higgs boson cross section is performed in a fiducial volume matching closely the phase space region sampled by the detector. The measured cross section can be *inclusive* for the whole fiducial volume or *differential* constructed by dividing the inclusive fiducial volume into several bins of a particular kinematic variable. This allows comparison of the shapes and rates of the events to the predictions, providing more information to test alternative theoretical models. The list of variables that have been studied and the motivation for their choice is as follows:

Inclusive variables

- $p_T^{\gamma\gamma}$, **the transverse momentum**, and $|y_{\gamma\gamma}|$, **the rapidity of the diphoton system** describe the fundamental kinematics of the Higgs boson. The low-p_T region of the Higgs boson transverse momentum spectrum exhibits a Sudakov peak due to initial state radiation [10]. Therefore, this region is very sensitive to resummation effects, as the logarithmic expansion develops with $\mathcal{O}(\alpha_s \ln^2 \frac{p_T^H}{M_H})$. Figure 7.2 shows as an example the effect of the resummation of terms of order up to N^3LL [11] on the Higgs boson p_T distribution. In addition, the low-p_T region can be used to set bounds on light-quark Yukawa couplings [12, 13], as shown in Fig. 7.3a. On the other hand, the high-p_T region of the spectrum is sensitive to the couplings between the Higgs boson and the heaviest quarks, i.e. the *top* quark. For example, in the region $p_T^H > m_t$, the top quark mass cannot be considered infinite, and a dependence on the top quark mass has to be considered. Also, boosted Higgs boson production can resolve loop effects from heavy BSM particles, yielding sensitivity of the high-p_T region to BSM physics [14, 15]. More details on this will be shown in Chap. 8. The rapidity of the Higgs boson, $|y_{\gamma\gamma}|$, is sensitive to PDFs, and similar to the Higgs boson p_T, its distribution can be used to probe the light-quark Yukawa couplings, as shown in Fig. 7.3b [13].

- $N_{jets}^{\geq 30Gev}$, **the jet multiplicity associated with the production of a Higgs boson**. The jet multiplicity is categorised in 4 bins : *0-jet, 1-jet, 2-jets, $\geq$ 3-jets*, with a jet p_T threshold of 30 GeV. The jet multiplicity can be used to probe the different Higgs boson production mechanisms. The 0-jet events are dominated by gluon-fusion. The VH and VBF production mechanisms become increasingly important for 1-jet and 2-jet events, making these bins sensitive to the relative strengths of the ggH effective coupling and the VVH couplings. Higgs bosons produced in association with a top-antitop quark pair are important for very large jet multiplicities, making the $\geq$ 3-jets bin sensitive to the relative strength of the ttH coupling. In addition, the $N_{jets}^{\geq 30Gev}$ distribution is also sensitive to the amount of QCD radiation. For example, gluon-fusion with second-order real emission corrections will produce two additional jets in the event.

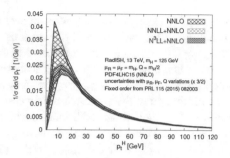

Fig. 7.2 Comparison between normalized distributions for the Higgs boson p_T at N^3LL logarithmic accuracy matched to fixed order QCD NNLO calculation [11]

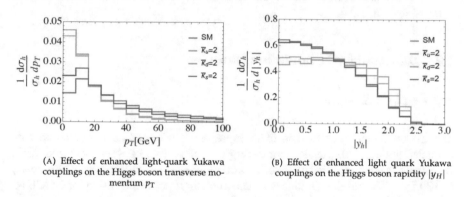

(A) Effect of enhanced light-quark Yukawa couplings on the Higgs boson transverse momentum p_T

(B) Effect of enhanced light quark Yukawa couplings on the Higgs boson rapidity $|y_H|$

Fig. 7.3 Comparison between normalized distributions of **A** the Higgs boson transverse momentum and **B** the Higgs boson rapidity between the Standard Model predictions (in blue) and the case in which the Yukawa couplings of the light quarks (s, u, d) are significantly enhanced (about 100 times for s and about 2000 times for u and d) so that they are twice larger than the SM b-quark Yukawa coupling ($\overline{k}_q = y_q/y_b^{SM}$) [13]

Jet kinematics variables

- **The leading jet transverse momentum,** $p_T^{j_1}$. The leading jet transverse momentum directly probes hard quark and gluon radiation in inclusive events. This variable predominantly tests fixed-order QCD calculations of gluon-fusion.
- **The dijet invariant mass,** m_{jj}. The invariant mass of the leading and sub-leading jets with jet p_T of at least 30 GeV. This variable is sensitive to the VBF production mode (which leads to events with large m_{jj}) and is useful in distinguishing it from gluon-fusion.
- **The dijet azimuthal angle difference,** $\Delta\phi_{jj,\text{signed}}$. The azimuthal angle difference of the leading and sub-leading jets with jet p_T of 30 GeV. The azimuthal angles of the jets are ordered according to the jet with the highest rapidity. This variable is sensitive to the charge conjugation and parity properties of the Higgs boson effective interactions with gluons and weak bosons in the gluon-fusion and VBF production channels respectively, making this variable sensitive to CP vio-

lation in the Higgs sector. For example, in gluon-fusion events with pure CP-even couplings, the distribution of this variable will exhibit a dip at $\pi/2$ and peaks at 0 and $\pm\pi$, whereas a pure CP-odd coupling would lead to the opposite behavior. VBF events, on the other hand, have a $\Delta\phi_{jj,\text{signed}}$ distribution which is approximately flat with a slight rise towards $\Delta\phi_{jj,\text{signed}} = \pm\pi$. Any additional anomalous CP-even or CP-odd contribution to the interaction between the Higgs boson and weak bosons would manifest itself as an additional oscillatory component, and any interference between the SM and anomalous couplings can produce distributions peaked at either $\Delta\phi_{jj,\text{signed}} = 0$ or $\Delta\phi_{jj,\text{signed}} = \pm\pi$ [16, 17]. More details on using $\Delta\phi_{jj,\text{signed}}$ to probe non-SM CP effects are given in Sect. 8.1.2.

The procedure used to choose the binning for each differential distribution is detailed in Sect. 7.3.4.

7.2   Data and MC Simulation Samples

7.2.1   Data Sample

The measurements presented in this chapter are performed on pp collision data collected with the ATLAS detector during the full Run-2, between 2015 and 2018, with a proton bunch spacing of 25 ns. This corresponds to a total integrated luminosity of 139 fb^{-1}. The baseline luminosity measurement is performed using the LUCID-2 detector [18]. The luminosity is measured with an uncertainty of 1.7% determined from a calibration of the luminosity scale using x-y beam-separation scans, following a methodology similar to that detailed in Ref. [19]. This method is known as the Van Der Meer scan [20]. The luminosity collected in each year, after data quality and trigger requirements have been applied, and its relative contribution to the full data set is summarized in Table 7.1.

Table 7.1   Summary of the Run-2 data set taken between 2015 and 2018 [21]

Year	Luminosity (fb^{-1})	Fraction (%)
2015	3.2	2.3
2016	33	23.5
2017	44.3	31.6
2018	59.9	42.7
Total	139 ± 1.7%	

7.2.2 Monte Carlo Simulation Samples

7.2.2.1 Higgs Boson Signal Default Simulation

Simulated event samples of Higgs bosons decaying to two photons are generated for the main production mechanisms: gluon-fusion ggF, vector boson fusion VBF, associated production with a vector boson WH and ZH, associated production with top anti-top pair $t\bar{t}H$, and associated production with bottom anti-bottom pair $b\bar{b}H$. These production mechanisms, described in Sect. 1.3.2, are used to study the shape of the signal $m_{\gamma\gamma}$ distribution discussed in Sect. 7.4.1, as well as to calculate the correction factors described in *Interlude B*. The hard scattering process leading to the production to the Higgs boson is generated using POWHEG (Positive Weight Hardest Emission Generator) [22–25], with the PDF4LHC15 PDF set [26]. All generated samples assume a Higgs boson of mass $m_H = 125$ GeV and width $\Gamma_H = 4.07$ MeV [27]. The Higgs boson decay to two photons, as well as the effects of the underlying event, parton showering, and hadronization, are modeled by interfacing the output of the $pp \rightarrow H + X$ event generation with PYTHIA 8 [28], using the AZNLO set of parameters that are tuned to data [29]. The stable particles created from these simulations (and defined in Sect. 7.3.2) are then passed to a GEANT4 simulation [30] of the response of the ATLAS detector, and the same reconstruction algorithms as those used for data are executed. The samples are normalized according to the most accurate theoretical predictions of the corresponding Higgs boson production cross sections, multiplied by the $H \rightarrow \gamma\gamma$ branching ratio of $0.227^{+0.006}_{-0.006}\%$ [31].

The details of the event generation for the different production modes, also summarized in Table 7.2, are:

- Gluon-fusion events are generated with POWHEG NNLOPS [22], which is a state of the art generator based on MINLO HJ and the POWHEG method, reaching NNLO+NNLL accuracy in its description of the Higgs boson p_T and rapidity distribution. The ggF sample is normalized to the total cross section calculated at N^3LO (QCD) with additional NLO electroweak corrections [32]. The ggF sample includes approximately $18M$ events.
- Vector Boson Fusion (VBF) events are generated with POWHEG- BOX [33] at NLO accuracy in QCD. The VBF sample is normalized to a cross section calculation with an approximate NNLO accuracy in QCD and includes NLO electroweak corrections [34–36]. The VBF sample includes approximately $7M$ events.
- Events from associated production of a Higgs boson and a vector boson (VH, with $V = W^{\pm}$, Z bosons) are generated with POWHEG- BOX. The samples include quark-initiated ($q\bar{q} \rightarrow VH$) and gluon-initiated ($gg \rightarrow ZH$) events. The samples are generated with NLO accuracy in QCD for quark-initiated production and LO for gluon-initiated production. The samples are normalized to a cross section calculation at NNLO accuracy in QCD using $VH@NNLO$ with electroweak NLO correction [37–39] for the quark initiated production and NLO accuracy in QCD for the gluon initiated production. The VH samples include approximately $5M$ events.

Table 7.2 Event generators and PDF sets used to model signal and background processes. The cross sections of Higgs production processes are reported for a center of mass energy of $\sqrt{s} = 13$ TeV and a SM Higgs with mass $m_H = 125.09$ GeV. The table reports the order of the ME generation and the order of the inclusive cross section calculation that is used to normalize the sample

Process	Generator	Showering	PDF set	Order of ME $\sqrt{s} = 13$ TeV	σ [pb]	Order of σ calculation
ggF	POWHEG NNLOPS	PYTHIA 8	PDF4LHC15	NNLO+NLL (QCD)	48.52	N^3LO(QCD)+NLO(EW)
VBF	POWHEG-BOX	PYTHIA 8	PDF4LHC15	NLO (QCD)	3.78	approximate-NNLO(QCD)+NLO(EW)
WH	POWHEG-BOX	PYTHIA 8	PDF4LHC15	NLO (QCD)	1.37	NNLO(QCD)+NLO(EW)
$q\bar{q} \to ZH$	POWHEG-BOX	PYTHIA 8	PDF4LHC15	NLO (QCD)	0.76	NNLO(QCD)+NLO(EW)
$gg \to ZH$	POWHEG-BOX	PYTHIA 8	PDF4LHC15	LO (QCD)	0.12	NLO(QCD)+NLO(EW)
$t\bar{t}H$	POWHEG-BOX	PYTHIA 8	PDF4LHC15	NLO (QCD)	0.51	NLO(QCD)+NLO(EW)
$b\bar{b}H$	POWHEG-BOX	PYTHIA 8	PDF4LHC15	NLO (QCD)	0.49	NNLO(QCD)+NLO(EW)
$\gamma\gamma$	SHERPA	SHERPA	CT10	NLO (QCD)	19.2×10^3	NLO (QCD)

- Associated production with a top quark pair ($t\bar{t}H$) samples are generated with POWHEG- BOX at NLO accuracy in QCD. The samples are normalised to a cross section calculation at NLO accuracy in QCD with electroweak NLO correction [40–42]. The $t\bar{t}H$ sample includes approximately $8M$ events.
- Associated production with bottom quark pair ($b\bar{b}H$) samples are generated with POWHEG- BOX with NLO accuracy in QCD. The samples are normalised cross section calculation at to NNLO accuracy in QCD with electroweak NLO correction [43, 44]. The $b\bar{b}H$ sample includes approximately $400k$ events.

The uncertainty on the predicted cross section of the different production modes is computed from uncertainties from QCD, PDF, and α_S as detailed in Sect. 1.3.

7.2.2.2 Background Simulation

Non-resonant diphoton events $pp \rightarrow \gamma\gamma + X$ (i.e. irreducible background events) are generated using SHERPA 2.2.4 [45] and the CT10 PDF set. The matrix elements are calculated at next-to-leading order accuracy in the strong coupling constant α_s for the real emission of up to one additional parton, and at leading-order accuracy in α_s for the real emission of two or three additional partons. They are then merged with the Sherpa parton shower using the MEPS@NLO prescription. The background samples are passed through a fast parametric simulation of the ATLAS detector response [46]. The fast simulation is used instead of the full detector simulation in order to accelerate the production time due to the large number of events to be simulated ($\mathcal{O}(100M)$ events), given the SM diphoton cross section $\sigma = 19.2$ nb. The sample is then normalized to data using data sidebands, after having estimated the contribution to the yield in the same control region from reducible background events.

7.2.2.3 Pileup Simulation

Contributions due to *in-time* pileup are included by overlaying the simulated hard-scatter events with minimum-bias events generated with PYTHIA8 [28]. Pileup events are overlaid onto the hard scattering events during digitization. The number of included pileup events is obtained by randomly drawing a number from a Poisson distribution with a mean of μ, where μ is the average number of additional proton-proton interactions per bunch crossing. The simulation of pileup is done before data taking was complete and the actual information about the pileup is known. Therefore, each sample is generated with a broad range of pileup values, μ, in order to encompass all possible pileup conditions which may be experienced during data taking. This distribution will then be corrected after data taking to account for the actual pileup distribution in data. *Out-of-time* pileup is included by adding detector signals from previous bunch crossings, also simulated from PYTHIA8 minimum-bias events.

The frequency of these signals is modeled on the nominal bunch structure used by the LHC.

7.2.2.4 Additional SM Higgs Boson Signal Simulation

As detailed in the introduction, one of the main advantages of particle level fiducial cross section measurements is that they can be compared directly to different SM cross section calculations. Therefore, in addition to the nominal SM predictions detailed in Sect. 7.2.2.1, the measured cross sections will be compared to additional state-of-the-art SM predictions. Several of these additional theory predictions are computed for the inclusive phase space without the selection criteria defining the fiducial volume (detailed in Sect. 7.3.2). Therefore, in order to compare these theory predictions to the fiducial cross section we apply an acceptance factor, α_{fid} to map the inclusive phase space to the particle level fiducial volume defined as:

$$\alpha_{\text{fid}} = \frac{\sigma^{\text{fid}}(pp \to H \to \gamma\gamma, \text{particle level})}{\sigma^{\text{inc}}(pp \to H \to \gamma\gamma, \text{particle level})} \tag{7.1}$$

These acceptance corrections are computed using the default SM predictions and include the same theoretical uncertainties (QCD, PDF, and α_S) as the default predictions. The following predictions for the gluon-fusion production mode will be compared to the unfolded fiducial cross sections after combining the default non-gluon-fusion predictions:

- NNLOJET+SCET [47] calculation for the $p_T^{\gamma\gamma}$ distribution providing predictions using a N^3LL resummation matched to an NNLO fixed-order calculation. These calculations include as well the effect of the finite top-quark mass which provides an accurate description of the transverse momentum region above the top-quark mass $p_T^H > m_t$.
- The $|y_{\gamma\gamma}|$ distribution is compared to SCETLIB, which provides predictions for $|y_{\gamma\gamma}|$ at NNLO+NNLL$'_\phi$ accuracy, derived by applying a resummation of the virtual corrections to the gluon form factor [48, 49]. The subscript ϕ refers to the fact that the applied resummation is to the gluon form factor. The underlying NNLO predictions are obtained using the MCFME event generator [50, 51] with zero-jettiness subtractions [52, 53].
- The N_{jets} distribution is compared to the following predictions:
  - The perturbative JVE+N^3LO prediction of Ref. [54], which includes QCD NNLL resummation of the p_T of the leading jet which is matched to the N^3LO total cross section. This calculation was performed only for the inclusive one-jet cross section.
  - The perturbative STWZ- BLPTW predictions of Refs. [55, 56], which include NNLL$'$+NNLO QCD resummation for the p_T of the leading jet, combined with

a NLL′+NLO QCD resummation for the subleading jet.[1] The numerical predictions for $\sqrt{s} = 13$ TeV are taken from Ref. [31].

- The perturbative NNLOJET prediction of Refs. [57, 58]. This is a fixed-order NNLO QCD prediction inclusive $H + 1$-jet production.
- The perturbative GoSam prediction of Refs. [59, 60], which provides the fixed-order loop contributions accurate at NLO in QCD for the inclusive $H +$ zero-jet, $H +$ one-jet, $H +$ two-jet, and $H +$ three-jet regions. The real-emission contributions at fixed order in QCD are provided by Sherpa [45].
- The Sherpa (Meps@Nlo) prediction of Refs. [61–64] that is accurate to NLO in QCD for the inclusive $H +$ zero-jet, $H +$ one-jet, $H +$ two-jet, and $H +$ three-jet regions and includes top-quark finite mass effects.
- The MG5_aMC@NLO prediction of Refs. [65, 66], which includes up to two jets at NLO accuracy using the FxFx merging scheme [67]. The central merging scale is taken to be 30 GeV. These predictions are derived in the full phase space. The MG5_aMC@NLO predictions are shown for the jet multiplicity. Uncertainties are estimated by varying the merging scale between 20 and 50 GeV.

- The $p_T^{j_1}$ distribution is compared to:

  - The same NNLOJET prediction described above for the N_{jets} distribution.
  - SCETlib(STWZ) [49, 55], which provides predictions for $p_T^{j_1}$ at NNLL′+NNLO accuracy by applying a resummation in $p_T^{j_1}$.

- The m_{jj} and $\Delta\phi_{jj,signed}$ distributions are compared to Sherpa (Meps@Nlo) and GoSam described above for the N_{jets} distribution.

7.2.2.5 Corrections to Simulations

Several corrections are applied to the Monte-Carlo simulation samples. These correction factors are different for the samples with the full detector simulation and the fast simulation ones. The corrections are as follows:

- A reweighting procedure is used to correct the average number of interactions per bunch crossing, $\langle\mu\rangle$. The reweighting is applied to simulated events to ensure that the pileup distribution of the simulation matches that observed in data.
- A similar reweighting procedure is applied to match the z distribution of the primary vertex in the simulation to that observed in the data.
- The photon energy spectrum is smeared to match the resolution observed in data. Also, correction factors are used to correct shower shapes based on the comparison between data and simulation as detailed in Sect. 4.3.1.1.

---

[1] The prime superscript indicates that the leading contributions from N^3LL (resp. NNLL) are included along with with the full NNLL (resp. NLL) corrections.

- Correction factors are applied to simulated samples to account for the residual differences between data and the simulation for photon identification and (track and calorimeter) isolation efficiencies.

7.3   Event Selection

The event selection used in this analysis is described in this section. The selection criteria applied to data and the detector-level simulations are described in Sect. 7.3.1. The particle-level selection used to define the fiducial region is given in Sect. 7.3.2. More emphasis is given to the particle-level isolation selection in Sect. 7.3.3. The choice of the binning of the differential variables is detailed in Sect. 7.3.4.

7.3.1   Detector-Level Event Selection

7.3.1.1   Event Preselection

Events first go through a *preselection* based on data quality and trigger requirements. Only events collected when all detector subsystems were operating in nominal running conditions and with good data quality are kept. These requirements are applied only to data. Selected events are then required to have at least one reconstructed primary vertex (PV) candidate. The reconstructed vertices are required to be consistent with the x and y coordinates of the beam spot.

Events are collected using a high-level trigger that requires at least two reconstructed photons with E_T larger than 35 and 25 GeV. These photons are required to pass photon identification criteria based on the energy leakage in the hadronic compartment and on the shower shape in the different layers of the electromagnetic calorimeter. For data collected in 2015 and 2016 the photon candidates were required to pass *loose* photon identification criteria (detailed in Sect. 4.3.1.1). In the data collected in 2017 and 2018, due to the higher pileup, photon candidates were required to pass the *medium* photon identification criteria, in order to limit the rate of the diphoton trigger.

7.3.1.2   Object Selection

Photons Photon candidates are reconstructed as detailed in Sect. 4.3.1. These candidates are then calibrated using the procedure detailed in Chap. 5. The photon candidates used for the measurement of this chapter are then required to pass the following requirements:

- Kinematic requirements ($E_T > 25$ GeV, $|\eta| < 2.37$) are applied. Photons reconstructed in the transition region between the barrel and the end-cap $1.37 < |\eta| < 1.52$ are rejected. This is a result of the poor resolution in this region due to the large amount of material upstream of the electromagnetic calorimeter, as well as of the poor photon/hadron discrimination in this region due to the limited segmentation of the readout cells of the first layer of the EM calorimeter.
- An *ambiguity* requirement is required for photon candidates in order to reduce the number of electrons reconstructed as photons. The $e \rightarrow \gamma$ misidentification is a result of electrons and photons having similar signatures in the detector. Therefore, it may happen that the same particle can be reconstructed as both electron and photon. This may yield genuine electrons reconstructed as photons and passing the offline selection. The electron-to-photon fake rates are measured in electron-enriched control regions (such as the $Z \rightarrow ee$) and selecting events using photon reconstruction and identification criteria instead of electron ones. Electron-to-photon fake rate measurements, detailed in Ref. [68], showed that the measured fake rate ranges from $< 2\%$ in the barrel up to 7% in the end-caps. This fake rate can be mitigated by up to 50% [69] by checking the compatibility of reconstructed tracks with calorimeter energy deposits for reconstructed photon candidates.
- Loose off-line photon identification requirements are applied.

Diphoton primary vertex The two highest p_T photon candidates passing the previous requirements are used to reconstruct the diphoton candidate of the event and to identify the diphoton primary vertex. The latter is selected among all reconstructed PV candidates using a neural-network algorithm, detailed in Ref. [70], based on the following input quantities:

- The scalar sum $\sum p_T$ of the momenta of the tracks associated with each reconstructed vertex.
- The sum $\sum p_T^2$ of the squared momenta of the tracks associated with each reconstructed vertex.
- The difference in azimuthal angle $\Delta\phi$ between the vector sum of the track momenta and the transverse momentum vector of the diphoton system.
- The pointing based on the hybrid primary vertex variable $\frac{(z_{vertex} - z_{HPV})}{\sigma_{HPV}}$ which uses the flight direction of the photons as determined by the measurement using the longitudinal segmentation of the calorimeter and the conversion point or hits in the precision tracking devices.

The efficiency to select a reconstructed PV within 0.3 mm of the true interaction point, studied with simulated $H \rightarrow \gamma\gamma$ samples, was found to be significantly higher for this neural-network algorithm when compared to the default algorithm used in ATLAS, which chooses the PV candidate with the largest $\sum p_T^2$ of the associated tracks ("hardest vertex"), as shown in Fig. 7.4a. The performance of the diphoton primary vertex neural-network algorithm in data and simulation is studied by selecting $Z \rightarrow ee$ candidate events, removing electron tracks from the list of reconstructed tracks, and thus treating the electron candidates as photons. The results of the validation are shown in Fig. 7.4b, demonstrating good agreement between data and the simulation.

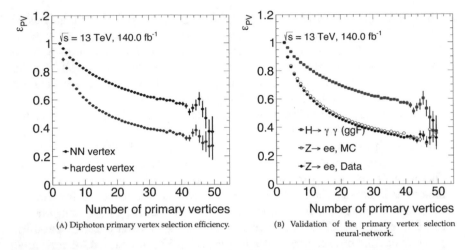

(A) Diphoton primary vertex selection efficiency.   (B) Validation of the primary vertex selection neural-network.

Fig. 7.4 **A** A comparison of the vertex selection efficiency using the neural network algorithm (black) or the highest Σp_T vertex (red). **B** Validation of the neural-network algorithm using $Z \to ee$ events in data and the simulation [6]

Jets Jets are reconstructed using the anti-k_t algorithm [71] with the distance parameter $R = 0.4$. Jets are required to pass the kinematic requirements $|\eta| < 4.4$ and $p_T > 25$ GeV. In order to reduce the number of reconstructed jets produced by the additional pileup interactions, a selection based on the *Jet Vertex Tagger* (*JVT*, detailed in Sect. 4.3.2) [72] is used. The JVT, computed for jets with $|\eta| < 2.5$ and $p_T < 120$ GeV, estimates the probability of a jet to originate from pileup or from a hard scattering, based on the number of tracks associated to the primary vertex. Selected jets are required to have JVT value larger than 0.59.

Overlap removal Particles produced in the pp collisions and interacting with the detector may be reconstructed as more than one candidate by different object reconstruction algorithms. In particular, photons can also be reconstructed as jets. An overlap removal procedure is implemented that removes jets candidates that are within $\Delta R(jet, \gamma) < 0.4$ of a reconstructed photon passing tight identification requirements.

Diphoton system and final event selection The Higgs boson candidate is formed from the two highest p_T photon candidates. The diphoton candidate must also satisfy the following requirements:

- The photon candidates are required to pass the *tight* identification criteria, as defined in Sect. 4.3.1.1.
- The photon candidates are required to pass the following isolation criteria:

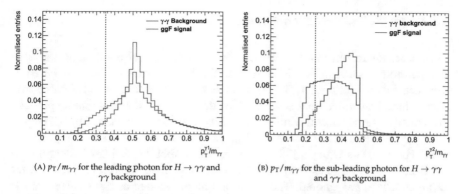

(A) $p_T/m_{\gamma\gamma}$ for the leading photon for $H \rightarrow \gamma\gamma$ and $\gamma\gamma$ background

(B) $p_T/m_{\gamma\gamma}$ for the sub-leading photon for $H \rightarrow \gamma\gamma$ and $\gamma\gamma$ background

Fig. 7.5 Distribution of $p_T/m_{\gamma\gamma}$ for gluon-fusion Higgs boson signal and $\gamma\gamma$ background for the **A** leading photon and **B** sub-leading photon. Selected events are required to have $p_T/m_{\gamma\gamma} > 0.35\ (0.25)$ for the leading (sub-leading) photons

- The track isolation variable `ptcone20`, calculated using only the tracks from the diphoton primary vertex, must satisfy `ptcone20 < 0.05 × $p_T$`.
- The calorimeter isolation variable `topoetcone20` must satisfy `topoetcone20 < 0.065 × $p_T$`

• The diphoton invariant mass must be in the region $m_{\gamma\gamma} \in [105, 160]$ GeV. The diphoton invariant mass is calculated as:

$$m_{\gamma\gamma} = \sqrt{2E_1 E_2 (1 - \cos(\theta))}, \qquad (7.2)$$

where E_1 and E_2 are the leading and sub-leading photon energies and θ is the angle between them. The angle between the selected photons is estimated using the neural-network selected diphoton vertex, as the η position of the photons (determined from the barycenter of the energy cluster in the calorimeter) is corrected to point to the selected diphoton PV.

• The leading (sub-leading) photon must satisfy a relative-p_T requirement defined as

$$p_T/m_{\gamma\gamma} > 0.35\ (0.25).$$

The $p_T/m_{\gamma\gamma}$ distribution is shown in Fig. 7.5 for the leading and sub-leading photons using gluon-fusion signal and $\gamma\gamma$ background. This relative-p_T cut has no significant impact on the signal shape, but it simplifies the analytical modeling of the invariant mass shape of the background [73].

7.3.2   Particle-Level Event Selection

To compare the results to other experiments or to theoretical predictions, the cross sections are measured in fiducial regions defined at the particle level, i.e. based on event generation information. The definition of the fiducial volume and of the observables is based on *stable* final state particles that enter the detector. *Stable* particles are defined as particles with lifetime $c\tau_0 > 10$ mm that are not created by the GEANT simulation of the detector response [74]. The type of each particle is represented by the Particle Data Group's [75] "Monte Carlo Numbering Scheme" identifier (or "PDG ID"), referred to as PdgId in the following. Particles originating from the pileup collisions are ignored. The selection criteria applied at particle-level are as follows:

Photons Particle-level photons are identified by PdgId $= 22$. They are required to have not originated during hadronization. This means that their parent should not have $|$PdgId$| \geq 111$. If the parent is a τ lepton ($|$PdgId$| = 15$) or a photon, corresponding to a final state radiative emission, then the PdgId of the grandparent is checked, and so on.

Particle-level photons are required to have generator-level $p_T > 25$ GeV and $|\eta| < 1.37$ or $1.52 < |\eta| < 2.37$. In addition, they are required to pass a particle-level isolation requirement, detailed in Sect. 7.3.3.

Jets Particle-level jets are clusters of stable particles obtained using the anti-k_t clustering algorithm with a radius parameter $R = 0.4$ [71]. Muons and neutrinos are not included in the clustering at the particle-level since they do not leave significant energy deposits in the calorimeters and so do not enter the detector-level jet finding. Selected particle-level jets are required to pass the kinematic selection $p_T > 25$ GeV and $|y| < 4.4$. Jets are rejected if they are reconstructed within $\Delta R < 0.4$ of a selected photon or within $\Delta R < 0.2$ of a selected electron.

Event selection The two photons with the highest p_T are chosen as the Higgs candidate. These photons are selected after the kinematic photon requirements have been applied. The diphoton invariant mass $m_{\gamma\gamma}$ is required to lie in the range $m_{\gamma\gamma} \in [105, 160]$ GeV. The two photon candidates are required to pass relative-p_T requirements of $p_T/m_{\gamma\gamma} > 0.35$ (0.25) for the leading (sub-leading) photon.

7.3.3   Particle-Level Isolation

As detailed in Sect. 7.1.1, the motivation for a measurement of the Higgs boson cross section in a fiducial region matching the experimental selection is to reduce model-dependent extrapolations of the acceptance and efficiency to the full phase space. However, a residual model dependence in the measured cross section can be present if the correction for detector effects (unfolding), described in *Interlude B*, depend on the production mode, as discussed in the following.

Let us consider the simplest of the unfolding methods, which is the bin-by-bin correction factor defined in Eq. (7.3):

$$\sigma_i \times BR = \frac{\nu_i^{sig}}{c_i \times \mathcal{L}_{int}}, \qquad (7.3)$$

where σ_i is the measured cross section in bin i, ν_i^{sig} is the measured signal yield, $\mathcal{L}_{int}$ is the integrated luminosity, and

$$c_i = \frac{n_i^{det}}{n_i^{ptcl}} \qquad (7.4)$$

is the correction factor. The correction factor is determined from the expected detector and particle level yields, n_i^{det} and n_i^{ptcl} respectively, as estimated from the simulations. If different Higgs boson production modes yield different values of the correction factor, when these factors are combined into a single value used to unfold the measurement, model dependence is introduced via the weights of the combination, which are determined as the relative fractions of the Standard Model cross sections for each process:

$$c_i = \frac{\Sigma_s w_s n_{i,s}^{det}}{\Sigma_s w_s n_{i,s}^{ptcl}}, \qquad (7.5)$$

where $s \in \{ggF, VBF, WH, ZH, ttH, bbH\}$. The larger the disagreement between the different $c_{i,s}$ (the correction factor for each production mode), the bigger the model dependence is.

7.3.3.1 Particle-Level Photon Isolation

As shown in Sect. 7.3.2, the kinematic selection requirements are similar between the detector-level and the particle level. The detector-level photon isolation requirement, however, requires further attention.

The detector-level photon isolation requirement is imposed in order to separate photons originating from the Higgs boson decay from photons emitted from hadronic jets. For example, collimated photons from boosted $\pi^0 \to \gamma\gamma$ decays can mimic the signature of single photons in the EM calorimeter. These photons, nevertheless, will be surrounded by hadronic activity, unlike photons from $H \to \gamma\gamma$.

Without a corresponding particle-level isolation requirement, this would lead to correction factors c_i that are highly dependent on the Higgs boson production mechanism (as shown in Fig. 7.9). This would cause a significant model dependence of the final measured cross section. To reduce the model dependence of the measurement, a photon isolation requirement was included in the particle-level selection, with the

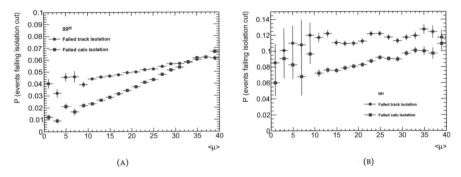

Fig. 7.6 The probability for an event to fail the detector-level calorimeter or track isolation requirements for **A** ggF and **B** ttH. The figures show that the track isolation is much stronger in rejecting events than the calorimeter isolation at zero pileup, which is the region relevant for the particle level quantities. The effect of pileup is more evident for ggF where more pileup results in an increased rejection from the calorimeter isolation, whereas for ttH the existence of hard scatter jets makes the event rejection almost pileup independent

drawback that theoretical predictions become more complicated [76]. A different solution to this problem, based on a veto on ΔR between photons and nearby jets, is shown in Appendix.

The particle-level isolation requirement was chosen to mimic the efficiency of the experimental one. Two particle-level isolation quantities are defined, to mimic the calorimeter-based and track-based detector-level isolation: the particle-level (calorimeter) isolation, `truthcalocone20`, defined as the sum of transverse energies of stable particles (with $p_T > 1$ GeV) in a cone of radius 0.2 around the photon, and the particle-level track isolation, `ptcone20`$_{particle}$, defined in a similar way but considering only charged particles [74].

Since pileup is not considered at particle-level, the matching with the detector-level must be done in the limit of zero $\langle\mu\rangle$. The inefficiency of the track- and calorimeter-based isolation selection for gluon-fusion and ttH events as a function of $\langle\mu\rangle$ is shown in Fig. 7.6. The figure shows that for $\langle\mu\rangle = 0$, the track-isolation requirement removes more events than the calorimeter isolation one. For this reason, at the particle-level, the isolation requirement will be made only on track isolation. This simplifies theory calculation without compromising the overall isolation requirement.

To find the particle-level isolation requirement to be applied in order to select a phase space closely matching the one selected by the detector-level isolation requirement, the following mapping procedure between the detector-level and particle-level isolation variables is followed:

- at *detector-level*, apply the full cutflow including only the weakest of the two isolation requirements (i.e. the calorimeter isolation), rejecting events without a reconstructed diphoton system;
- at *particle-level*, select stable photons with $p_T > 20$ GeV that are not originating from hadrons;

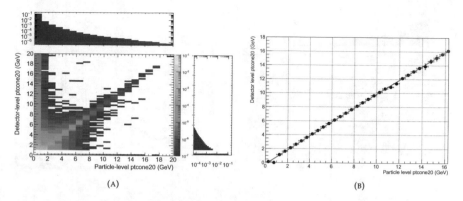

Fig. 7.7 **A** Detector-level versus particle-level isolation track-isolation for matched diphotons for the different Higgs boson production mechanisms weighted by their cross sections. The plot shows also the 1D projections on top of the corresponding axes. **B** Profile of the 2D histogram using the median and error computed from bootstrapping with 500 toys with the linear fit to determine detector-level isolation as function of the particle level isolation

- match detector-level and particle-level diphotons within $\Delta R < 0.1$;
- for each of the two photons, separately for each Higgs boson production mode, fill a 2D histogram with their detector-level and particle-level track-isolation variables;
- combine the 2D histograms of different Higgs boson production modes weighting them by their Standard Model cross sections. The combined 2D histogram is shown in Fig. 7.7a;
- profile the combined 2D histogram by finding, for each bin of the particle-level isolation, the median of the detector-level isolation distribution, and its uncertainty through a bootstrap technique based on the RMS of 500 pseudo datasets. The median is used instead of the mean in order to account for the peak at zero coming from true isolated photons. The resulting profile histogram is shown in Fig. 7.7b.
- perform a linear fit of the profile to find the relation between detector-level and particle-level track isolation. The fit is modeled as

$$\texttt{ptcone20}_{detector} = p_0 + p_1 \times \texttt{ptcone20}_{particle};$$

- use the previous formula to find the $\texttt{ptcone20}_{particle}$ threshold corresponding to the detector-level isolation requirement $\texttt{ptcone20}_{detector} < 0.05 \times p_T$.

The resulting mapping fit shows that the detector-level and particle level isolation variables are equivalent since the fit parameters are

$$p_1 = 0.997 \pm 0.009 \;, \; p_0 = -0.121 \pm 0.095,$$

and therefore the particle-level track isolation requirement will be

$$\texttt{ptcone20}_{particle} < 0.05 \times p_T$$

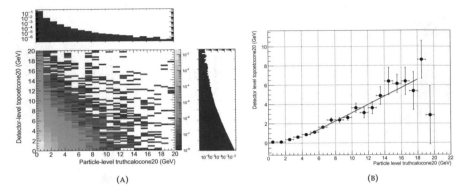

(A)   (B)

Fig. 7.8 **A** 2D histogram with the detector and particle-level isolation calorimeter isolation variables for matched diphotons for the different Higgs production mechanisms weighted with their cross sections. The plot shows also the 1D projections on top of the corresponding axes. **B** Profile of the 2D histogram using the median and error computed from bootstrapping with 500 toys

A similar procedure was performed to map the calorimeter isolation (applying the full detector-level selection, including the track-isolation requirement but excluding the calorimeter-isolation one). The results are shown in Fig. 7.8. The fit was performed in the range corresponding to the events failing the detector-level calorimeter isolation (`topoetcone20` $\succ$ 1.5 GeV). The resulting fit parameters are

$$p_1 = 0.431 \pm 0.047 \,, \; p_0 = -1.202 \pm 0.394,$$

which results into a particle-level calorimeter requirement of

$$\texttt{truthcalocone20} < 0.15 \times p_T + 2.789 \,(\text{GeV}).$$

7.3.3.2 Correction Factors Dependence on the Higgs Boson Production Mode

Figure 7.9 shows the effect of applying the particle-level isolation requirement on the correction factors. The plot shows that applying the particle-level track isolation significantly reduces the dependence of the correction factors on the Higgs boson production mode. This is more evident for the $t\bar{t}H$ production mode as the jet activity causes more events to fail detector-level isolation criteria and hence results in a smaller correction factor. The plot also shows that the effect of applying a particle-level calorimeter isolation requirement in addition to the particle-level track isolation one is minimal, as expected from the studies of Sect. 7.3.3.1 and Fig. 7.6. The residual dependence of the correction factors on the Higgs boson production mode after the particle-level isolation requirement and the corresponding model dependence introduced in the cross section measurement will be estimated in Sect. 7.5.3.1.

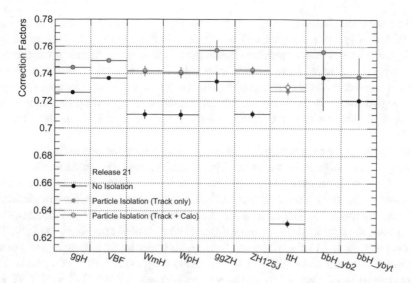

Fig. 7.9 Correction factors for the inclusive fiducial volume. Black points show the case when no particle-level isolation is applied and displays a significant dependence on the production process. Green points show the case when a particle-level track isolation requirement is applied. The red points are for the case of particle-level track and calorimeter isolation. Production-mode dependence is reduced when the isolation required is used

7.3.4 Binning of the Differential Variables

The choice of the binning for the differential variables is performed before unblinding the data in the $m_{\gamma\gamma}$ signal region, i.e. the subset of events passing the full detector-level selection and with the additional requirement $m_{\gamma\gamma} \in [121 - 129]$ GeV. The optimal binning is chosen in order to:

- increase the expected signal significance in each bin, to avoid having bins with low statistics as they can cause instabilities in the fitting procedure.
- increase the migration purity and the reconstruction efficiency in each bin in order to reduce uncertainties from the unfolding procedure. The migration purity P is defined as $P = n^{ptcl\&det}/n^{det}$, where $n^{ptcl\&det}$ is the event yield at both detector-level and particle-level bin, and n^{det} is the event yield at a given detector-level bin. The purity is sensitive to migrations due to resolution effects, in addition to fake events wrongly reconstructed in the detector-level fiducial volume. The reconstruction efficiency ϵ is defined as $\epsilon = n^{ptcl\&det}/n^{ptcl}$, with n^{ptcl} the event yield in a given particle-level bin.

The following criteria are adopted for the choice of the binning:

- Expected significance close to or greater than 2σ.
- Migration purity and reconstruction efficiency close to or higher than 50%,

The expected significance in each bin is computed as $s/\sqrt{b}$, where s is the number of SM Higgs boson signal events expected in that bin, and b is the number of background events under the Higgs boson peak. The number of background events, b, is roughly estimated from a linear extrapolation of the events counted in data in the $m_{\gamma\gamma}$ side-bands around the signal region. Using the full Run-2 dataset, we expect 6247 signal events in the inclusive fiducial region, corresponding to a significance $s/\sqrt{b} \approx 12$. The purity in the inclusive fiducial region is 98.2%, and the efficiency is 72%.

Using the above requirements, a simple algorithm is developed to find the optimal binning for the differential variables. The algorithm uses the same bin edges of the $H \rightarrow ZZ^* \rightarrow 4\ell$ analysis to facilitate the combination of the two measurements. A summary of the chosen binning with purities, efficiencies, and expected significance is shown in Table 7.3. The table shows large changes in expected significance and purity between adjacent bins for some differential variables, such as $p_T^{\gamma\gamma}$ and $p_T^{j_1}$. This is generally due to the change in bin width, with wider bins yielding higher purities due to smaller migrations. The lower efficiencies for jet variables are the result of the worse energy resolution of jets and hence jets with true $p_T^{\text{particle−level}} > 30$ GeV may not pass the detector-level requirement $p_T^{\text{detector−level}} > 30$ GeV. In addition, jets from pileup collisions decrease the purity as they increase the number of detector-level jets without a corresponding increase in particle-level jets. More details on pileup jets are given in Sect. 7.5.2.6.

7.3.4.1   Signal Composition and Migration Matrix for the Differential Variables

The relative fractions of the selected events from the different Higgs boson production modes as a function of the bins of the differential variables are shown in Figs. 7.10 and 7.11. The figures also show the detector migration matrix, which gives the probability for an event to be reconstructed in a given bin i of the detector-level observable x giving it was generated in a bin j of the corresponding particle-level variable y. The migration matrix is a measure of the purities and bin migrations. More details on this are provided in *Interlude B*. From the figures, one can observe that:

- Large m_{jj} events are dominated by VBF.
- Events with 3 jets or more receive an important contribution from the $t\bar{t}H$ production mode.
- Events with large $p_T^{\gamma\gamma}$ and $p_T^{j_1}$ receive important contributions from VBF, VH and $t\bar{t}H$.

7.4   Signal and Background Invariant Mass Models

The total Higgs boson event yield and the Higgs boson yield in each bin of the differential distributions are determined using a fit to the diphoton invariant mass,

Table 7.3 Summary of selected binning for the different differential variables, along expected signifcances, migration purities, and reconstruction efficiencies. Bold indicates: efficiency<0.5, purity<0.5, expected significance<2. Bins with significance <2, typically overflow bins, are not included in the measurement

$p_T^{\gamma\gamma}$ GeV	Bin	Purity = $\frac{n^{reco.\&ptcl.}}{n^{reco.}}$	Efficiency = $\frac{n^{reco.\&ptcl.}}{n^{ptcl.}}$	Exp. SM yield	Exp. SM unc.	Exp. SM Sig.
	0–5	0.84	0.57	275.4	105.2	2.6
	5–10	0.81	0.56	588.7	158.9	3.7
	10–15	0.80	0.55	633.1	177.3	3.5
	15–20	0.80	0.55	582.6	165.3	3.5
	20–25	0.80	0.55	510.4	149.7	3.4
	25–30	0.80	0.55	439.5	135.4	3.2
	30–35	0.79	0.55	375.8	133.3	2.8
	35–45	0.88	0.61	598.0	155.6	3.8
	45–60	0.91	0.62	626.6	151.4	4.1
	60–80	0.93	0.63	527.9	134.2	3.9
	80–100	0.92	0.64	327.8	100.3	3.3
	100–120	0.92	0.65	212.3	60.5	3.5
	120–140	0.91	0.66	147.4	43.6	3.4
	140–170	0.93	0.69	146.7	38.3	3.8
	170–200	0.93	0.69	90.6	25.8	3.5
	200–250	0.95	0.72	82.6	22.3	3.7
	250–350	0.96	0.74	58.3	14.4	4.0
	350–∞	0.98	0.76	23.4	8.7	2.7
$|y_{\gamma\gamma}|$	Bin	Purity = $\frac{n^{reco.\&ptcl.}}{n^{reco.}}$	Efficiency = $\frac{n^{reco.\&ptcl.}}{n^{ptcl.}}$	Exp. SM yield	Exp. SM unc.	Exp. SM Sig.
	0.00–0.15	0.97	0.70	662.8	142.1	4.7
	0.15–0.30	0.96	0.69	652.2	141.4	4.6
	0.30–0.45	0.96	0.69	634.2	134.5	4.7
	0.45–0.60	0.95	0.69	600.3	134.9	4.4
	0.60–0.75	0.95	0.68	564.7	137.1	4.1
	0.75–0.90	0.95	0.68	524.3	141.9	3.7
	0.90–1.20	0.96	0.67	919.6	201.7	4.6
	1.20–1.60	0.96	0.66	948.1	234.6	4.0
	1.60–2.40	0.97	0.65	741.0	203.4	3.6
$N_{jets}[p_T^j >$ 30]	Bin	Purity = $\frac{n^{reco.\&ptcl.}}{n^{reco.}}$	Efficiency = $\frac{n^{reco.\&ptcl.}}{n^{ptcl.}}$	Exp. SM yield	Exp. SM unc.	Exp. SM Sig.
	=0	0.91	0.56	3306.1	407.3	8.1
	=1	0.64	**0.48**	1783.8	252.6	7.1
	=2	0.58	**0.46**	806.9	155.2	5.2
	≥3	0.58	0.56	373.9	104.9	3.6

(continued)

Table 7.3 (continued)

p_T^{j1} GeV	Bin	Purity = $\frac{n^{reco.\&ptcl.}}{n^{reco.}}$	Efficiency = $\frac{n^{reco.\&ptcl.}}{n^{ptcl.}}$	Exp. SM yield	Exp. SM unc.	Exp. SM Sig.
	=0 jet	0.91	0.56	3306.1	407.3	8.1
	30–60	0.60	0.50	1511.3	239.0	6.3
	60–90	0.69	**0.48**	511.7	132.4	3.9
	90–120	0.67	**0.46**	532.0	117.5	4.5
	120–350	0.88	0.65	391.7	84.0	4.6
	350–∞	0.89	0.67	17.9	14.5	**1.1**
$\Delta\Phi_{jj}$	Bin	Purity = $\frac{n^{reco.\&ptcl.}}{n^{reco.}}$	Efficiency = $\frac{n^{reco.\&ptcl.}}{n^{ptcl.}}$	Exp. SM yield	Exp. SM unc.	Exp. SM Sig.
	<2 jets	0.95	0.63	5089.9	474.2	10.7
	$-\pi--\pi/2$	0.62	0.50	310.1	113.8	2.7
	$-\pi/2-0.0$	0.62	0.53	280.0	70.0	4.0
	$0.0-\pi/2$	0.62	0.53	280.0	70.0	4.0
	$\pi/2-\pi$	0.62	0.50	310.1	113.8	2.7
m_{jj} GeV	Bin	Purity = $\frac{n^{reco.\&ptcl.}}{n^{reco.}}$	Efficiency = $\frac{n^{reco.\&ptcl.}}{n^{ptcl.}}$	Exp. SM yield	Exp. SM unc.	Exp. SM Sig.
	<2 jets	0.95	0.63	5089.9	474.2	10.7
	0–170	0.53	**0.47**	548.9	133.5	4.1
	170–500	0.65	0.52	379.6	89.6	3.8
	500–1500	0.65	0.54	208.2	59.3	3.5
	1500–∞	0.69	0.61	43.9	17.6	2.4

$m_{\gamma\gamma}$. The fitted model (detailed in Sect. 7.6) is the sum of two analytic functions that model the signal and background components. The details of the signal model are shown in Sect. 7.4.1, while those of the background model are provided in Sect. 7.4.2.

7.4.1 Signal Model

The shape of the invariant mass distribution for the signal is studied using simulated signal events. The Higgs boson decay in the diphoton channel $H \rightarrow \gamma\gamma$ is resonant. Therefore, the signal is expected to follow a Breit-Wigner distribution, peaking at the Higgs mass, m_H. The effect of the interference between the diphoton background and the $H \rightarrow \gamma\gamma$ signal on the shift of the Higgs mass was found to be negligible with respect to the experimental uncertainty [77]. The Higgs boson has a narrow width of 4.07 MeV [27]. This width is much smaller than the measured energy resolution of photons (typically ≥ 1 GeV, as shown in Fig. 5.40). Therefore, the observed distribution will be dominated by the smearing induced by the finite resolution of

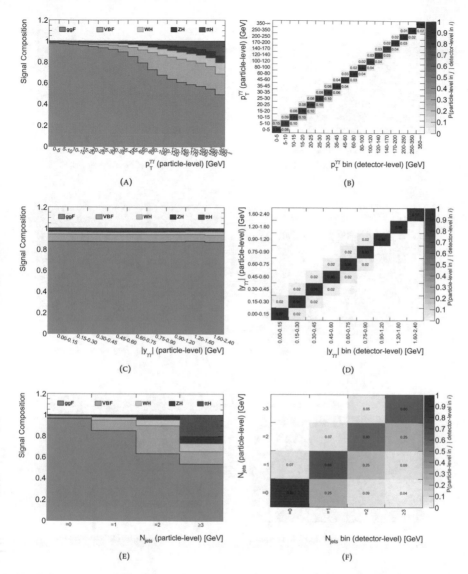

Fig. 7.10 Variations of the relative contributions to the total signal yield from the different Higgs boson production mode with the bins of **A** $p_T^{\gamma\gamma}$ **C** $|y_{\gamma\gamma}|$ and **E** N_{jets}. The migrations matrices for **B** $p_T^{\gamma\gamma}$ **D** $|y_{\gamma\gamma}|$ and **F** N_{jets}, the matrices have the bins in the detector-level in the x-axis and the particle-level in the y-axis. The bins show the probability of an event generated in particle-level bin j given it was reconstructed in detector-level bin i

the detector. In this study, the $m_{\gamma\gamma}$ distribution is modeled by a double-sided Crystal Ball function (DSCB), shown in Eq. (7.6), which is a function chosen empirically and consisting of a Gaussian core with power-law tails:

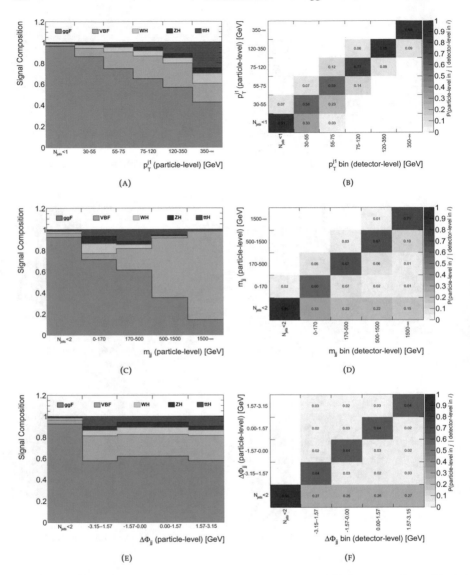

Fig. 7.11 Variations of the relative contributions to the total signal yield from the different Higgs boson production mode in bins of **A** p_T^{j1}, **C** m_{jj}, **E** $\Delta\Phi_{jj}$. The migrations matrices for **B** p_T^{j1}, **D** m_{jj}, **F** $\Delta\Phi_{jj}$, the matrices have the bins in the detector-level in the x-axis and the particle-level in the y-axis. The bins show the probability of an event generated in particle-level bin j given that it was reconstructed in detector-level bin i

Fig. 7.12 Example of a double-sided Crystal Ball function

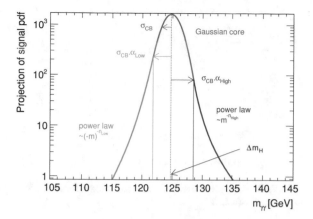

$$
CB(m_{\gamma\gamma}) = N \times
\begin{cases}
e^{-t^2/2}, & \text{if } -\alpha_{\text{low}} \leq t \leq \alpha_{\text{high}}. \\[2mm]
e^{-\frac{1}{2}\alpha_{\text{low}}^2}\left[\dfrac{1}{R_{\text{low}}}\left(R_{\text{low}} - \alpha_{\text{low}} - t\right)\right]^{-n_{\text{low}}}, & \text{if } t < -\alpha_{\text{low}}. \\[2mm]
e^{-\frac{1}{2}\alpha_{\text{high}}^2}\left[\dfrac{1}{R_{\text{high}}}\left(R_{\text{high}} - \alpha_{\text{high}} - t\right)\right]^{-n_{\text{high}}}, & \text{if } t > \alpha_{\text{high}}.
\end{cases}
$$

(7.6)

where $t = (m_{\gamma\gamma} - \mu_{\text{CB}})/\sigma_{\text{CB}}$, $R_{\text{low}} = \frac{\alpha_{\text{low}}}{n_{\text{low}}}$, and $R_{\text{high}} = \frac{\alpha_{\text{high}}}{n_{\text{high}}}$. Here N is a normalization parameter, μ_{CB} and σ_{CB} are the mean and the width of the Gaussian distribution, α_{low} and α_{high} are the positions of the transitions from the Gaussian core to the exponential tails (in units of σ_{CB} on the low and high mass sides, and n_{low} and n_{low} are the exponents of the low and high mass tails. An illustration of this function is shown in Fig. 7.12.

The different shape parameters of the DSCB function are estimated from a fit to the simulated $H \rightarrow \gamma\gamma$ samples. The fit is performed using the combination of the different Higgs boson production modes samples detailed in Sect. 7.2.2.1 weighted by their SM cross sections. This parameterization is derived separately for each bin of the differential variables providing the nominal signal template. The full signal model will then include shape changes due to systematic uncertainties on μ_{CB} and σ_{CB}, detailed in Sect. 7.5.1.1, as nuisance parameters constrained from the photon energy scale and resolution measurements. The nominal signal model parameters are fixed in the final signal extraction fit to data. The resulting values of μ_{CB} in each bin are shifted by $+90\,\text{MeV}$ to match the measured Higgs boson mass of 125.09 GeV [78]. Examples of these fits are shown in Fig. 7.13 for the inclusive fiducial region and for the bin $p_{\text{T}}^{\gamma\gamma} \in [60\text{–}80]$ GeV.

7.4.2 Background Model

The estimation of the Higgs boson signal yield requires the precise determination of the background in the selected data. The main backgrounds affecting this anal-

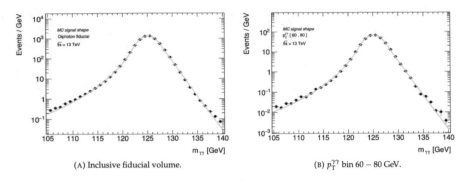

(A) Inclusive fiducial volume.   (B) $p_T^{\gamma\gamma}$ bin $60 - 80$ GeV.

Fig. 7.13 Examples of $m_{\gamma\gamma}$ pf the signal model for **A** the inclusive fiducial volume and **B** the bin $p_T^{\gamma\gamma} \in [60–80]$ GeV

ysis can be categorized into *irreducible* and *reducible* components. The irreducible component comes from non-resonant prompt photon pairs produced in processes such as $q\bar{q} \rightarrow \gamma\gamma$, $qg \rightarrow \gamma\gamma$, or $gg \rightarrow \gamma\gamma$. The photons from these processes are largely indistinguishable from Higgs boson photons, hence the name "irreducible". The reducible component, on the other hand, arises from events with γ-jet and jet-jet final states where jets of hadrons are misidentified as photons. The photon identification and isolation requirement reject most of these events, thanks to their jet rejection factor between 5000 and 10000. Hence the reducible component of the background after the selection is sub-dominant, and the jet-jet component of the reducible background is negligible (as shown in Fig. 7.14).

The background $m_{\gamma\gamma}$ distributions in each bin of the differential variables and the inclusive sample are smoothly falling and can be described using analytical functions chosen empirically. The analytical functions are determined using templates of the background that are built for each bin of the differential variables and the inclusive sample. These templates include both the reducible and irreducible components. The procedures for obtaining these templates and for choosing the analytical functions are detailed in the next sections.

7.4.2.1   Background $m_{\gamma\gamma}$ Templates

The background templates are built from the sum of the reducible and irreducible components, as follows:

- **Irreducible $\gamma\gamma$ background**. The $m_{\gamma\gamma}$ distribution of this background component is obtained from large ($\mathcal{O}(100M)$ events) simulated $\gamma\gamma$ event samples generated with SHERPA (Sect. 7.2.2.2). The events from this sample are required to pass the nominal event selection detailed in Sect. 7.3.
- **Reducible background**. The $m_{\gamma\gamma}$ distribution of the reducible γ-jet component is determined from data control regions, defined by inverting the tight photon identification criteria on any of the two photon candidates while keeping all the

other nominal event selection requirements. These control regions have a small (typically ~10–20%) contamination from $\gamma\gamma$ events, due to the inefficiency of the photon identification algorithm. The $\gamma\gamma$ contamination in each bin is estimated from the SHERPA diphoton sample and is subtracted from the data control regions. An additional data control region is built by inverting both the photon identification and isolation requirements of the nominal selection. This procedure provides higher statistics, resulting in a more accurate estimate of the shape of the reducible component. This is more evident for bins where the default control region is poorly populated (e.g. the high $p_T^{\gamma\gamma}$ or the $N_{\text{jets}} \geq 3$ bins). A smoothing procedure is then performed in order to suppress statistical fluctuations since the final $m_{\gamma\gamma}$ template requires fine binning (0.25 GeV/bin). The smoothing is performed by fitting the ratio between the irreducible template from data control regions to the high statistics simulated $\gamma\gamma$ sample with a second-order polynomial function. The polynomial function is then used to reweight the shape of the high statistics $\gamma\gamma$ simulated event sample to match that of the reducible template.

The final background templates are built by adding the smoothed reducible and irreducible components after weighting them with their relative fraction in the signal region. The relative fraction of the reducible and irreducible components is estimated using a double two-dimensional side-band method ($2 \times 2D$), detailed in Refs. [79, 80]. The $2 \times 2D$ side-band method computes the background fraction in the signal region using the yields in data in background-enriched control regions, populated by events in which at least one of the two photon candidates that either fail the identification or isolation requirements or both (i.e. not passing the tight identification and the isolation).

The fractions of the irreducible and reducible background components were found to be $\left(75^{+3}_{-4}\right)\%$ and $\left(25^{+3.5}_{-2.4}\right)\%$ respectively in the inclusive fiducial region. Figure 7.14 shows the relative fractions of the reducible and irreducible backgrounds as a function of $p_T^{\gamma\gamma}$ and N_{jets} bin. The background fractions were also checked with pileup. The measured fraction was found to be resilient against additional activity from higher pileup. The contribution of $\gamma\gamma$ events is reduced by less than 4% from $\mu = 15$ to $\mu = 50$ as shown in Fig. 7.15.

The templates are normalized to the event yield in data in the $m_{\gamma\gamma}$ side-bands. The final background template for the inclusive fiducial region and that for one particular $p_T^{\gamma\gamma}$ bin are shown in Fig. 7.16.

7.4.2.2 Determination of the Background Model

Several functional forms were considered for the parametrization of the background $m_{\gamma\gamma}$ spectrum. They include:

- A power-law (Pow) function.
- Exponentiated polynomials of first (Exp), second (ExpPoly2) or third (ExpPoly3) degree. As an example, the exponentiated second-order polynomial has the form:

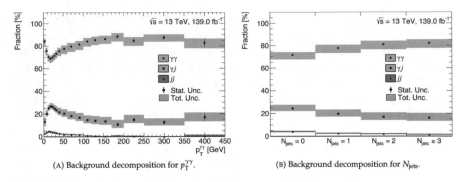

(A) Background decomposition for $p_T^{\gamma\gamma}$.　　　(B) Background decomposition for N_{jets}.

Fig. 7.14 Background fractions as a function of $p_T^{\gamma\gamma}$ and N_{jets} [6]. The composition is estimated using the $2 \times 2D$ method detailed in Refs. [79, 80]

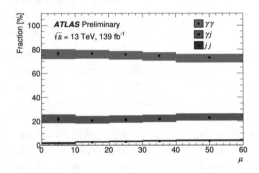

Fig. 7.15 Fraction of $\gamma\gamma$, γ-jet, and jet-jet events in data measured in bins of μ (the average number of interactions per bunch crossing) using the background decomposition. The error bars on the markers represent the statistical uncertainty and the filled rectangles the total uncertainty, which is dominated by systematic sources [81]

$$\mathcal{B}\left(m_{\gamma\gamma}; \alpha^{\mathbf{bkg}}\right) = N\left(\alpha^{\mathbf{bkg}}\right) \cdot \exp\left(-\frac{m_{\gamma\gamma}}{\alpha_1^{\text{bkg}}} - \frac{m_{\gamma\gamma}^2}{\alpha_2^{\text{bkg}}}\right), \tag{7.7}$$

where α_1^{bkg}, α_2^{bkg} are nuisance parameters and $N\left(\alpha^{\mathbf{bkg}}\right)$ normalises the function to unity.

- Bernstein polynomials of degree 3 (Bern3), 4 (Bern4), and 5 (Bern5).

An inappropriate choice of the analytical function describing the background can underestimate (or overestimate) the number of background events under the signal peak, thus biasing the measured signal events in the signal+background fit to the data. This bias is called the *spurious signal* and is detailed as follows.

The spurious signal for each functional form can be used as a measure of the bias in the signal yield and is used as a tool to select the background function. The spurious signal is estimated by fitting the background templates detailed in the previous section using a signal + background parameterization. In an ideal case, where the analytical function describes perfectly the background template, the number of fitted spurious

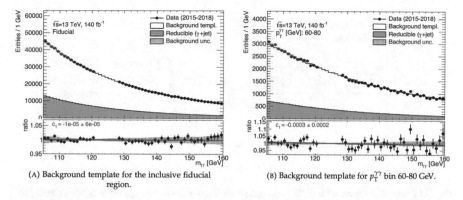

(A) Background template for the inclusive fiducial region.

(B) Background template for $p_T^{\gamma\gamma}$ bin 60-80 GeV.

Fig. 7.16 Background templates for the inclusive fiducial region and $p_T^{\gamma\gamma} \in [60\text{–}80]$ GeV. The figure compares the templates to data sidebands. The reducible component which is part of the template is shown in yellow. The ratio between the templates and data sidebands is shown, and a linear fit is performed to assess the agreement between the two. The slope of the fit for each bin was found to be compatible with zero, confirming that the final templates describe well the shape of the data sidebands [6]

signal will be zero on average. We look for a background function that minimizes the spurious signal while limiting the number of degrees of freedom of the model to avoid increasing too much the statistical uncertainty on the signal yield. More concretely, the selected function must satisfy the following requirements on the number of spurious signal events (N_{sp}) and its statistical uncertainty (Δ_{sp}):

- $N_{sp} \pm 2\Delta_{sp}$ is less than 20% of the background uncertainty, which is an estimate of the statistical uncertainty on the signal δS.
- $N_{sp} \pm 2\Delta_{sp}$ is less than 10% of the number of expected signal events N_{ref} for each bin.

The estimation of the spurious signal is performed in the region 121–129 GeV around the expected signal. This means that the signal+background fit is repeated for S with masses between 121 and 129 GeV. There is an additional loose goodness-of-fit test which requires the χ^2 probability to be higher than 1%, in order to make sure that the chosen functions can describe reasonably well the background templates.

The main challenge for the estimation of the spurious signal arises from the statistical fluctuations in the background template. These statistical fluctuations can artificially induce spurious signal, and make the tested model fail one of the two requirements previously described. This is most likely the case for bins with low Monte Carlo statistics. In order to deal with such problem, the MC simulated sample that is used to build the background template is required to be very large in size, $\mathcal{O}(100)$ times the data, in order to provide smooth $m_{\gamma\gamma}$ distributions even in the bins with a low number of events. Furthermore, an envelope around the spurious signal of $2\Delta_{sp}$ is considered in order to reduce the effect of the fluctuations, i.e. avoiding the situation where the spurious-signal criteria rejects a background model when it

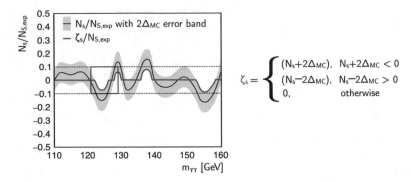

Fig. 7.17 An illustration of the effect of statistical fluctuation on the spurious signal criteria [68]. The illustration shows the ratio between the fitted spurious signal N_s and the expected Higgs boson signal N_{ref}. The light blue band represents double the statistical uncertainty on the fitted spurious signal $2\Delta_{sp}$. The dashed horizontal lines represent the 10% criteria on N_s/N_{ref}

has no statistical power to do so. This is known as the relaxed spurious signal criteria, and it is considered when the nominal spurious signal criteria ($N_{sp} < 20\% \ \delta S$ or $N_{sp} < 10\% \ N_{ref}$) are not satisfied. An illustration of the effect of the statistical fluctuations is shown in Fig. 7.17.

Smoothing using Gaussian processes The number of events in the $\gamma\gamma$ simulation samples for the full Run-2 do not satisfy the requirement to be $\mathcal{O}(100)$ times the data. This problem manifests itself more evidently in the lower statistics bins of the kinematic distributions (such as high $p_T^{\gamma\gamma}$). In such bins, the statistical fluctuations of the background template will be misidentified as spurious signal. An illustration of such cases is shown in Fig. 7.18. In this figure, a low statistics pseudo-dataset was generated from an exponential background-only function and fitted with a PDF composed of the same background function and a Gaussian signal. The signal and background fit resulted in a non-zero N_{signal} which in the context of spurious signal test will be identified as spurious signal or uncertainty on the background modeling despite that the analytical function of the background was known.

To solve this problem, a smoothing procedure using Gaussian Processes regression (GPR) is used [83]. The goal of this technique is to suppress statistical fluctuations in the background template without biasing the shape of wider features in the background template. In order to do this, a prior has to be defined which is specified in a kernel object. For the presented studies a combination of a Gibbs kernel [84] and an additional custom error kernel is used.

The Gibbs kernel is motivated for smoothly-falling spectra where the length scale increases linearly with x. The Gibbs kernel has two hyper-parameters:

- a length scale λ, with larger λ meaning more smoothing.
- the slope of the length scale b_λ.

Fig. 7.18 An illustration of the limitations of the spurious signal test on a low statistics background template. A known analytical function is used to generate a pseudo dataset that is then fit with signal+background PDF resulting in non-zero N_{signal}. This is a result of the statistical fluctuations of the background template. Scheme based on Ref. [82]

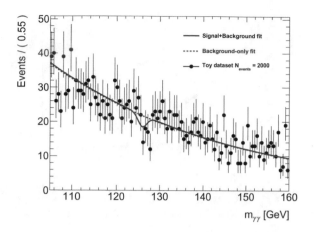

In order to account for the statistical uncertainties in the original background templates, an additional noise kernel is used. The custom error kernel is a kernel similar to a white noise kernel but with an error magnitude that decreases linearly with x to mimic Poissonian statistical uncertainties. The error kernel has two hyperparameters: ϵ and b_ϵ. These parameters are highly constrained to approximate the error bars from the original background template.

The hyperparameters of the Gibbs kernel are optimized to smooth out statistical fluctuations without removing any real features of the background. The optimization is performed by scanning a 2D space of the hyper-parameters (λ, b_λ) for a background template with injected narrow and wide signals. The injected signal is of magnitude set to 1% of the total background integral. The narrow signal width is in the order of the bin width, whereas the wide signal is in the order of the expected signal width. The difference between the smooth shape (GPR fit without injected signal) and the GPR fit under the injected signal is compared, and the criteria for the hyperparameters are defined:

- Smooth out at least 33% of narrow injected features defined as with width equal to half the bin width.
- Smooth out less than 25% of the wide injected features defined as with width equal to that of the expected signal.

Using these criteria, GPR smoothing was performed. Examples of the results of smoothing are shown in Fig. 7.19 for the inclusive selection and one low statistics bin in $p_T^{\gamma\gamma}$.

As expected, there is no visible change for the inclusive selection, but for the $p_T^{\gamma\gamma}$ bin with fewer statistics, the smoothed template reduces the statistical fluctuation which would otherwise appear a spurious signal uncertainty. Studies using pseudo-datasets from known analytical functions were performed in order to check whether the GPR smoothing can introduce a bias to the analysis based on the inclusive selection. These studies compared the difference between the GPR smoothed templates and the true known shape for each of the generated pseudo-datasets. These tests found

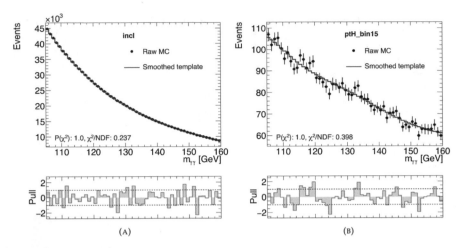

Fig. 7.19 The raw template (black dots) and the smoothed one (red line) for **A** the inclusive selection **B** and the 15th $p_T^{\gamma\gamma}$ bin $p_T^{\gamma\gamma} \in$ [200–250] GeV

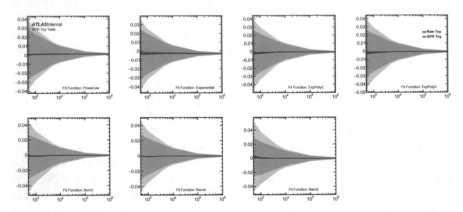

Fig. 7.20 Comparison of the spurious signal mean and width for the non-smoothed toys (red) and the GPR-smoothed toys (blue) as function of the number of toy events [6]

excellent agreement between the GPR smoothed and the true shape within 1% [6]. A similar test was performed comparing the estimated spurious signal between the GPR smoothed templates and the non-smoothed templates from known analytical functions. The results of such studies are shown in Fig. 7.20 for different background shapes showing the mean and standard deviation of the number of spurious signal for the different signal shapes. The figure shows a smaller width for the GPR smoothed templates with a mean consistent with the non-smoothed templates confirming that the smoothing does not bias the background shape but only reduces statistical fluctuations.

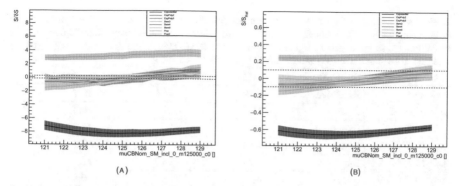

Fig. 7.21 Results of the spurious signal test for the inclusive fiducial region using the GPR smoothed templates. **A** the fitted spurious signal S relative to the background uncertainties δS **B** the fitted spurious signal relative to the expected signal yield S/S_{ref}

Results The spurious signal test is performed using the different analytical functions described before as potential background models, using the GPR smoothed background templates. An example of such tests is shown in Fig. 7.21, where the distribution of the number of the fitted spurious signal relative to the statistical uncertainty δS and expected signal N_{ref} is shown. In case multiple functions satisfy the above spurious signal requirements, the one with fewest degrees of freedom is be chosen, in order to reduce the statistical uncertainty on the extracted signal yield. The final selected functions for each bin are summarized in Table 7.4. Using GPR smoothed templates achieved a significant reduction of the spurious signal uncertainty, 20% on average. The improvement is more significant for the low-statistics bins, reaching more than 70% reduction in some bins.

7.4.2.3 F-Test

After the background model is selected based on the spurious signal criteria, the selected function is checked against data in the sidebands of the $m_{\gamma\gamma}$ distribution, to test if a higher-order function is needed to model the data better. This is because data might contain features not found in the MC simulation.

This test is performed by comparing the fit χ^2 and the number of degrees of freedom for each function and a higher-order function in which additional degrees of freedom have been introduced. The idea is to use the higher-order model instead of the simpler one if it gives a significant improvement in the agreement between data and the model. To quantify the significance of this improvement, the following test statistic is computed:

$$F_{1,2} = \frac{\frac{\chi_1^2 - \chi_2^2}{p_2 - p_1}}{\frac{\chi_2^2}{n - p_2}}, \tag{7.8}$$

Table 7.4 Summary of spurious signal studies for the smoothed background template for the different bins of the differential observables. The table shows the maximum number of fitted spurious signal relative to the expected signal in a given bin ($\max(S)/S_{\mathrm{ref}}$), the maximum number of the fitted spurious signal $\max(S)$, and the total background uncertainty

Selection	Sel. Func.	$\max(S)/S_{\mathrm{ref}}$ (%)	$\max(S)$	σ_{tot} (%)
Inclusive	ExpPoly2	−6.79	−367	10.7
$p_{\mathrm{T}}^{\gamma\gamma}$ 0–5 GeV	Pow	6.16	13.7	46.8
$p_{\mathrm{T}}^{\gamma\gamma}$ 5–10 GeV	ExpPoly2	2.69	13	33.2
$p_{\mathrm{T}}^{\gamma\gamma}$ 10–15 GeV	Bern4	−27.4	−143	43.5
$p_{\mathrm{T}}^{\gamma\gamma}$ 15–20 GeV	ExpPoly2	−3.24	−15.7	31.6
$p_{\mathrm{T}}^{\gamma\gamma}$ 20–25 GeV	Bern3	3.67	15.8	36.4
$p_{\mathrm{T}}^{\gamma\gamma}$ 25–30 GeV	ExpPoly2	1.73	5.56	36.3
$p_{\mathrm{T}}^{\gamma\gamma}$ 30–35 GeV	Bern4	−1.89	−4.1	40.4
$p_{\mathrm{T}}^{\gamma\gamma}$ 35–45 GeV	ExpPoly2	−8.46	−43.6	30.7
$p_{\mathrm{T}}^{\gamma\gamma}$ 45–60 GeV	ExpPoly2	−9.81	−54.2	29
$p_{\mathrm{T}}^{\gamma\gamma}$ 60–80 GeV	Exponential	5.21	25.1	25.7
$p_{\mathrm{T}}^{\gamma\gamma}$ 80–100 GeV	Bern3	8.06	24.4	33.3
$p_{\mathrm{T}}^{\gamma\gamma}$ 100–120 GeV	Exponential	9.56	19.2	30.7
$p_{\mathrm{T}}^{\gamma\gamma}$ 120–140 GeV	Exponential	−4.13	−5.87	29.3
$p_{\mathrm{T}}^{\gamma\gamma}$ 140–170 GeV	Pow	−1.65	−2.12	24.4
$p_{\mathrm{T}}^{\gamma\gamma}$ 170–200 GeV	Pow	5.76	5.13	26.6
$p_{\mathrm{T}}^{\gamma\gamma}$ 200–250 GeV	Exponential	2.75	2.25	23.3
$p_{\mathrm{T}}^{\gamma\gamma}$ 250–350 GeV	Exponential	1.43	0.755	23
$p_{\mathrm{T}}^{\gamma\gamma}$ 350–∞ GeV	Pow	−6.52	−1.53	28.1
$\lvert y_{\gamma\gamma}\rvert$ 0–0.15	ExpPoly2	−7.12	−41	20.5
$\lvert y_{\gamma\gamma}\rvert$ 0.15–0.30	ExpPoly2	3.93	22.3	24.3
$\lvert y_{\gamma\gamma}\rvert$ 0.30–0.45	ExpPoly2	−7.55	−41.4	25.6
$\lvert y_{\gamma\gamma}\rvert$ 0.45–0.60	ExpPoly2	3.25	17	23.9
$\lvert y_{\gamma\gamma}\rvert$ 0.60–0.75	ExpPoly2	14.2	69.7	30.5
$\lvert y_{\gamma\gamma}\rvert$ 0.75–0.90	ExpPoly2	−9.54	−43.3	30.8
$\lvert y_{\gamma\gamma}\rvert$ 0.90–1.20	ExpPoly2	−2.65	−20.9	24.2
$\lvert y_{\gamma\gamma}\rvert$ 1.20–1.60	Bern4	−9.96	−81.6	28.4
$\lvert y_{\gamma\gamma}\rvert$ 1.60–2.40	Bern4	−22.7	−147	39.6
$N_{\mathrm{jets}} = 0$	Bern4	9.37	273	16.1
$N_{\mathrm{jets}} = 1$	ExpPoly2	−4.23	−64.2	18.5
$N_{\mathrm{jets}} = 2$	ExpPoly2	3.15	20.3	23.3
$N_{\mathrm{jets}} \geq 3$	ExpPoly2	12.3	35	37.9

(continued)

Table 7.4 (continued)

Selection	Sel. Func.	max$(S)/S_{\text{ref}}$ (%)	max(S)	σ_{tot} (%)
$p_{\text{T}}^{j_1}$ 30–60 GeV	ExpPoly2	7.45	83	23.3
$p_{\text{T}}^{j_1}$ 60–90 GeV	ExpPoly2	−5.22	−23.5	45.1
$p_{\text{T}}^{j_1}$ 90–120 GeV	Exponential	−6.72	−33.4	23.8
$p_{\text{T}}^{j_1}$ 120–350 GeV	ExpPoly2	−4.75	−18	21.9
$p_{\text{T}}^{j_1}$ 350–∞ GeV	Pow	−4.68	−0.787	78
$m_{jj}/\Delta\phi_{jj,\text{signed}}$ underflow	ExpPoly2	−8.66	−390	12.9
$\Delta\phi_{jj,\text{signed}}$ $-\pi--\pi/2$	Bern3	−5.25	−13.2	46
$\Delta\phi_{jj,\text{signed}}$ $-\pi/2-0$	Exponential	−7.86	−16.9	31.8
$\Delta\phi_{jj,\text{signed}}$ $0-\pi/2$	Exponential	−7.86	−16.9	31.8
$\Delta\phi_{jj,\text{signed}}$ $\pi/2-\pi$	Bern3	−5.25	−13.2	46
m_{jj} 0–170 GeV	ExpPoly2	8.85	39.3	31
m_{jj} 170–500 GeV	Bern3	−7.98	−24.4	38.5
m_{jj} 500–1500 GeV	ExpPoly2	−2.35	−3.07	39.4

where χ_1^2 and χ_2^2 are the χ^2 of the two fits, p_1 and p_2 are the number of degrees of freedom for the functions $f_1(x)$ and $f_2(x)$, and n is the number of bins used in the fit.

The distribution of the test statistic $F_{1,2}$ is computed using 1000 pseudo-datasets for each bin of the differential distributions. The pseudo-datasets are drawn from the function with fewer degrees of freedom, after fixing the shape parameters to the values obtained from a fit to the data $m_{\gamma\gamma}$ sidebands.

The F-test is then performed on the functions chosen using the spurious signal tests. The background function with more degrees-of-freedom is then chosen instead of the lower-degree one if $P(F' \geq F) < 0.05$, where P is the probability of observing a F' value higher than the observed one if randomly picked from the test-statistic distribution calculated with pseudo-data. If this is the case, the background function with one more degree of freedom is used, and the F-test is repeated. All selected background functions passed the F-test criteria with the exception of four bins detailed in Table 7.5 in which the function with a higher degree of freedom was chosen.

Table 7.5 F-Test results for analysis bins that required increasing the number of degrees of freedom

Var	Bin	Func.	S/S_{ref} (%)	New Func.	new S/S_{ref} (%)
$p_T^{\gamma\gamma}$	20–25 GeV	Bern3	3.67	Bern4	16.9
$p_T^{\gamma\gamma}$	60–80 GeV	Exponential	5.21	ExpPoly2	8.52
$|y_{\gamma\gamma}|$	0.75–0.9	ExpPoly2	−9.54	ExpPoly3	−9.53
$\Delta\phi_{jj}$	0-±1.57	Exponential	−7.86	ExpPoly2	5.9

7.5   Systematic Uncertainties

In this section, we will review the different sources of systematic uncertainty affecting the cross section measurement. According to the formula $\sigma_i = \frac{\nu_i^{sig}}{c_i \times \mathcal{L}_{int}}$, the uncertainties can be categorized into:

- Uncertainties affecting the signal yield ν_i^{sig}. These are the systematic uncertainties related to the signal and background models. They are described in Sect. 7.5.1. The inclusion of these uncertainties in the signal extraction fit is described in Sect. 7.6.
- Uncertainties affecting the correction factors c_i and the luminosity $\mathcal{L}_{int}$. Luminosity and experimental uncertainties on the correction factors are discussed in Sect. 7.5.2. The theoretical uncertainties on the correction factors are discussed in Sect. 7.5.3.

7.5.1   Signal and Background Model Uncertainties

The sources of signal and background model uncertainties are described in this section.

7.5.1.1   Signal Model Uncertainties

The main uncertainties in the signal diphoton invariant mass shape arise from the photon energy scale and resolution uncertainties. The details of these uncertainties are given in Chap. 5. The photon energy scale uncertainty affects μ_{CB}, resulting in a shift of the position of the peak, as shown in Fig. 7.22a. On the other hand, the photon energy resolution uncertainty affects σ_{CB}, broadening or narrowing the width of the signal, as shown in Fig. 7.22b. The effects of these uncertainties will be included in the final signal extraction fits as nuisance parameters, as detailed in Sect. 7.6.

The signal shape uncertainties are computed using the full decorrelation scheme (detailed in Chap. 5). This scheme includes 9 photon energy resolution uncertainties and 39 photon energy scale uncertainties. The combined effect of these uncertainties,

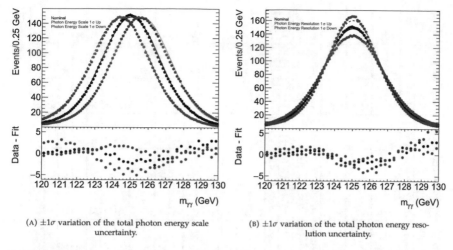

(A) ±1σ variation of the total photon energy scale uncertainty.

(B) ±1σ variation of the total photon energy resolution uncertainty.

Fig. 7.22 Effect of varying the total **A** the photon energy scale and **B** the photon energy resolution on the Higgs boson signal shape

depicted in Figs. 5.40 and 5.39, is shown in Table 7.6 for reference; in the fit, all 9+39 sources are considered.

In addition to the previous uncertainties, the following uncertainties affect the signal shape:

- The uncertainty on the measured Higgs boson mass, 125.09 ± 0.24 GeV [78], is taken into account as an additional nuisance parameter in the final signal extraction fit. It affects the position of the peak of the signal model and is fully correlated among all bins.
- Signal composition uncertainties arise from assuming SM relative cross sections for the different Higgs boson production modes. The topologies of the different production mechanisms can result in changes in resolution, as the resolution varies as a function of the pseudorapidity and the photon energy. This uncertainty is estimated by varying the relative weights of each production mechanism within their measured uncertainty using the procedure detailed in Sect. 7.5.3.1. These uncertainties were found to be negligible ($\ll 1\%$) with respect to the uncertainties on the resolution, and hence will not be used for the final signal extraction.

7.5.1.2 Background Model Uncertainties

The uncertainty in the fitted signal yield induced by the choice of the analytical model of the background invariant mass distribution is the spurious signal, described in Sect. 7.4.2.2.

Table 7.6 Relative change on the μ_{CB} and σ_{CB} of the signal model, Eq. (7.6), due to the total photon energy resolution and photon energy scale variations. These total variations do not enter into the final signal extraction fit, but are shown for reference on their impact

Variable	Bin 1	Bin 2	Bin 3	Bin 4	Bin 5	Bin 6	Bin 7	Bin 8	Bin 9	Bin 10	Bin 11	Bin 12	Bin 13	Bin 14	Bin 15	Bin 16	Bin 17	Bin 18
Inclusive Scale UP	+0.2%	X	X	X	X	X	X	X	X	X	X	X	X	X	X	X	X	X
Inclusive Scale down	−0.2%	X	X	X	X	X	X	X	X	X	X	X	X	X	X	X	X	X
Inclusive Resolution UP	+8.9%	X	X	X	X	X	X	X	X	X	X	X	X	X	X	X	X	X
Inclusive Resolution down	−7.6%	X	X	X	X	X	X	X	X	X	X	X	X	X	X	X	X	X
$p_T^{\gamma\gamma}$ Scale UP	+0.2%	+0.2%	+0.2%	+0.2%	+0.2%	+0.2%	+0.2%	+0.2%	+0.2%	+0.2%	+0.3%	+0.3%	+0.3%	+0.3%	+0.3%	+0.4%	+0.4%	+0.5%
$p_T^{\gamma\gamma}$ Scale down	−0.2%	−0.2%	−0.2%	−0.2%	−0.2%	−0.2%	−0.2%	−0.2%	−0.2%	−0.2%	−0.3%	−0.3%	−0.3%	−0.3%	−0.3%	−0.4%	−0.4%	−0.5%
$p_T^{\gamma\gamma}$ Resolution UP	+8.0%	+8.2%	+8.2%	+8.4%	+8.4%	+8.3%	+8.3%	+8.5%	+8.7%	+9.0%	+9.4%	+10.0%	+11.2%	+12.0%	+13.5%	+14.9%	+17.1%	+22.2%
$p_T^{\gamma\gamma}$ Resolution down	−6.9%	−6.8%	−6.9%	−6.9%	−7.1%	−7.1%	−7.1%	−7.3%	−7.4%	−7.6%	−8.2%	−8.9%	−9.9%	−10.8%	−11.9%	−13.6%	−15.7%	−21.2%
$|y_{\gamma\gamma}|$ Scale UP	+0.2%	+0.2%	+0.2%	+0.2%	+0.2%	+0.3%	+0.3%	+0.3%	+0.4%	X	X	X	X	X	X	X	X	X
$|y_{\gamma\gamma}|$ Scale down	−0.2%	−0.2%	−0.2%	−0.2%	−0.2%	−0.3%	−0.3%	−0.3%	−0.4%	X	X	X	X	X	X	X	X	X
$|y_{\gamma\gamma}|$ Resolution UP	+7.7%	+7.7%	+7.8%	+7.9%	+8.2%	+8.5%	+9.1%	+10.4%	+13.1%	X	X	X	X	X	X	X	X	X
$|y_{\gamma\gamma}|$ Resolution down	−6.3%	−6.3%	−6.3%	−6.6%	−6.9%	−7.2%	−7.7%	−9.2%	−12.1%	X	X	X	X	X	X	X	X	X
N_{jets} Scale UP	+0.2%	+0.2%	+0.3%	+0.3%	X	X	X	X	X	X	X	X	X	X	X	X	X	X
N_{jets} Scale down	−0.2%	−0.2%	−0.3%	−0.3%	X	X	X	X	X	X	X	X	X	X	X	X	X	X
N_{jets} Resolution UP	+8.4%	+9.0%	+10.0%	+10.5%	X	X	X	X	X	X	X	X	X	X	X	X	X	X
N_{jets} Resolution down	−7.0%	−7.7%	−8.5%	−9.3%	X	X	X	X	X	X	X	X	X	X	X	X	X	X
p_T^{j1} Scale UP	+0.2%	+0.2%	+0.3%	+0.3%	+0.3%	+0.4%	X	X	X	X	X	X	X	X	X	X	X	X
p_T^{j1} Scale down	−0.2%	−0.2%	−0.2%	−0.3%	−0.3%	−0.4%	X	X	X	X	X	X	X	X	X	X	X	X
p_T^{j1} Resolution UP	+8.5%	+8.6%	+9.2%	+10.2%	+12.5%	+19.0%	X	X	X	X	X	X	X	X	X	X	X	X
p_T^{j1} Resolution down	−7.1%	−7.3%	−7.5%	−8.8%	−11.1%	−17.6%	X	X	X	X	X	X	X	X	X	X	X	X
m_{jj} Scale UP	+0.2%	+0.3%	+03%	+0.3%	+0.3%	X	X	X	X	X	X	X	X	X	X	X	X	X
m_{jj} Scale down	−0.2%	−0.3%	−0.3%	−0.3%	−0.3%	X	X	X	X	X	X	X	X	X	X	X	X	X
m_{jj} Resolution UP	+8.7%	+9.9%	+10.3%	+10.2%	+10.1%	X	X	X	X	X	X	X	X	X	X	X	X	X

(continued)

Table 7.6 (continued)

Variable	Bin 1	Bin 2	Bin 3	Bin 4	Bin 5	Bin 6	Bin 7	Bin 8	Bin 9	Bin 10	Bin 11	Bin 12	Bin 13	Bin 14	Bin 15	Bin 16	Bin 17	Bin 18
m_{jj} Resolution down	−7.2%	−8.5%	−8.8%	−9.1%	−9.8%	X	X	X	X	X	X	X	X	X	X	X	X	X
$\Delta\phi_{jj}$ Scale UP	+0.2%	+0.3%	+0.3%	+0.3%	+0.3%	X	X	X	X	X	X	X	X	X	X	X	X	X
$\Delta\phi_{jj}$ Scale down	−0.2%	−0.3%	−0.3%	−0.3%	−0.3%	X	X	X	X	X	X	X	X	X	X	X	X	X
$\Delta\phi_{jj}$ Resolution UP	+8.7%	+9.2%	+10.9%	+10.7%	+9.4%	X	X	X	X	X	X	X	X	X	X	X	X	X
$\Delta\phi_{jj}$ Resolution down	−7.2%	−8.2%	−9.5%	−9.5%	8.0%	X	X	X	X	X	X	X	X	X	X	X	X	X

7.5.2   Experimental Uncertainties

7.5.2.1   Luminosity

The relative uncertainty on the full Run-2 dataset luminosity is 1.7% [21]. This uncertainty is derived from a calibration of the luminosity scale using Van Der Meer scans [20]. The estimation of the uncertainties is detailed in Ref. [85].

7.5.2.2   Trigger Efficiency

The efficiency of the diphoton trigger is measured in data using a bootstrap method, detailed in Ref. [86], measuring the trigger efficiency turn-on curve in events collected with prescaled, lower-threshold triggers. This results in an efficiency of $99.16^{+0.23}_{-0.49}$ (stat)$^{+0.34}_{-0.52}$ (syst)%, in agreement with predictions from the simulation. The corresponding uncertainty is taken as an uncertainty on the correction factors.

7.5.2.3   Vertex Selection Efficiency

As detailed in Sect. 7.3.1.2, the vertex selection uses pointing information from the diphoton system to choose the primary vertex candidate of the event. The difference in the efficiency of selecting the primary vertex between data and the simulation can affect the correction factors (since a reconstructed primary vertex is a requirement). Therefore, an uncertainty on the correction factor is assigned from the difference in the vertex selection efficiency between data and the simulation using $Z \rightarrow ee$ events after ignoring the electron tracks. The ratio of the efficiency is used to increase the weights of events in the simulation with $|z_{reconstructed} - z_{true}| > 0.3$ mm. The uncertainty on the efficiency of this selection is generally found to be $<0.3\%$.

7.5.2.4   Photon Selection

Photon identification efficiency As detailed in Sect. 7.3.1.2, selected photons are required to pass *tight* identification criteria. The efficiency of this selection $\epsilon_{\text{tight ID}}$ is measured in data using the methods detailed in Sect. 4.3.1.1. A per-photon correction factor is then extracted to account for differences in efficiency between data and the simulation. The uncertainties in these correction factors are translated to the unfolding correction factors by varying the photon identification efficiency correction factors within their uncertainties. This variation will cause events to migrate across the boundaries of the fiducial region. The difference in the correction factor between these variations and the nominal one is taken as the uncertainty. They are summarized in Table 7.7.

Photon isolation efficiency Similar to the photon identification efficiency, the uncertainty in the simulation-to-data isolation efficiency corrections, measured using the methods detailed in Sect. 4.3.1.2, translates into an uncertainty in the correction factor. The uncertainties are obtained for track and calorimeter isolation and are added in quadrature. They are summarized in Table 7.7.

Photon energy scale and resolution In addition to affecting the signal invariant mass distribution, as detailed in Sect. 7.5.1.1, the photon energy scale and resolution uncertainties will also have an effect on the correction factors due to migrations across the boundaries of the fiducial region. The combined effect of all scale and resolution uncertainties on the correction factors is shown in Table 7.7. Given the small size of the total effect of these uncertainties, the combined uncertainty scheme (1 NP) is used. This uncertainty on the correction factor is in addition to, and uncorrelated with, the uncertainties on the signal yield from the full decorrelation model (39 PES + 9 PER NPs).

7.5.2.5 Jet Selection

Jet energy scale and resolution The correction factors for jet observables (such as N_{jets}) are affected by the uncertainties in the jet energy scale and resolution (similarly to photons). These uncertainties reflect the remaining difference in jet energy scale and resolution between data and the simulation as estimated through the transverse momentum balance technique in Z+jets, γ+jet, and dijet events, as detailed in [87]. The impact of these uncertainty sources (31 in total) on the correction factors was estimated, and their combined effect is summarized in Table 7.8.

Jet vertex tagging efficiency for jets from the hard scattering The jet vertex tagging (JVT) algorithm is used to suppress pileup jets (jets originating from additional vertices in the event) [72]. Jets in data and simulation are required to pass a selection based on the JVT variable (detailed in Sect. 7.3.1.2). The difference in the efficiency of this requirement between data and the simulation for jets from the hard scatter is used to derive an uncertainty in the correction factor. This uncertainty is very small, given the small uncertainties on the correction factor for data and simulation differences in the JVT selection efficiencies shown in Fig. 7.28b. The computed uncertainty is, at maximum, 0.3%.

7.5.2.6 Pileup

Modeling of inelastic cross section As detailed in Sect. 7.2.2.1, a *pileup reweighting* procedure is performed in which the distribution of the average number of interactions per bunch crossing, $\langle \mu \rangle$, in the simulation is reweighted to match that of the data. The modeling of pileup in the simulation is based on simulations of inelastic pp collisions. The pileup reweighting is varied to cover the uncertainty in the ratio between the predicted and measured inelastic cross section in the fiducial volume defined

Table 7.7 Magnitude of photon experimental uncertainties on the bin-by-bin correction factors for the fiducial inclusive and the differential observables. The label PES/PER refers to the combined effects of photon energy scale and resolution uncertainties, whereas PhotonEff refers to the combined effect of photon identification and isolation efficiencies

Inclusive	PES/PER up	PES/PER dn	PhotonEff up	PhotonEff dn
–	0.06%	0.06%	1.63%	1.64%
$p_T^{\gamma\gamma}$	PES/PER up	PES/PER dn	PhotonEff up	PhotonEff dn
0–5 GeV	0.96%	0.86%	1.74%	1.76%
5–10 GeV	0.6%	0.58%	1.76%	1.77%
10–15 GeV	0.36%	0.36%	1.77%	1.78%
15–20 GeV	0.22%	0.32%	1.76%	1.78%
20–25 GeV	0.2%	0.06%	1.75%	1.77%
25–30 GeV	0.02%	0.12%	1.74%	1.75%
30–35 GeV	0.12%	0.09%	1.72%	1.73%
35–45 GeV	0.13%	0.14%	1.68%	1.7%
45–60 GeV	0.25%	0.26%	1.63%	1.64%
60–80 GeV	0.39%	0.34%	1.56%	1.57%
80–100 GeV	0.48%	0.54%	1.48%	1.49%
100–120 GeV	0.46%	0.46%	1.39%	1.4%
120–140 GeV	0.62%	0.54%	1.31%	1.32%
140–170 GeV	0.77%	0.66%	1.22%	1.23%
170–200 GeV	0.85%	1.08%	1.13%	1.14%
200–250 GeV	1.42%	1.28%	1.04%	1.05%
250–350 GeV	1.9%	1.99%	1%	1.01%
350–∞ GeV	2.92%	2.87%	1.13%	1.14%
$|y_{\gamma\gamma}|$	PES/PER up	PES/PER dn	PhotonEff up	PhotonEff dn
0.00–0.15	<0.01%	0.01%	1.58%	1.59%
0.15–0.30	<0.01%	0.01%	1.57%	1.59%
0.30–0.45	0.04%	0.02%	1.56%	1.58%
0.45–0.60	0.02%	0.04%	1.55%	1.57%
0.60–0.75	0.03%	0.01%	1.57%	1.58%
0.75–0.90	0.03%	0.06%	1.59%	1.6%
0.90–1.20	0.11%	0.1%	1.66%	1.68%
1.20–1.60	0.25%	0.26%	1.74%	1.75%
1.60–2.40	0.13%	0.12%	1.75%	1.77%
$N_{jets}, p_T^j \geq$ 30GeV	PES/PER up	PES/PER dn	PhotonEff up	PhotonEff dn
= 0 jet	0.08%	0.08%	2.5%	2.53%
= 1 jet	0.06%	0.06%	2.33%	2.35%
= 2 jets	0.03%	0.04%	2.05%	2.07%
≥ 3 jets	0.04%	0.03%	1.81%	1.83%

(continued)

Table 7.7 (continued)

p_T^{j1}	PES/PER up	PES/PER dn	PhotonEff up	PhotonEff dn
Underflow	0.08%	0.08%	2.5%	2.53%
30–55 GeV	0.08%	0.08%	2.4%	2.42%
55–75 GeV	0.03%	0.04%	2.2%	2.22%
75–120 GeV	0.02%	0.01%	1.98%	2%
120–350 GeV	0.01%	0.01%	1.67%	1.68%
350–∞ GeV	0.02%	0.03%	1.51%	1.52%
$m_{jj}, p_T^j \geq$ 30GeV	PES/PER up	PES/PER dn	PhotonEff up	PhotonEff dn
Underflow	0.08%	0.07%	2.44%	2.46%
0–170 GeV	0.04%	0.04%	2.12%	2.13%
170–500 GeV	0.03%	0.03%	1.96%	1.98%
500–1500 GeV	0.03%	0.05%	1.74%	1.76%
1500–∞ GeV	0.04%	0.02%	1.48%	1.49%
$\Delta\phi_{jj}, p_T^j \geq$ 30GeV	PES/PER up	PES/PER dn	PhotonEff up	PhotonEff dn
$-\pi--\pi/2$	0.05%	0.05%	2.03%	2.04%
$-\pi/2-0$	0.02%	0.03%	1.92%	1.94%
$0.-\pi/2$	0.03%	0.02%	1.92%	1.94%
$\pi/2-\pi$	0.05%	0.06%	2.03%	2.04%

by $m > 13$ GeV, where m is the mass of the non-diffractive hadronic system [88]. This is achieved by shifting the data μ distribution by $\pm 3\%$ before reweighting the simulated samples. An example of such variations is shown in Fig. 7.23. The effect of such variations on the correction factors are summarized in Table 7.11.

Jet vertex tagging efficiency for pileup Jets Pileup jets originate from additional inelastic pp collisions in the event. They are not present in the particle-level fiducial volume, and as such, they reduce the purity of the bins of the jet-related differential observables. To reduce the fraction of pileup jets in the selected sample, thus improving the purity and reducing related model uncertainties, the JVT algorithm is applied for central jets with $|\eta| < 2.5$ [72]. The residual pileup jet contamination is corrected for in the unfolding procedure. However, any difference in the efficiency of the JVT requirement in rejecting fake jets from pileup between data and simulation can bias the correction factor. For example, having more pileup jets passing the JVT requirement in the simulation than in data can cause large migrations between bins of the jet differential distributions. This is most important for the large jet multiplicity distribution bins where pileup jets will be (wrongly) counted as jets from the hard scatter. Therefore, an uncertainty in the correction factors due to the uncertainty in the JVT efficiency for pileup jets is computed. The JVT efficiency for pileup jets is defined as (Table 7.9):

Table 7.8 Magnitude of the impact of jet energy scale and resolution uncertainties on the correction factors

N_{jets}, $p_T^j \geq 30\text{GeV}$	JES/JER up	JES/JER dn
$= 0$ jet	6.5%	4.57%
$= 1$ jet	2.33%	3.61%
$= 2$ jets	6.66%	8.45%
≥ 3 jets	12.6%	19.1%
p_T^{j1}	JES/JER up	JES/JER dn
Underflow	6.5%	4.57%
30–55 GeV	6.37%	10.8%
55–75 GeV	3.69%	4.36%
75–120 GeV	3.67%	3.84%
120–350 GeV	3.38%	3.57%
350–∞ GeV	3.04%	3.59%
m_{jj}, $p_T^j \geq 30\text{GeV}$	JES/JER up	JES/JER dn
Underflow	2.86%	2.08%
0–170 GeV	8.95%	12.1%
170–500 GeV	7.39%	10.2%
500–1500 GeV	9.26%	13.5%
1500–∞ GeV	11.6%	16.5%
$\Delta\phi_{jj}$, $p_T^j \geq 30\text{GeV}$	JES/JER up	JES/JER dn
$-\pi - -\pi/2$	8.24%	11%
$-\pi/2 - 0$	8.99%	12.6%
$0 - \pi/2$	8.99%	12.6%
$\pi/2 - \pi$	8.14%	11.4%

$$\epsilon_{\text{PU central jet, JVT}} = \frac{N_{\text{PU central jets}}^{\text{pass JVT req.}}}{N_{\text{All PU central jets}}} \qquad (7.9)$$

In the simulation, a jet is defined as originating from pileup if it has $p_T > 10$ GeV and is not matched to a particle-level jet within $\Delta R < 0.2$. The fraction of such jets passing the JVT requirement gives the efficiency in the simulation. In data, the only quantity we can measure is the JVT efficiency for all jets (hard-scatter and pileup)

$$\epsilon_{\text{Central jets, JVT}} = \frac{N_{\text{Central jets}}^{\text{pass JVT req.}}}{N_{\text{All central jets}}} \equiv \frac{N_{\text{HS central jets}}^{\text{pass JVT req.}} + N_{\text{PU central jets}}^{\text{pass JVT req.}}}{N_{\text{All central jets}}} \qquad (7.10)$$

Hard-scatter and pileup jets have very different JVT efficiency, as shown in Fig. 7.24. Therefore, the combined JVT efficiency can not be translated directly to the JVT efficiency of pileup jets, and the number of jets from the hard-scatter must be subtracted

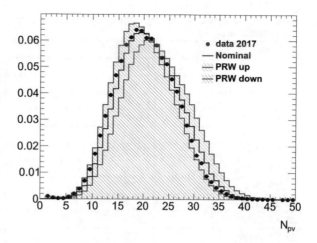

Fig. 7.23 Effect of the pileup reweighting uncertainty variations on the distribution of the number of primary vertices, N_{pv}

from Eq. (7.10). This number is estimated from the simulation after normalizing the total number of jets in the simulation to the total number of jets in data. The number of hard-scatter jets from the simulation can be used to correct the data, as the JVT correction factors between data and simulation were found to be close to unity [72]. The final pileup jet JVT efficiency in data and the simulation is summarized in Table 7.10.

The final uncertainty in the correction factors due to the JVT efficiency for pileup jets is then estimated by randomly removing a fraction of pileup jets in the simulation corresponding to

$$\epsilon_{PU\,jet,\,JVT}^{Data}/\epsilon_{PU\,jet,\,JVT}^{MC} - 1 = 10\%,$$

re-computing the correction factors, and taking their difference to the nominal correction factors as an uncertainty. This uncertainty is shown in Table 7.11.

7.5.3 Theoretical Uncertainties

7.5.3.1 Physics Modelling Uncertainties

The model dependence of the correction factors and its corresponding uncertainty is evaluated from the following variations:

1. The relative contributions of the different Higgs production mechanisms are varied within their corresponding experimental bounds.
2. The bias of the unfolding method, as detailed in *Interlude B*.
3. The impact of the modeling of the underlying event.
4. The modeling of Dalitz events, $H \rightarrow \gamma\gamma^* \rightarrow \gamma f \bar{f}$

Signal composition uncertainty As detailed in Sect. 7.3.3, the final correction factor for each bin of the differential distributions is the combination of the correction factors

Table 7.9 Magnitude of the impact of the pileup reweighting uncertainties on the bin-by-bin correction factors

Inclusive	PRW up	PRW dn
–	1.55%	1.3%
$p_T^{\gamma\gamma}$	PRW up	PRW dn
0–5 GeV	1.6%	1.36%
5–10 GeV	1.63%	1.42%
10–15 GeV	1.72%	1.37%
15–20 GeV	1.62%	1.4%
20–25 GeV	1.61%	1.36%
25–30 GeV	1.72%	1.4%
30–35 GeV	1.57%	1.35%
35–45 GeV	1.65%	1.34%
45–60 GeV	1.63%	1.36%
60–80 GeV	1.61%	1.34%
80–100 GeV	1.48%	1.27%
100–120 GeV	1.42%	1.11%
120–140 GeV	1.11%	0.9%
140–170 GeV	1.09%	0.93%
170–200 GeV	0.96%	0.85%
200–250 GeV	0.77%	0.69%
250–350 GeV	0.56%	0.56%
350–∞ GeV	0.27%	0.44%
$|y_{\gamma\gamma}|$	PRW up	PRW dn
0.00–0.15	1.48%	1.24%
0.15–0.30	1.53%	1.22%
0.30–0.45	1.45%	1.23%
0.45–0.60	1.39%	1.19%
0.60–0.75	1.49%	1.18%
0.75–0.90	1.47%	1.23%
0.90–1.20	1.6%	1.29%
1.20–1.60	1.66%	1.43%
1.60–2.40	1.77%	1.56%
$N_{jets}, p_T^j \geq 30$ GeV	PRW up	PRW dn
= 0 jet	2.87%	2.36%
= 1 jet	0.94%	0.73%
= 2 jets	0.11%	0.24%
≥ 3 jets	2.13%	3%
p_T^{j1}	PRW up	PRW dn
Underflow	2.87%	2.36%
30–55 GeV	0.63%	0.8%
55–75 GeV	1.05%	0.93%

(continued)

Table 7.9 (continued)

	PRW up	PRW dn
75–120 GeV	1.26%	1.06%
120–350 GeV	0.92%	0.84%
350–∞ GeV	0.63%	0.62%
m_{jj}, $p_T^j \geq 30$ GeV	PRW up	PRW dn
Underflow	2.2%	1.79%
0–170 GeV	0.93%	1.34%
170–500 GeV	0.34%	0.58%
500–1500 GeV	1.06%	1.53%
1500–∞ GeV	0.64%	0.99%
$\Delta\phi_{jj}$, $p_T^j \geq 30$ GeV	PRW up	PRW dn
$-\pi - -\pi/2$	0.64%	1.04%
$-\pi/2 - 0$	0.92%	1.29%
$0. - \pi/2$	0.83%	1.15%
$\pi/2 - \pi$	0.65%	1.01%

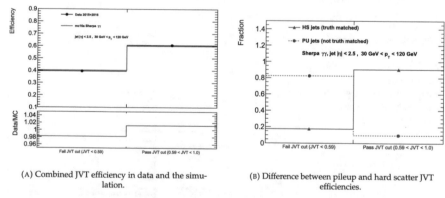

(A) Combined JVT efficiency in data and the simulation.

(B) Difference between pileup and hard scatter JVT efficiencies.

Fig. 7.24 **A** Combined JVT efficiency for data and simulation. This efficiency is defined as $\epsilon_{\text{All jet, JVT}} = \frac{N_{\text{jets}}^{\text{pass JVT req.}}}{N_{\text{All jets}}}$. **B** JVT efficiency for pileup and hard-scattering jets

from each Higgs boson production mode, weighted by its SM cross section. This will introduce model dependence if the correction factors vary significantly among the different production modes. The uncertainty induced by this model dependence is estimated by varying the cross sections of the different production modes within their experimental bounds. These variations are taken from the Run-1 ATLAS+CMS combined Higgs boson couplings measurement [89]. This combination resulted in the measurement of the signal strength of the different production modes, μ_i, defined as $\mu_i = \frac{\sigma_i}{\sigma_i^{SM}}$. The measured signal strength of each production mode is shown in Fig. 7.25.

Table 7.10 JVT efficiency for all jets and pileup jets computed using Eq. (7.10) and Eq. (7.9), respectively for data and simulation. The pileup jet JVT efficiency in data is computed by subtracting the hard scatter jet contribution, as detailed in the text

	Data	Simulation	Ratio
All jets JVT eff. $\epsilon_{\text{All jet, JVT}}$	63%	63%	1.012
PU jets JVT eff. $\epsilon_{\text{PU jet, JVT}}$	17%	15%	1.09

Table 7.11 Relative uncertainties on the correction factors due to the JVT pileup jet uncertainty

	Bin1	Bin2	Bin3	Bin4	Bin5
$N_{\text{jets}}^{\geq 30\text{Gev}}$	+1.1%	−0.4%	−1.5%	−3.8%	−
p_T^{j1}	−2.0%	−0.4%	−0.1%	−0.1%	−0.02%
m_{jj}	−2.5%	−1.9%	−2.0%	−1.2%	−
$\Delta\phi_{jj,\text{signed}}$	−2.6%	−1.8%	−1.8%	−2.6%	−

Fig. 7.25 Best-fit results for the production signal strengths for the combination of ATLAS and CMS data. The results from each experiment are also shown. The error bars indicate the 1σ (thick lines) and 2σ (thin lines) intervals. The measurements of the global signal strength μ are also shown [89]

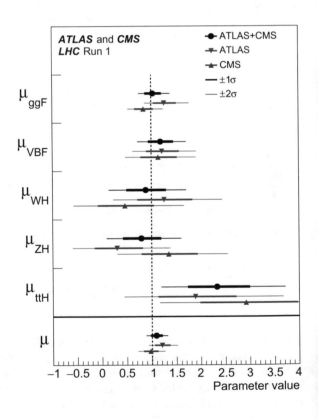

An updated measurement of the $t\bar{t}H$ production mode signal strength was performed using 80 fb^{-1} of Run-2 data, resulting in a measurement of $\mu_{\text{ttH}} = 1.32^{+0.28}_{-0.26}$ [90]. Similarly, the uncertainty on the measured signal strength from the observation of the Higgs boson decay to bottom quarks is used to derive variations for the $b\bar{b}H$ production mode [91]. Using all these measurement, one can obtain the following up and down orthogonal variations to the different Higgs boson production modes:

$$
\vec{\lambda}^{+1\sigma} = \begin{bmatrix} \mu_{\text{ggH}} \\ \mu_{\text{VBF}} \\ \mu_{\text{WH}} \\ \mu_{\text{ZH}} \\ \mu_{\text{ttH}} \\ \mu_{\text{bbH}} \end{bmatrix}, \begin{bmatrix} 1.145 \\ 1 \\ 1 \\ 1 \\ 1 \\ 1 \end{bmatrix}, \begin{bmatrix} 1 \\ 1.203 \\ 1 \\ 1 \\ 1 \\ 1 \end{bmatrix}, \begin{bmatrix} 1 \\ 1 \\ 1.433 \\ 1 \\ 1 \\ 1 \end{bmatrix}, \begin{bmatrix} 1 \\ 1 \\ 1 \\ 1.468 \\ 1 \\ 1 \end{bmatrix}, \begin{bmatrix} 1 \\ 1 \\ 1 \\ 1 \\ 1.5 \\ 1 \end{bmatrix}, \begin{bmatrix} 1 \\ 1 \\ 1 \\ 1 \\ 1 \\ 1.22 \end{bmatrix}
$$

$$(7.11)$$

$$
\vec{\lambda}^{-1\sigma} = \begin{bmatrix} \mu_{\text{ggH}} \\ \mu_{\text{VBF}} \\ \mu_{\text{WH}} \\ \mu_{\text{ZH}} \\ \mu_{\text{ttH}} \\ \mu_{\text{bbH}} \end{bmatrix}, \begin{bmatrix} 0.855 \\ 1 \\ 1 \\ 1 \\ 1 \\ 1 \end{bmatrix}, \begin{bmatrix} 1 \\ 0.797 \\ 1 \\ 1 \\ 1 \\ 1 \end{bmatrix}, \begin{bmatrix} 1 \\ 1 \\ 0.567 \\ 1 \\ 1 \\ 1 \end{bmatrix}, \begin{bmatrix} 1 \\ 1 \\ 1 \\ 0.532 \\ 1 \\ 1 \end{bmatrix}, \begin{bmatrix} 1 \\ 1 \\ 1 \\ 1 \\ 0.7 \\ 1 \end{bmatrix}, \begin{bmatrix} 1 \\ 1 \\ 1 \\ 1 \\ 1 \\ 0.78 \end{bmatrix}
$$

$$(7.12)$$

Each of these variations will then be used to vary the cross section of its respective production mode, and the correction factor is re-calculated. The difference between the cross section from the varied composition and the nominal one defines the uncertainty on the correction factor. These uncertainties are summarized in Table 7.12.

Unfolding bias As detailed in *Interlude B*, the bias from the unfolding method is taken into account as a systematic uncertainty in the measured cross section. The systematic uncertainty from the unfolding should reflect the difference between the data and the underlying physics model used to compute the correction factors. Therefore, the unfolded $p_T^{\gamma\gamma}$ and $|y_{\gamma\gamma}|$ distributions will be used to derive a reweighting function. These observables are used since they are, to a good approximation, uncorrelated for diphotons.

The reweighting function is derived by smoothing the ratio between the data and the simulation using a Gaussian kernel. The smoothed reweighting functions are then applied to the simulation to make it more closely reflect the observed distributions in data. The $p_T^{\gamma\gamma}$ and $|y_{\gamma\gamma}|$ distributions after the reweighting are shown in Fig. 7.26, along with data. The reweighted simulation samples are then used to compute the correction factors, and the difference to the nominal correction factors is taken as a systematic uncertainty. The uncertainty due to unfolding bias is summarized in Table 7.13.

Table 7.12 Impact of production mode variations on the correction factors, given in %

Variable	Bin1	Bin2	Bin3	Bin4	Bin5	Bin6	Bin7	Bin8	Bin9	Bin10	Bin11	Bin12	Bin13	Bin14	Bin15	Bin16	Bin17
Inclusive	0.03	–	–	–	–	–	–	–	–	–	–	–	–	–	–	–	–
$p_T^{\gamma\gamma}$	0.01	<0.01	<0.01	0.01	<0.01	0.01	0.01	0.02	0.03	0.04	0.06	0.06	0.08	0.08	0.07	0.08	0.09
$|y_{\gamma\gamma}|$	0.04	0.04	0.03	0.03	0.04	0.04	0.04	0.05	0.02	–	–	–	–	–	–	–	–
p_T^{j1}	0.2	0.6	0.3	0.1	0.5	1.0	–	–	–	–	–	–	–	–	–	–	–
$N_{\text{jets}}^{\geq 30 \text{Gev}}$	0.2	0.1	1.3	1.5	–	–	–	–	–	–	–	–	–	–	–	–	–
$\Delta\phi_{jj}$	0.2	1.1	0.9	1.0	1.1	–	–	–	–	–	–	–	–	–	–	–	–
m_{jj}	0.2	0.6	0.9	2.1	1.6	–	–	–	–	–	–	–	–	–	–	–	–

Table 7.13 Impact of the modeling uncertainties on the correction factor, in %, evaluated by reweighting the Higgs p_T and y distributions in MC to the observed spectra in data

Variable	Bin1	Bin2	Bin3	Bin4	Bin5	Bin6	Bin7	Bin8	Bin9	Bin10	Bin11	Bin12	Bin13	Bin14	Bin15	Bin16	Bin17
Inclusive	+0.2 −0.05	–	–	–	–	–	–	–	–	–	–	–	–	–	–	–	–
$p_T^{\gamma\gamma}$	+2.9 −0.2	+0.2 −0.8	+0.2 −0.6	+0.2 −0.2	+0.2 <−0.01	+0.3 −0.060	+0.3 −0.1	+0.3 −0.1	+0.3 −0.1	+0.3 −0.03	+0.2 −0.02	+0.3 −0.02	+0.2 <−0.01	+0.2 −0.01	+0.2 −0.02	+0.2 −0.07	+0.1 −0.3
$\lvert y_{\gamma\gamma}\rvert$	+0.01 −0.05	<+0.01 −0.05	<+0.01 −0.05	<+0.01 −0.02	<+0.01 −0.05	<+0.01 −0.08	<+0.01 −0.08	+0.01 −0.07	+0.1 −0.02	–	–	–	–	–	–	–	–
$N_{jets}^{\geq 30Gev}$	+0.2 −0.5	+2.6 −0.3	+1.1 −0.3	+0.6 −0.2	–	–	–	–	–	–	–	–	–	–	–	–	–
$\Delta\phi_{jj}$	+0.3 −0.06	+0.9 −0.3	+1.2 −0.3	+1.1 −0.2	+0.8 −0.3	–	–	–	–	–	–	–	–	–	–	–	–
m_{jj}	+0.3 −0.0	+1.4 −0.3	+0.6 −0.2	+0.7 −0.1	+0.5 −0.04	–	–	–	–	–	–	–	–	–	–	–	–
p_T^{j1}	+0.2 −0.5	+2.6 −0.3	+1.5 −0.3	+0.8 −0.3	+0.1 −0.1	+0.05 −2.6	–	–	–	–	–	–	–	–	–	–	–

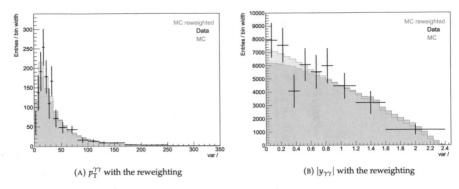

(A) $p_T^{\gamma\gamma}$ with the reweighting   (B) $|y_{\gamma\gamma}|$ with the reweighting

Fig. 7.26   Distribution of $p_T^{\gamma\gamma}$ and $|y_{\gamma\gamma}|$ from the simulation after applying a reweighting from data. The nominal distribution is shown in blue, the reweighted one in orange and the data points are shown

Underlying event modeling   As detailed in Sect. 7.2.2.1, the different simulation samples that are used in the analysis are interfaced with showering and hadronisation models. These models convert the inclusive parton-level cross sections into exclusive particle-level ones. In addition, these models also model the underlying events (i.e. all particles from the same not coming from the hard scatter vertex). The modeling of such processes can have an effect on the correction factors, as they can cause migrations in or out of the fiducial region, and hence will be sources of uncertainty in the correction factors. In order to estimate these uncertainties, the correction factors are computed using two samples with the same matrix element calculation but interfaced with different hadronization models.

For this study, we use PYTHIA 8 [28] (used for the nominal MC samples) and HERWIG 7 [92]. They use different algorithms for the parton shower and for the modeling of non-perturbative effects (hadronization and multi-parton interactions). The difference between the correction factors using the POWHEG gluon-fusion sample interfaced with either model is taken as a systematic uncertainty. The uncertainty is summarized in Table 7.14.

Dalitz events   The PYTHIA 8 generator that is used to model the Higgs boson decay and the parton showering includes by default the *Dalitz* decay channel $H \rightarrow \gamma\gamma^* \rightarrow \gamma f \bar{f}$, where γ^* is an off-shell photon, and f is any charged fermion. These events represent around 6% of the generated events. However, they are not considered as part of the fiducial volume, as they do not have stable diphoton final states and they are removed at the particle-level. The remaining events are reweighted to maintain the correct normalization.

At the detector-level, around 0.3% of all events that pass the fiducial selections in the simulated signal samples are Dalitz events. This fraction was estimated for the bins of the different distributions and was found to be constant. These events are not removed at the reconstruction level, and therefore, they will be corrected for when unfolding to particle-level. However, the Dalitz decay branching ratio is poorly

Table 7.14 Impact of underlying event modelling and parton shower uncertainties on the correction factors given in %

Variable	Bin1	Bin2	Bin3	Bin4	Bin5	Bin6	Bin7	Bin8	Bin9	Bin10	Bin11	Bin12	Bin13	Bin14	Bin15	Bin16	Bin17	Bin18
Inclusive	0.094	–	–	–	–	–	–	–	–	–	–	–	–	–	–	–	–	–
$p_T^{\gamma\gamma}$	−0.086	−0.086	−0.086	−0.086	−0.086	−0.086	−0.086	−0.086	−0.086	−0.086	−0.086	−0.086	−0.086	−0.086	−0.086	−0.086	−0.086	−0.086
$p_T^{j_1}$	2.23	2.23	2.23	2.23	2.23	2.23	–	–	–	–	–	–	–	–	–	–	–	–
$|y_{\gamma\gamma}|$	−0.11	−0.11	−0.11	−0.11	−0.11	−0.11	−0.11	−0.11	−0.11	–	–	–	–	–	–	–	–	–
N_{jets}	1.78	1.23	−1.563	−1.85	–	–	–	–	–	–	–	–	–	–	–	–	–	–
m_{jj}	2.19	2.19	2.19	2.19	2.19	–	–	–	–	–	–	–	–	–	–	–	–	–
$\Delta\phi_{jj,\text{signed}}$	1.71	1.71	1.71	1.71	1.71	–	–	–	–	–	–	–	–	–	–	–	–	–

Table 7.15 The breakdown of uncertainties on the inclusive diphoton fiducial cross section measurement. The uncertainties from the statistics of the data and the systematic sources affecting the signal extraction are shown. The remaining uncertainties are associated with the unfolding correction factor and luminosity

Source	Uncertainty (%)
Statistics	6.9
Signal extraction syst.	7.9
Photon energy scale & resolution	4.6
Background modelling (spurious signal)	6.4
Correction factor	2.6
Pile-up modelling	2.0
Photon identification efficiency	1.2
Photon isolation efficiency	1.1
Trigger efficiency	0.5
Theoretical modelling	0.5
Photon energy scale & resolution	0.1
Luminosity	1.7
Total	11.0

known, and different generators produce different results. Therefore, a conservative 100% uncertainty is assigned to the Dalitz contribution. This results in an uncertainty of $\approx 0.3\%$ on the bin-by-bin correction factor for the different bins of the differential variables.

7.5.4   Summary of Uncertainties

A summary of the different theoretical and experimental uncertainties in the correction factors is shown in Table 7.15 for the inclusive fiducial region. The leading sources of uncertainties are shown in Figs. 7.27 and 7.28 for the differential cross section measurements.

7.6   Signal Extraction

The Higgs boson signal yield ν^{sig}, in the inclusive fiducial region and each bin of the differential variables, is extracted using an unbinned extended maximum likelihood fit to data. The fits are done simultaneously to $m_{\gamma\gamma}$ distributions for all the bins of a given differential variable or fiducial region that we are interested in. The total likelihood function is the product of the per-bin likelihood $\mathcal{L}_i$:

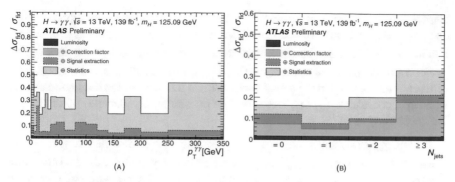

Fig. 7.27 Breakdown of the statistical and systematic uncertainties on the cross sections for measurement in bins of **A** $p_T^{\gamma\gamma}$ and **B** N_{jets}

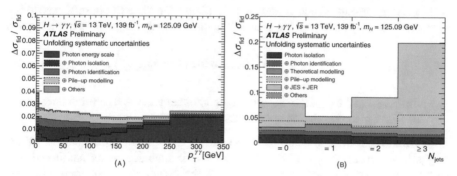

Fig. 7.28 Breakdown of leading systematic uncertainties due to unfolding on the correction factors for **A** $p_T^{\gamma\gamma}$ and **B** N_{jets}

$$\mathcal{L}_i(m_{\gamma\gamma}|\nu_i^{sig}, \nu_i^{bkg}) = \frac{e^{-\nu_i}}{n_i!} \prod_j^n \left[\nu_i^{sig} \, \mathcal{S}(m_{\gamma\gamma}^j|\overrightarrow{\theta^{sig}}_i) + \nu_i^{bkg} \, \mathcal{B}(m_{\gamma\gamma}^j|\overrightarrow{\theta^{bkg}}_i) \right] \quad (7.13)$$

where ν_i^{sig} and ν_i^{bkg} are the fitted number of signal and background events, respectively. In a given bin i, $\nu_i = \nu_i^{sig} + \nu_i^{bkg}$ is the mean value of the underlying Poisson distribution for the n_i events in that bin. For a given event j, $m_{\gamma\gamma}^j$ is the diphoton invariant mass. $\mathcal{S}(m_{\gamma\gamma}^j|\overrightarrow{\theta^{sig}})$ is the signal PDF, and it is a function of the vector of nuisance parameters, $\overrightarrow{\theta^{sig}}$, describing the signal shape and the different uncertainties affecting it, as detailed in Sect. 7.5.1.1. $\mathcal{B}(m_{\gamma\gamma}^j|\overrightarrow{\theta^{bkg}})$ is the background PDF, and it is a function of a vector of nuisance parameters $\overrightarrow{\theta^{bkg}}$, describing the different shape and normalization parameters of the background function detailed in Sect. 7.4.2.

Signal shape nuisance parameters In the simultaneous fit of the diphoton mass spectrum, we consider two sources of uncertainty on the signal shape: uncertainties in the resolution due photon energy resolution $\vec{\theta}_{\text{EnRes}}$ and uncertainties in the peak position due to photon energy scale $\vec{\theta}_{\text{EnScale}}$. As detailed in *Interlude A*, these different uncertainties in the signal shape will be included in the likelihood as multiplicative constraints $\prod_k G_k$. The choice of constraint, G_k, depends on the type of the uncertainty as follows:

- The energy scale uncertainties are implemented via a Gaussian constraint. For each uncertainty in $\vec{\theta}_{\text{EnScale}}$, the pulls are treated asymmetrically. The constraint term $G_{\text{EnScale}}(\theta_{\text{EnScale}}; 0, 1)$ has a mean of $\theta_{\text{EnScale}} = 0$ and a width of 1, ensuring that values of $\theta_{\text{EnScale}} = \pm 1$ correspond to $\pm 1\sigma$ pulls on the nuisance parameters and correctly penalise the likelihood accordingly.
- The energy resolution uncertainties are implemented via a log-normal constraint, as they can only have positive values. The log-normal constraint is implemented as:

$$\Theta_{\text{EnRes}} = \exp\left(\theta_{\text{EnRes}}\sqrt{\log\left(1 + \delta_{\text{EnRes}}^2\right)}\right), \qquad (7.14)$$

where θ_{EnRes} and δ_{EnRes} are the nuisance parameter and the resolution uncertainty values for a given resolution uncertainty source.

Higgs mass nuisance parameters The uncertainty on the measurement of the Higgs boson mass, $m_H = 125.09 \pm 0.24$ GeV, is included as an additional nuisance parameter that allows for a shift of the signal peak position. The nuisance parameter is included in the likelihood as a symmetric Gaussian constraint $G_{\text{Mass uncert}}(\theta_{\text{Mass uncert}}; 0, 1)$.

The nuisance parameters are correlated between all bins in a given distribution. The nuisance parameters are allowed to float for the fit to data, and any deviations from zero will be penalized by a reduction in the likelihood.

Other uncertainties that do not affect the shape of the diphoton mass spectrum are not included in the fit and are dealt with as part of the correction for detector effects. The spurious signal, described in Sect. 7.4.2.2, is not included in the fit. It is added in quadrature to the uncertainty in the signal yield from the fit.

The total number of nuisance parameters in the fit is thus 49+N = 39 (photon energy scale uncertainties) + 9 (photon energy resolution uncertainties) + N (background function parameters) + 1 (Higgs mass uncertainty). The free parameters on the fit are therefore θ_{EnScale}, θ_{EnRes}, $\theta_{\text{Background}}$, ν^{sig} and ν^{bkg}, for each of the bins considered. The fitted number of signal events is not constrained to be positive. The results of the fit to the data for the extraction of the signal yields for the inclusive fiducial region, and in one $p_{\text{T}}^{\gamma\gamma}$ bin are shown in Fig. 7.29. The signal extraction fits for all the bins of the differential observables are shown in Appendix. The final results for the extracted signal yield and its total error are shown in Table 7.16. The fitted signal yield is then used to estimate the final spurious signal uncertainty by scaling SS/S_{ref} with the ratio $S_{\text{ref}}/S_{\text{obs}}$.

Table 7.16 Fitted signal yield in the inclusive fiducial and different bins of the kinematic distributions. The uncertainties shown account for the data statistics and the different systematic uncertainties that enter the signal extraction fit (i.e. photon energy scale and resolution, and Higgs mass uncertainty)

Variable	Bin 1	Bin 2	Bin 3	Bin 4	Bin 5	Bin 6	Bin 7	Bin 8	Bin 9	Bin 10	Bin 11	Bin 12	Bin 13	Bin 14	Bin 15	Bin 16	Bin 17	Bin 18
Inclusive fit yield	6546.1 ± 533.4	X	X	X	X	X	X	X	X	X	X	X	X	X	X	X	X	X
$p_T^{\gamma\gamma}$ fit yield	213.2 ± 105.2	495.7 ± 158.9	692.0 ± 177.3	914.2 ± 165.3	641.1 ± 149.7	395.0 ± 135.4	599.2 ± 133.3	510.4 ± 155.6	517.8 ± 151.4	611.1 ± 134.2	229.8 ± 100.3	197.3 ± 60.5	132.9 ± 43.6	200.6 ± 38.3	80.4 ± 25.8	115.3 ± 22.4	33.4 ± 14.4	25.9 ± 8.7
$|y_{\gamma\gamma}|$ fit yield	866.5 ± 142.0	825.9 ± 141.4	448.8 ± 134.54	669.5 ± 134.9	609 ± 137.1	655.4 ± 141.9	965.6 ± 201.7	902.5 ± 234.6	646.6 ± 203.4	X	X	X	X	X	X	X	X	X
N_{jets} fit yield	3573.4 ± 407.2	1773 ± 252.6	879.1 ± 155.2	421.4 ± 104.9	X	X	X	X	X	X	X	X	X	X	X	X	X	X
p_T^{j1} fit yield	3573.4 ± 407.2	1726.9 ± 239.0	392.1 ± 132.4	619.4 ± 117.5	350.4 ± 84.0	5.1 ± 14.6	X	X	X	X	X	X	X	X	X	X	X	X
m_{jj} fit yield	5268.8 ± 476.0	593.1 ± 133.5	357.9 ± 115.8	288.1 ± 59.3	45.3 ± 17.6	X	X	X	X	X	X	X	X	X	X	X	X	X
$\Delta\phi_{jj}$ fit yield	5268.8 ± 476.0	409.0 ± 113.8	216.3 ± 69.9	265.9 ± 72.6	386.7 ± 115.1	X	X	X	X	X	X	X	X	X	X	X	X	X

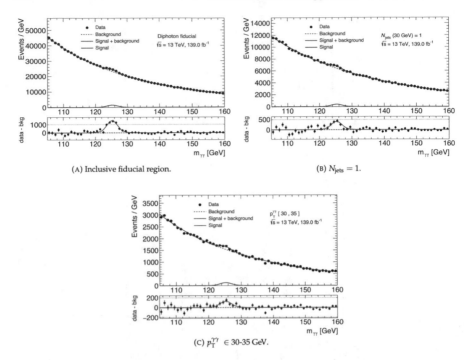

Fig. 7.29 Signal+background fit to data for **A** the inclusive fiducial region and examples from the fit to the bins of the differential distributions **B** $N_{\text{jets}} = 1$ and **C** $p_{\text{T}}^{\gamma\gamma}$ [30,35] GeV

7.6.1 Nuisance Parameter Rankings

In order to check if some of the nuisance parameters (NP) are overconstrained, or significantly pulled after the fit from their initial best estimate, the pulls and the impacts (defined in *Interlude A*) of the different nuisance parameters are inspected. As an example, Fig. 7.30 shows the NPs for the inclusive fiducial region. These plots show the pre-fit and post-fit values of the nuisance parameters, and their effect on the expected and observed uncertainties on the extracted signal yield. The nuisance parameters are ranked from top to bottom according to their impact.

The *pre-fit* impacts are determined by shifting a given nuisance parameter by $\pm 1\sigma$ from its nominal value and redoing the fit while letting the other nuisance parameters float, and then computing the variation in the fitted signal yield with respect to the nominal case. The *post-fit* impacts are determined by setting all the nuisance parameters to their best fit values and reporting the pulls from a fit to the Asimov dataset. The observed NP ranking plot shows no over-constraints for the different energy scale and resolution nuisance parameters. There are some small observed negative pulls for the resolution as the observed data signal is compatible with a smaller resolution of approximately 90 MeV. A similar pull is observed for the Higgs mass nuisance parameter, corresponding to a change of the Higgs mass

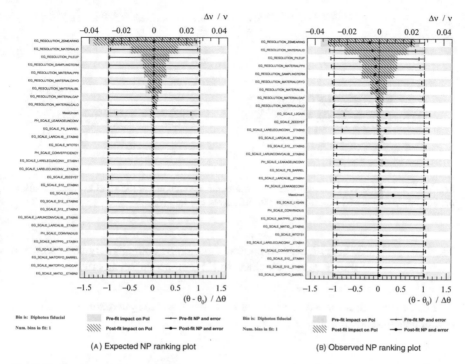

Fig. 7.30 Ranking of the pre-fit and post-fit pulls for the inclusive fiducial region for an Asimov dataset. The red (black) points are the pre(post)-fit NPs with their associated uncertainty. The yellow (blue hatched) band is the pre(post)-fit uncertainty, showing the expected and observed impact of a given uncertainty on the extracted signal yield

of about $+32.5$ MeV, which is compatible with the uncertainty from the combined ATLAS+CMS mass measurement of ± 240 MeV.

7.6.2 Break-Down of Uncertainties

The fit to data leads to an estimate of the signal yield together with its total uncertainty. The following procedure is then followed in order to separate the total uncertainty into its statistical and systematic components:

1. Obtain the best-fit values of the nuisance parameters (that were float) in the signal extraction fit.
2. Construct an Asimov dataset using the observed signal yield and nuisance parameters.
3. Perform a fit on the Asimov dataset, fixing the nuisance parameters in the likelihood to the observed best-fit values. The statistical uncertainty is obtained from the uncertainty in the fitted yield.

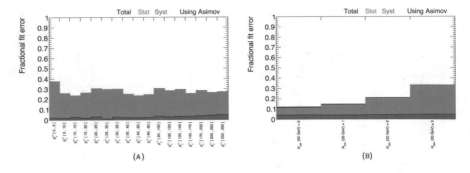

Fig. 7.31 Breakdowns of the uncertainty components as fractions of the yield using the Asimov method based on the MC expected yields for inclusive $p_T^{\gamma\gamma}$ and N_{jets}

4. The systematic uncertainty component can then be computed by subtraction in quadrature of the total error obtained in (1) and the statistical error obtained in (3)

The expected decomposition of the total uncertainty into systematic and statistical components is shown in Fig. 7.31.

7.6.3   Signal Yield Cross Checks

In the previous sections, the Higgs boson signal yield in each category has been extracted from a simultaneous fit to all the bins of a differential distribution. The underflow and overflow bins are included, such that the sum of the yields in all bins should be equivalent to the yields in the inclusive case. This can be used as a cross-check between the different distributions, comparing the integrals of these distributions to check that they are consistent. This is shown in Fig. 7.32.

The uncertainty on the integral is calculated as $\sqrt{\sum_{ij} Cov_{ij}^{\text{stat}}}$, where Cov_{ij}^{stat} is the statistical covariance of the different bins. We take into account only the statistical uncertainties since the systematic uncertainties are largely correlated between the different distributions. The statistical covariance is computed by the product of the correlations between bins ρ_{ij} and the uncertainties. The statistical correlation between bins of the same distribution is zero since events are not shared between bins of the same distributions.

Fig. 7.32 A comparison of the integrals of the fitted signal yields including the under- and over-flow bins for the different distributions. The integrals are found to be consistent with the fiducial integrated signal yield. The spurious signal of the integrated fiducial is shown for reference

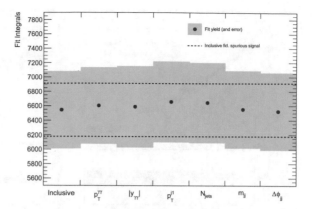

7.7 Fiducial Inclusive and Differential Cross Sections

7.7.1 Closure Test with an Asimov Dataset

A cross-check procedure is performed before the actual signal extraction on data by comparing the particle-level cross sections from simulations with the expected unfolded cross sections obtained using an Asimov dataset. The Asimov data-set is built using the signal parameterization from simulation, and background parameterisations from data sidebands ($m_{\gamma\gamma} \in [105 - 120] \cup [130 - 160]$GeV). The results are shown in Fig. 7.33. The results show that no bias is observed. The results show the expected statistical and systematic uncertainty (from the fitting and unfolding). The statistical uncertainty is reduced by almost a factor of 2 compared to the cross section measurement using the 36 fb^{-1} dataset [9], whereas the experimental systematic errors are overall of the same order as those reported 36 fb^{-1} with larger contribution from spurious signal uncertainty and jet systematics due to the higher pileup conditions. The experimental uncertainty is largely reduced with respect to the measurement with the 80 fb^{-1} dataset that was performed using preliminary calibrations [93].

7.7.2 Results

The measured cross sections are obtained using the signal yields estimated in Sect. 7.6, and the bin-by-bin correction factors, as detailed in *Interlude B*. The cross section for $pp \to H \to \gamma\gamma$ measured in the baseline fiducial region is:

$$\sigma_{\text{fid}} = 65.2 \pm 4.5 \,(\text{stat.}) \pm 5.6 \,(\text{exp.}) \pm 0.3 \,(\text{theory}) = 65.2 \pm 7.1 \,\text{fb}, \quad (7.15)$$

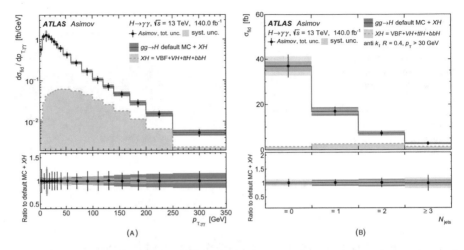

Fig. 7.33 Cross-check of the differential cross section estimated from an Asimov dataset compared with the particle-level simulated cross section

which is to be compared with the default Standard Model prediction of 63.5 ± 3.3 fb [32]. The measured cross section is found to be within one standard deviation of the default SM prediction, detailed in Sect. 7.2.2.1. The uncertainty on the measured cross section is dominated by systematic uncertainties the signal extraction: photon energy resolution and background modeling (spurious signal).

Figure 7.34 shows the measured unfolded differential cross sections as a function of the diphoton kinematics, $p_T^{\gamma\gamma}$ and $|y_{\gamma\gamma}|$. The unfolded differential distributions are compared to the default MC prediction for ggF and XH and to the additional theory predictions detailed in Sect. 7.2.2.4. In general, the observed distributions from data are in excellent agreement with the default SM predictions over the full rapidity range. The $p_T^{\gamma\gamma}$ distribution extends up to 350 GeV, a region where top mass effects start to become sizeable. The statistical errors for the last bin prevent any conclusive statement about the presence of such effects in the data. The inclusive cross section for $p_T^{\gamma\gamma} > 350$ GeV is measured to be 0.23 ± 0.14 fb, with the uncertainty being predominantly statistical, and is in good agreement with the default prediction of about 0.21 ± 0.04 fb.

Figures 7.35 and 7.36 show the results for the jet-related observables, N_{jets}, $p_T^{j_1}$, m_{jj} and $\Delta\phi_{jj,\text{signed}}$. The N_{jets} distribution includes both the exclusive and inclusive jet multiplicities. The inclusive jet multiplicities are computed as follows:

- The $N_{\text{jets}} \geq 0$ corresponds to the fiducial integrated cross section.
- The $N_{\text{jets}} \geq 1, 2$ are computed by summing the observed cross section from the exclusive $N_{\text{jets}} = 1, 2$ to the inclusive $N_{\text{jets}} \geq 3$. The uncertainties are propagated including the full experimental covariance of the exclusive N_{jets} bins.

Good agreement is observed between the measured N_{jets} distributions and all predictions with precision better than NLO. The predictions of SHERPA and

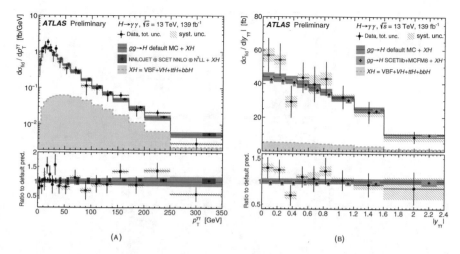

Fig. 7.34 Cross sections measured as a function of the diphoton kinematics: **A** $p_T^{\gamma\gamma}$, **B** $|y_{\gamma\gamma}|$. The cross section as a function of $p_T^{\gamma\gamma}$ is shown in the range 0–350 GeV, while for $p_T^{\gamma\gamma} > 350$ GeV it is measured to be 0.23 ± 0.14 fb with the uncertainty being predominantly statistical. The measurement for $p_T^{\gamma\gamma} > 350$ GeV agrees with the default prediction within less than one standard deviation. All measurements are compared to the default MC prediction in additional predictions for different ggF components added to the same XH prediction

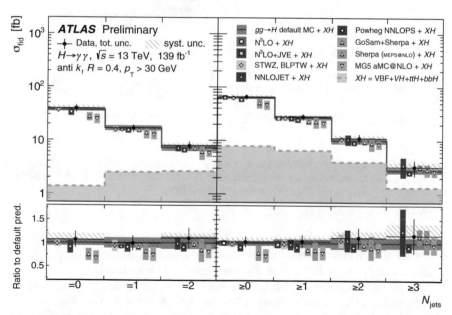

Fig. 7.35 Cross sections measured as a function of the jet multiplicity, N_{jets}, in exclusive and inclusive bins. All measurements are compared to the default MC prediction and additional predictions for different ggF components added to the same XH prediction

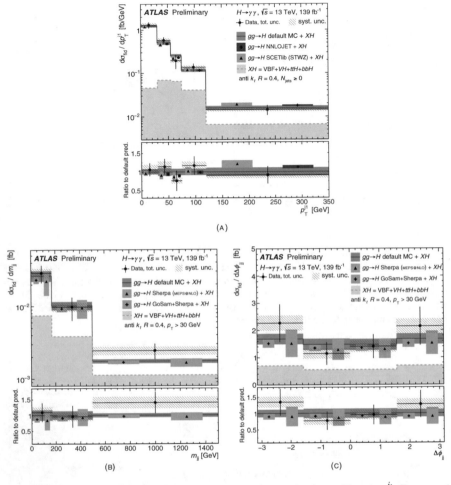

Fig. 7.36 Cross sections measured as a function of jet kinematic observables: **A** $p_T^{j_1}$, **B** m_{jj} and **C** $\Delta\phi_{jj,\text{signed}}$. All measurements are compared to the default MC prediction in additional predictions for different ggF components added to the same XH prediction

MG5_aMC@NLO underestimate the inclusive and zero-jet cross section, but still, give a reasonable description of the shape of the measured distributions. An overall scaling to the N³LO prediction yields better agreement with the data.

The $p_T^{j_1}$ distribution covers the same kinematic range as the Higgs boson $p_T^{\gamma\gamma}$ measurement, with coarser binning at low p_T. All predictions agree well with the data, with the NNLOJET prediction providing the best precision in the high $p_T^{j_1}$ region. The first bin of the $p_T^{j_1}$ distribution represents events that do not contain a jet passing the corresponding fiducial selections.

The m_{jj} and $\Delta\phi_{jj,\text{signed}}$ distributions are compared to SHERPA (MEPS@NLO) and GOSAM predictions that are of NLO accuracy for this jet multiplicity ($N_{\text{jets}} \geq 2$).

Table 7.17 Probabilities from a χ^2 compatibility test comparing data and the default SM prediction for each differential distribution. The χ^2 is computed using the covariance matrix constructed from the full set of uncertainties on the data measurements and the theory uncertainties on the SM prediction

Observables	$p(\chi^2)$ with Default MC Prediction (%)
$p_T^{\gamma\gamma}$	44
$\lvert y_{\gamma\gamma} \rvert$	68
$p_T^{j_1}$	77
N_{jets}	96
$\Delta\phi_{jj,\text{signed}}$	82
m_{jj}	75

Table 7.18 List of event selection requirements which define the fiducial phase space for the cross section measurement in the $H \rightarrow 4\ell$ channel. SFOS lepton pairs are same-flavour opposite-sign lepton pairs [94]

Leptons and jets	
Leptons	$p_T > 5$ GeV, $\lvert\eta\rvert < 2.7$
Jets	$p_T > 30$ GeV, $\lvert y \rvert < 4.4$
remove jets with	$\Delta R(\text{jet}, \ell) < 0.1$
Lepton selection and pairing	
Lepton kinematics	$p_T > 20, 15, 10$ GeV
Leading pair (m_{12})	SFOS lepton pair with smallest $\lvert m_Z - m_{\ell\ell} \rvert$
Subleading pair (m_{34})	remaining SFOS lepton pair with smallest $\lvert m_Z - m_{\ell\ell} \rvert$
Event selection (at most one quadruplet per event)	
Mass requirements	50 GeV $< m_{12} < 106$ GeV and 12 GeV $< m_{34} < 115$ GeV
Lepton separation	$\Delta R(\ell_i, \ell_j) > 0.1$
J/ψ veto	$m(\ell_i, \ell_j) > 5$ GeV for all SFOS lepton pairs
Mass window	105 GeV $< m_{4\ell} < 160$ GeV
If extra leptons with $p_T > 12$ GeV	Quadruplet with the largest ME

Good agreement is seen between the data and the predictions, including that of the default MC that is of LO accuracy for this jet multiplicity. In the highest m_{jj} bin that is the most sensitive to VBF production, the data are in agreement within the prediction within the uncertainty of the measurement. The $\Delta\phi_{jj,\text{signed}}$ distribution that is sensitive to the CP properties of the Higgs boson is in good agreement with the expected shape in the SM.

Compatibility of measured distributions with the Standard Model The compatibility between the measured differential distributions and the default SM prediction is assessed using a χ^2 test. The χ^2 is computed using the covariance matrix constructed

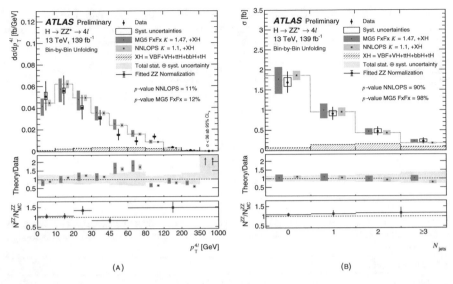

Fig. 7.37 Differential fiducial cross sections for **A** p_T^{4l} and **B** the number of jets, N_{jets}. The measured cross sections are compared to ggF predictions by POWHEG NNLOPS and MG5_aMC@NLO - FxFx normalized to the N^3LO total cross section with the listed K-factors. MC-based predictions for all other Higgs boson production modes XH are normalized to the SM predictions. The error bars on the data points show the total uncertainties, while the systematic uncertainties are indicated by the boxes. The shaded bands on the expected cross sections indicate the PDF and scale uncertainties. The p-values indicating the compatibility of the measurement and the SM prediction are shown as well. The p-values do not include the systematic uncertainty in the theoretical predictions. The central panels of **A** and **B** show the ratio of different predictions to the data. The grey area represents the total uncertainty of the measurement. The bottom panels of **A** and **B** show the fitted values of the ZZ normalisation factors [94]

from the full set of uncertainties on the data measurements, taking into account correlations between bins, as well as the theoretical uncertainties on the SM prediction. Table 7.17 reports the p-values of the χ^2 between data and the default MC prediction for all differential distributions. For all observables, the compatibility between the data and the SM prediction is excellent.

7.7.3   Combination with the $H \to 4\ell$ Channel

A measurement of the Higgs boson fiducial integrated and differential cross section was performed as well in the $H \to ZZ^* \to 4\ell$ channel using the full Run-2 data. A summary of the fiducial selection in the $H \to 4\ell$ is shown in Table 7.18. The results of this measurement are shown in Fig. 7.37 for $p_T^{4\ell}$ and N_{jets}. The results confirm the excellent agreement with the Standard Model, as observed in the $H \to \gamma\gamma$ channel.

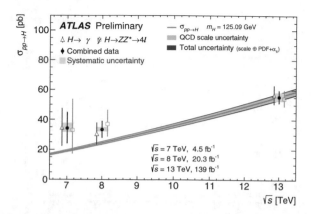

Fig. 7.38 Total $pp \to H + X$ cross sections measured at centre-of-mass energies of 7, 8, and 13 TeV, compared to Standard Model predictions. The $H \to \gamma\gamma$ channel (red triangles), $H \to ZZ^* \to 4l$ channel (green squares) and combined (black dots) measurements are shown. The individual channel results are offset along the x-axis for display purposes. The grey bands on the combined measurements represent the systematic uncertainty, while the error bars show the total uncertainty. The light blue band shows the estimated uncertainty due to missing higher-order corrections, and the dark blue band indicates the total uncertainty [31]. The total theoretical uncertainty corresponds to the higher-order-correction uncertainty summed in quadrature with the PDF and α_S uncertainties, and is partially correlated across center-of-mass energy values [95]

The results from both channels are then combined to provide a measurement of the inclusive and differential full phase space Higgs boson production cross sections. The combination is performed from an extrapolation of measurement in the fiducial region for each channel to the inclusive phase space using acceptance corrections (defined in Sect. 7.2.2.4). The inclusive acceptance corrections are approximately 50% for the $H \to \gamma\gamma$ and 49% for $H \to 4\ell$ for the full phase space. In the differential measurement they vary from about 45 (50)% at low p_T^H to 65 (75)% at high p_T^H for the $H \to 4\ell$ ($H \to \gamma\gamma$) channel. The chosen binning of the combination matches the coarser of the two measurements (i.e the $H \to 4\ell$ one) as both measurements have consistent bin boundaries.

The total Higgs boson production section measured using the $H \to \gamma\gamma$ is $58.6^{+6.7}_{-6.5}$ pb, and in the $H \to ZZ^* \to 4\ell$ is $54.4^{+5.6}_{-5.3}$ pb. Combining the two channels results in the following cross Sect. [95]:

$$56.1^{+4.5}_{-4.3} \left(\pm 3.2(\text{stat.}) \, {}^{+3.1}_{-2.8}(\text{sys.}) \right) \text{pb}. \tag{7.16}$$

All measured total cross sections are in agreement with the SM predictions of 55.6 ± 2.5 pb. The results are shown in Fig. 7.38 along with the cross sections measured using Run-1 data at $\sqrt{s} = 7$ and 8 TeV [96]. In addition, the results of the combined differential cross section as function of p_T^H are shown in Fig. 7.39 and compared to SM predictions (including QCD, PDF, and α_S uncertainties). The resulting combined

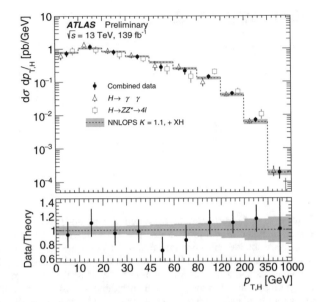

Fig. 7.39 The differential cross section as a function of Higgs boson transverse momentum p_T^H in the full phase space, as measured with the $H \rightarrow \gamma\gamma$ (red triangles) and $H \rightarrow ZZ^* \rightarrow 4l$ (green squares) decay channels, as well as the combined measurement (black dots). The blue dashed line shows the central value of the sum of the NNLOPS ggF prediction, scaled to the inclusive N^3LO prediction with a K-factor of 1.1, and the contribution of the other Higgs boson production modes XH. The SM prediction is overlaid with uncertainty bands that include PDF and α_S uncertainties as well as those due to missing higher-order corrections. For a better visibility, all bins are shown as having the same size, independent of their numerical width. The panel on the bottom shows the ratio of the combined measurement to the prediction [95]

distribution is dominated by statisical uncertainties with dominant systematic sources from the $H \rightarrow \gamma\gamma$ mainly being due to the background modeling and photon energy resolution.

References

1. Aad G et al (2012) Observation of a new particle in the search for the Standard Model Higgs boson with the ATLAS detector at the LHC. Phys Lett B716:1–29. https://doi.org/10.1016/j.physletb.2012.08.020, arXiv:1207.7214 [hep-ex]
2. Chatrchyan S et al (2012) Observation of a new boson at a mass of 125 GeV with the CMS experiment at the LHC. Phys Lett B716:30-61. https://doi.org/10.1016/j.physletb.2012.08.021, arXiv:1207.7235 [hep-ex]
3. Aad G et al (2013) Measurements of Higgs boson production and couplings in diboson final states with the ATLAS detector at the LHC. Phys Lett B726:88–119. https://doi.org/10.1016/j.physletb.2014.05.011, https://doi.org/10.1016/j.physletb.2013.08.010, arXiv:1307.1427 [hep-ex]

4. Aad G et al (2016) Measurements of the Higgs boson production and decay rates and coupling strengths using pp collision data at $\sqrt{s} = 7$ and 8 TeV in the ATLAS experiment. Eur Phys J C76(1):6. https://doi.org/10.1140/epjc/s10052-015-3769-y, arXiv:1507.04548 [hep-ex]
5. Tackmann F et al (2016) Simplified template cross sections. https://cds.cern.ch/record/2138079
6. Adelman J et al (2019) Measurement of fiducial and differential cross sections in the $H \rightarrow \gamma\gamma$ decay channel with 140fb^{-1} of 13 TeV proton-proton collision data with the ATLAS detector. Tech. rep. ATLCOM-PHYS-2019-035. Geneva, CERN. https://cds.cern.ch/record/2654897
7. Boudjema F et al (2013) On the presentation of the LHC Higgs Results. In: Workshop on likelihoods for the LHC searches Geneva, Switzerland, January 21–23, 2013. arXiv:1307.5865 [hep-ph]
8. Aad G et al (2014) Measurements of fiducial and differential cross sections for Higgs boson production in the diphoton decay channel at $\sqrt{s} = 8$ TeV with ATLAS. JHEP 09:112. https://doi.org/10.1007/JHEP09(2014)112, arXiv:1407.4222 [hep-ex]
9. Aaboud M et al (2018) Measurements of Higgs boson properties in the diphoton decay channel with 36 fb^{-1} of pp collision data at $\sqrt{s} = 13$ TeV with the ATLAS detector. Phys Rev D98:052005. https://doi.org/10.1103/PhysRevD.98.052005, arXiv:1802.04146 [hep-ex]
10. Collins JC, Soper DE, Sterman GF (1984) Transverse momentum distribution in Drell-Yan pair and W and Z boson production. Nucl Phys B 250.CERN-TH-3923, 199–224. 37p. https://cds.cern.ch/record/153460
11. Bizon W et al (2018) Momentum-space resummation for transverse observables and the Higgs p_T at N^{3}LL+NNLO. JHEP 02:108. https://doi.org/10.1007/JHEP02(2018)108, arXiv:1705.09127 [hep-ph]
12. Bishara F et al (2017) Constraining light-quark Yukawa couplings from Higgs distributions. Phys Rev Lett 118(12):121801. https://doi.org/10.1103/PhysRevLett.118.121801, arXiv:1606.09253 [hep-ph]
13. Soreq Y, Zhu HX, Zupan J (2016) Light quark Yukawa couplings from Higgs kinematics. JHEP 12:045. https://doi.org/10.1007/JHEP12(2016)045, arXiv:1606.09621 [hep-ph]
14. Grojean C et al (2014) Very boosted Higgs in gluon fusion. JHEP 05:022. https://doi.org/10.1007/JHEP05(2014)022, arXiv:1312.3317 [hep-ph]
15. Grazzini M et al (2017) Modeling BSM effects on the Higgs transverse-momentum spectrum in an EFT approach. JHEP 03:115. https://doi.org/10.1007/JHEP03(2017)115, arXiv:1612.00283 [hep-ph]
16. Klamke G, Zeppenfeld D (2007) Higgs plus two jet production via gluon fusion as a signal at the CERN LHC. JHEP 04:052. https://doi.org/10.1088/1126-6708/2007/04/052, arXiv:hep-ph/0703202 [hep-ph]
17. Andersen JR, Arnold K, Zeppenfeld D (2010) Azimuthal angle correlations for Higgs Boson plus multi-jet events. JHEP 06:091. https://doi.org/10.1007/JHEP06(2010)091, arXiv:1001.3822 [hep-ph]
18. Avoni G et al (2018) The new LUCID-2 detector for luminosity measurement and monitoring in ATLAS. JINST 13(07):P07017. https://doi.org/10.1088/1748-0221/13/07/P07017
19. Aaboud M et al (2016) Luminosity determination in pp collisions at $\sqrt{s} = 8$ TeV using the ATLAS detector at the LHC. Eur Phys J C76(12):653. https://doi.org/10.1140/epjc/s10052-016-4466-1, arXiv:1608.03953 [hep-ex]
20. van der Meer S (1968) Calibration of the effective beam height in the ISR. Tech. rep. CERN-ISR-PO-68-31. ISR-PO-68-31. Geneva, CERN, https://cds.cern.ch/record/296752
21. Luminosity determination in pp collisions at $\sqrt{s} = 13$ TeV using the ATLAS detector at the LHC. Tech. rep. ATLAS-CONF-2019-021. Geneva, CERN (2019). https://cds.cern.ch/record/2677054
22. Hamilton K et al (2013) NNLOPS simulation of Higgs boson production. JHEP 10:222. https://doi.org/10.1007/JHEP10(2013)222, arXiv:1309.0017 [hep-ph]
23. Nason P (2004) A New method for combining NLO QCD with shower Monte Carlo algorithms. JHEP 11:040. https://doi.org/10.1088/1126-6708/2004/11/040, arXiv:hep-ph/0409146
24. Frixione S, Nason P, Oleari C (2007) Matching NLO QCD computations with parton shower simulations: the POWHEG method. JHEP 11:070. https://doi.org/10.1088/1126-6708/2007/11/070, arXiv:0709.2092 [hep-ph]

25. Alioli S et al (2010) A general framework for implementing NLO calculations in shower Monte Carlo programs: the POWHEG BOX. JHEP 06:043. https://doi.org/10.1007/JHEP06(2010)043, arXiv:1002.2581 [hep-ph]

26. Butterworth J et al (2016) PDF4LHC recommendations for LHC Run II. J Phys G 43:023001. https://doi.org/10.1088/0954-3899/43/2/023001, arXiv:1510.03865 [hep-ph]

27. Andersen JR et al (2013) Handbook of LHC Higgs cross sections: 3. Higgs properties. In: Heinemeyer S et al (eds). https://doi.org/10.5170/CERN-2013-004, arXiv:1307.1347 [hep-ph]

28. Sjöstrand T, Mrenna S, Skands PZ (2008) A brief introduction to PYTHIA 8.1. Comput Phys Commun 178:852–867. https://doi.org/10.1016/j.cpc.2008.01.036, arXiv:0710.3820 [hep-ph]

29. ATLAS Collaboration (2014) Measurement of the Z/γ^* boson transverse momentum distribution in pp collisions at $\sqrt{s} = 7$ TeV with the ATLAS detector. JHEP 09:145. https://doi.org/10.1007/JHEP09(2014)145, arXiv:1406.3660 [hep-ex]

30. GEANT4 Collaboration (2003) GEANT4: a simulation toolkit. Nucl Instrum Meth A 506:250–303. https://doi.org/10.1016/S0168-9002(03)01368-8

31. de Florian D et al (2016) Handbook of LHC Higgs cross sections: 4. Deciphering the nature of the Higgs sector. https://doi.org/10.23731/CYRM-2017-002, arXiv:1610.07922 [hep-ph]

32. Anastasiou C et al (2015) Higgs boson gluon-fusion production in QCD at three loops. Phys Rev Lett 114:212001. https://doi.org/10.1103/PhysRevLett.114.212001, arXiv:1503.06056 [hep-ph]

33. Nason P, Oleari C (2010) NLO Higgs boson production via vector-boson fusion matched with shower in POWHEG. JHEP 02:037. https://doi.org/10.1007/JHEP02(2010)037, arXiv:0911.5299 [hep-ph]

34. Bolzoni P et al (2010) Higgs production via vector-boson fusion at NNLO in QCD. Phys Rev Lett 105:011801. https://doi.org/10.1103/PhysRevLett.105.011801, arXiv:1003.4451 [hep-ph]

35. Ciccolini M, Denner A, Dittmaier S (2008) Electroweak and QCD corrections to Higgs production via vector-boson fusion at the LHC. Phys Rev D77:013002. https://doi.org/10.1103/PhysRevD.77.013002, arXiv:0710.4749 [hep-ph]

36. Ciccolini M, Denner A, Dittmaier S (2007) Strong and electroweak corrections to the production of Higgs + 2jets via weak interactions at the LHC. Phys Rev Lett 99:161803. https://doi.org/10.1103/PhysRevLett.99.161803, arXiv:0707.0381 [hep-ph]

37. Mimasu K, Sanz V, Williams C (2016) Higher order QCD predictions for associated Higgs production with anomalous couplings to gauge bosons. JHEP 08:039. https://doi.org/10.1007/JHEP08(2016)039, arXiv:1512.02572 [hep-ph]

38. Campbell JM et al (2012) NLO Higgs boson production plus one and two jets using the POWHEG BOX, MadGraph4 and MCFM. JHEP 07:092. https://doi.org/10.1007/JHEP07(2012)092, arXiv:1202.5475 [hep-ph]

39. Luisoni G et al (2013) HW$^\pm$/HZ+0 and 1 jet at NLO with the POWHEG BOX interfaced to GoSam and their merging within MiNLO. JHEP 10:083. https://doi.org/10.1007/JHEP10(2013)083, arXiv:1306.2542 [hep-ph]

40. Zhang Y et al (2014) QCD NLO and EW NLO corrections to $t\bar{t}H$ production with top quark decays at hadron collider. Phys Lett B738:1–5. https://doi.org/10.1016/j.physletb.2014.09.022, arXiv:1407.1110 [hep-ph]

41. Dawson S et al (2003) Associated Higgs production with top quarks at the large hadron collider: NLO QCD corrections. Phys Rev D68:034022. https://doi.org/10.1103/PhysRevD.68.034022, arXiv:hep-ph/0305087 [hep-ph]

42. Beenakker W et al (2003) NLO QCD corrections to t anti-t H production in hadron collisions. Nucl Phys B653:151–203. https://doi.org/10.1016/S0550-3213(03)00044-0, arXiv:hep-ph/0211352 [hep-ph]

43. Dawson S et al (2004) Exclusive Higgs boson production with bottom quarks at hadron colliders. Phys Rev D69:074027. https://doi.org/10.1103/PhysRevD.69.074027, arXiv:hep-ph/0311067 [hep-ph]

44. Dittmaier S, Krämer M, Spira M (2004) Higgs radiation off bottom quarks at the Tevatron and the CERN LHC. Phys Rev D70:074010. https://doi.org/10.1103/PhysRevD.70.074010, arXiv:hep-ph/0309204 [hep-ph]
45. Gleisberg T et al (2009) Event generation with SHERPA 1.1. JHEP 02:007. https://doi.org/10.1088/1126-6708/2009/02/007, arXiv:0811.4622 [hep-ph]
46. ATLAS Collaboration (2010) The ATLAS simulation infrastructure. Eur Phys J C 70:823–874. https://doi.org/10.1140/epjc/s10052-010-1429-9, arXiv:1005.4568 [physics.ins-det]
47. Chen X et al (2019) Precise QCD description of the Higgs boson transverse momentum spectrum. Phys Lett B 788:425–430. https://doi.org/10.1016/j.physletb.2018.11.037, arXiv:1805.00736 [hep-ph]
48. Ebert MA, Michel JKL, Tackmann FJ (2017) Resummation improved rapidity spectrum for gluon fusion Higgs production. JHEP 05:088. https://doi.org/10.1007/JHEP05(2017)088, arXiv:1702.00794 [hep-ph]
49. Ebert MA et al. SCETlib: a C++ package for numerical calculations in QCD and soft-collinear effective theory. DESY-17-099. http://scetlib.desy.de
50. Campbell JM, Ellis RK (1999) An Update on vector boson pair production at hadron colliders. Phys Rev D60:113006. https://doi.org/10.1103/PhysRevD.60.113006, arXiv:hepph/9905386 [hep-ph]
51. Campbell JM, Ellis RK, Williams C (2011) Vector boson pair production at the LHC. JHEP 07:018. https://doi.org/10.1007/JHEP07(2011)018, arXiv:1105.0020 [hep-ph]
52. Boughezal R et al (2017) Color singlet production at NNLO in MCFM. Eur Phys J C 77(1):7. https://doi.org/10.1140/epjc/s10052-016-4558-y, arXiv:1605.08011 [hep-ph]
53. Gaunt J et al (2015) N-jettiness subtractions for NNLO QCD calculations. JHEP 09:058. https://doi.org/10.1007/JHEP09(2015)058, arXiv:1505.04794 [hep-ph]
54. Banfi A et al (2016) Jet-vetoed Higgs cross section in gluon fusion at N^3LO+NNLL with small-R resummation. JHEP 04:049. https://doi.org/10.1007/JHEP04(2016)049, arXiv:1511.02886 [hep-ph]
55. Stewart IW et al (2014) Jet p_T Resummation in Higgs Production at NNLL′+NNLO. Phys Rev D 89(5):054001. https://doi.org/10.1103/PhysRevD.89.054001. arXiv:1307.1808 [hep-ph]
56. Boughezal R et al (2014) Combining resummed Higgs predictions across jet bins. Phys Rev D 89:074044. https://doi.org/10.1103/PhysRevD.89.074044, arXiv:1312.4535 [hep-ph]
57. Chen X et al (2015) Precise QCD predictions for the production of Higgs + jet final states. Phys Lett B 740:147–150. https://doi.org/10.1016/j.physletb.2014.11.021, arXiv:1408.5325 [hep-ph]
58. Chen X et al (2016) NNLO QCD corrections to Higgs boson production at large transverse momentum. JHEP 10:066. https://doi.org/10.1007/JHEP10(2016)066. arXiv:1607.08817 [hep-ph]
59. Cullen G et al (2012) Automated one-loop calculations with GoSam. Eur Phys J C 72:1889. https://doi.org/10.1140/epjc/s10052-012-1889-1, arXiv:1111.2034 [hep-ph]
60. Cullen G et al (2014) GOSAM-2.0: a tool for automated one-loop calculations within the Standard Model and beyond. Eur Phys J C 74(8):3001. https://doi.org/10.1140/epjc/s10052-014-3001-5. arXiv:1404.7096 [hep-ph]
61. Bothmann E et al (2019) Event generation with Sherpa 2.2. arXiv:1905.09127 [hep-ph]
62. Hche S, Krauss F, Schnherr M (2014) Uncertainties in MEPS@NLO calculations of h+jets. Phys Rev D 90(1):014012. https://doi.org/10.1103/PhysRevD.90.014012, arXiv:1401.7971 [hep-ph]
63. Buschmann M et al (2015) Mass effects in the Higgs-Gluon coupling: boosted vs off-shell production. JHEP 02:038. https://doi.org/10.1007/JHEP02(2015)038, arXiv:1410.5806 [hep-ph]
64. Höche S et al (2013) QCD matrix elements + parton showers: the NLO case. JHEP 04:027. https://doi.org/10.1007/JHEP04(2013)027, arXiv:1207.5030 [hep-ph]
65. Alwall J et al (2014) The automated computation of tree-level and next-to-leading order differential cross sections, and their matching to parton shower simulations. JHEP 07:079 https://doi.org/10.1007/JHEP07(2014)079, arXiv:1405.0301 [hep-ph]

66. Frederix R et al (2016) Heavy-quark mass effects in Higgs plus jets production. JHEP 08:006. https://doi.org/10.1007/JHEP08(2016)006, arXiv:1604.03017 [hep-ph]
67. Frederix R, Frixione S (2012) Merging meets matching in MC@NLO. JHEP 12:061. https://doi.org/10.1007/JHEP12(2012)061, arXiv:1209.6215 [hep-ph]
68. ATLAS Collaboration (2017) Supporting note: selection and performance for the $H \rightarrow \gamma\gamma$ and $H \rightarrow Z_\gamma$ analyses, Spring 2017. Tech. rep. ATL-COM-PHYS-2017-357. Geneva, CERN. https://cds.cern.ch/record/2258158
69. Zecchinelli AG et al (2016) Electron-to-photon fake rate measurements: supporting documentation for the Photon identification in 2015 ATLAS data. Tech. rep. ATL-COM-PHYS-2016-575. Geneva, CERN. https://cds.cern.ch/record/2154427
70. Aad G et al (2014) Measurement of Higgs boson production in the diphoton decay channel in pp collisions at center-of-mass energies of 7 and 8 TeV with the ATLAS detector. Phys Rev D90(11): 112015. https://doi.org/10.1103/PhysRevD.90.112015, arXiv:1408.7084 [hep-ex]
71. Cacciari M, Salam GP, Soyez G (2008) The Anti-k(t) jet clustering algorithm. JHEP 0804:063. https://doi.org/10.1088/1126-6708/2008/04/063, arXiv:0802.1189 [hep-ph]
72. Tagging and suppression of pileup jets with the ATLAS detector. In: ATLAS-CONF-2014-018 (2014). https://cds.cern.ch/record/1700870
73. The HSG1 Group (2013) Measurement of the spin of the new particle observed at a mass of 126 GeV in the diphoton decay channel. Tech. rep. ATL-COM-PHYS-2013-107. Geneva, CERN. https://cds.cern.ch/record/1510537
74. Proposal for truth particle observable definitions in physics measurements. Tech. rep. ATL-PHYS-PUB-2015-013. Geneva, CERN (2015). https://cds.cern.ch/record/2022743
75. Tanabashi M et al (2018) Review of particle physics. Phys Rev D 98(3):030001. https://doi.org/10.1103/PhysRevD.98.030001, https://link.aps.org/doi/10.1103/PhysRevD.98.030001
76. Gordon LE (1997) Isolated photons at hadron colliders at O $(\alpha\alpha s2)$ (I): Spin-averaged case. Nuclear Physics B 501:175–196. https://doi.org/10.1016/S0550-3213(97)00339-8, arxiv:9611391 [hep-ph]
77. Estimate of the m_H shift due to interference between signal and background processes in the $H \rightarrow \gamma\gamma$ channel, for the $\sqrt{s} = 8$ TeV dataset recorded by ATLAS. Tech. rep. ATL-PHYS-PUB-2016-009. Geneva, CERN (2016), https://cds.cern.ch/record/2146386
78. Aad G et al (2015) Combined measurement of the Higgs Boson Mass in pp Collisions at $\sqrt{s} =$ 7 and 8 TeV with the ATLAS and CMS experiments. Phys Rev Lett 114:191803. https://doi.org/10.1103/PhysRevLett.114.191803, arXiv:1503.07589 [hep-ex]
79. Aad G et al (2013) Measurement of isolated-photon pair production in pp collisions at $\sqrt{s} =$ 7 TeV with the ATLAS detector. JHEP 01:086. https://doi.org/10.1007/JHEP01(2013)086, arXiv:1211.1913 [hep-ex]
80. Aad G et al (2012) Measurement of the isolated di-photon cross-section in pp collisions at $\sqrt{s} =$ 7 TeV with the ATLAS detector. Phys Rev D85:012003. https://doi.org/10.1103/PhysRevD. 85.012003, arXiv:1107.0581 [hep-ex]
81. Measurements and interpretations of Higgs-boson fiducial cross sections in the diphoton decay channel using 139 fb^{-1} of pp collision data at $\sqrt{s} = 13$ TeV with the ATLAS detector. Tech. rep. ATLAS-CONF-2019-029. Geneva, CERN (2019), https://cds.cern.ch/record/2682800
82. Hyneman R (2019) GaSBaG: GAussian Smoothing for BAckGrounds - ATLAS stat forum, https://indico.cern.ch/event/822461/contributions/3438712/attachments/1847895/3035531/GPR_BkgSmoothing_StatsForum_23May2019.pdf
83. Frate M et al (2017) Modeling smooth backgrounds and generic localized signals with Gaussian processes. arXiv:1709.05681 [physics.data-an]
84. Geman S, Geman D (1984) Stochastic relaxation, Gibbs distributions, and the Bayesian restoration of images. IEEE Trans Pattern Anal Mach Intell PAMI-6.6, 721–741. ISSN: 0162-8828. https://doi.org/10.1109/TPAMI.1984.4767596
85. Aad G et al (2013) Improved luminosity determination in pp collisions at $\sqrt{s} = 7$ TeV using the ATLAS detector at the LHC. Eur Phys J C73(8):2518. https://doi.org/10.1140/epjc/s10052-013-2518-3. arXiv:1302.4393 [hep-ex]

86. Monticelli F et al (2012) Performance of the electron and photon trigger in p-p collisions at $\sqrt{s} = 7$ TeV with the ATLAS detector at the LHC in 2011. Tech. rep. ATL-COM-DAQ-2012-008. Geneva, CERN. https://cds.cern.ch/record/1426717

87. Jet Calibration and Systematic Uncertainties for Jets Reconstructed in the ATLAS Detector at $\sqrt{s} = 13$ TeV. Tech. rep. ATL-PHYS-PUB-2015-015. Geneva, CERN (2015). https://cds.cern.ch/record/2037613

88. Aaboud M et al (2016) Measurement of the inelastic proton-proton cross section at $\sqrt{s} = 13$ TeV with the ATLAS Detector at the LHC. Phys Rev Lett 117(18):182002. https://doi.org/10.1103/PhysRevLett.117.182002, arXiv:1606.02625 [hep-ex]

89. Aad G et al (2016) Measurements of the Higgs boson production and decay rates and constraints on its couplings from a combined ATLAS and CMS analysis of the LHC pp collision data at $\sqrt{s} = 7$ and 8 TeV. JHEP 08:045. https://doi.org/10.1007/JHEP08(2016)045. arXiv:1606.02266 [hep-ex]

90. Aaboud M et al (2018) Observation of Higgs boson production in association with a top quark pair at the LHC with the ATLAS detector. Phys Lett B784:173–191. https://doi.org/10.1016/j.physletb.2018.07.035, arXiv:1806.00425 [hep-ex]

91. Aaboud M et al (2018) Observation of $H \rightarrow b\bar{b}$ decays and VH production with the ATLAS detector. Phys Lett B786:59–86. https://doi.org/10.1016/j.physletb.2018.09.013, arXiv:1808.08238 [hep-ex]

92. Bellm J et al (2016) Herwig 7.0/Herwig++ 3.0 release note. Eur Phys J C76(4):196. https://doi.org/10.1140/epjc/s10052-016-4018-8, arXiv:1512.01178 [hep-ph]

93. Measurements of Higgs boson properties in the diphoton decay channel using 80 fb^{-1} of pp collision data at $\sqrt{s} = 13$ TeV with the ATLAS detector. Tech. rep. ATLAS-CONF-2018-028. Geneva, CERN (2018). https://cds.cern.ch/record/2628771

94. ATLAS Collaboration (2017) Measurements of the Higgs boson inclusive, differential and production cross sections in the $4l$ decay channel at $\sqrt{s} = 13$ TeV with the ATLAS detector. Tech. rep. ATLAS-COMCONF-2019-045. Geneva, CERN, https://cds.cern.ch/record/2680220

95. ATLAS Collaboration (2019) Combined measurement of the total and differential cross sections in the $H \rightarrow \gamma\gamma$ and the $H \rightarrow ZZ^*4l$ decay channels at $\sqrt{s} = 13$ TeV with the ATLAS detector. Tech. rep. ATLASCOM-CONF-2019-049. Geneva, CERN, https://cds.cern.ch/record/2681143

96. Measurements of the total cross sections for Higgs boson production combining the $H \rightarrow \gamma\gamma$ and $H \rightarrow ZZ^*4l$ decay channels at 7, 8 and 13 TeV center-of-mass energies with the ATLAS detector. Tech. rep. ATLAS-CONF-2015-069. Geneva, CERN (2015), https://cds.cern.ch/record/2114841

Chapter 8
Higgs Boson Cross Section Interpretation Using the EFT Approach

As detailed in Chap. 7, the measurement of the Higgs boson cross section in a fiducial region results in a model-independent measurement. In addition, the measured cross sections are corrected for detector effects (unfolded), resulting in particle-level cross sections that can be compared directly to different theoretical predictions. In this chapter, we will use the effective field theory (EFT) framework, detailed in Chap. 2, as a tool to interpret our results in terms of constraints on anomalous Higgs boson interactions.

8.1 Strategy

In the EFT framework, deviations from SM predictions that can be probed using current ATLAS data arise from the non-zero values of the Wilson coefficients $c^{(6)}$ of dimension-6 operators $\mathcal{O}^{(6)}$ of the Lagrangian:

$$\mathcal{L}_{\text{EFT}} = \mathcal{L}_{\text{SM}} + \mathcal{L}_{D=6} \quad \text{where} \quad \mathcal{L}_{D=6} = \frac{c^{(6)}}{\Lambda_{\text{NP}}^2} \mathcal{O}^{(6)}. \tag{8.1}$$

The interpretation relies on comparisons between the observed differential cross sections and those predicted by simulated samples of the Higgs boson decays to diphotons that include the effect of dimension-6 operators. This is done using the FEYNRULES tool [1, 2]. FEYNRULES automatizes the computation of Feynman rules for a given Lagrangian, providing a universal output known as the *UFO* [3] file, that can be used as an input to different MC event generators. The event generation is performed with the MADGRAPH5 [4] generator at leading-order accuracy in QCD. Using these simulated samples, it is possible to generate leading-order SM

© The Editor(s) (if applicable) and The Author(s), under exclusive license
to Springer Nature Switzerland AG 2020
A. Tarek Abouelfadl Mohamed, *Measurement of Higgs Boson Production Cross Sections in the Diphoton Channel*, Springer Theses,
https://doi.org/10.1007/978-3-030-59516-6_8

and BSM predictions with non-zero Wilson coefficients. Using the generated events, one can obtain predictions for the fiducial inclusive and differential cross sections. The leading-order cross sections obtained with MADGRAPH5 are reweighted using the state-of-the-art SM Higgs boson signal simulations to account for higher-order QCD and electroweak corrections to the SM process, according to the formula:

$$
\frac{d\sigma}{dX} = \sum_j \left(\frac{d\sigma_j}{dX}\right)^{\text{SM MC}} \cdot \left(\frac{d\sigma_j}{dX}\right)^{\text{MG5}}_{c_i \neq 0} \Big/ \left(\frac{d\sigma_j}{dX}\right)^{\text{MG5}}_{c_i = 0},
\tag{8.2}
$$

where the summation is performed over the different Higgs boson production mechanisms j, $\left(\frac{d\sigma_j}{dX}\right)^{\text{SM MC}}$ denotes the SM differential cross section prediction for process j obtained from the state-of-the art SM simulations (detailed in Sect. 7.2.2.1), and $\left(\frac{d\sigma_j}{dX}\right)^{\text{MG5}}_{c_i \neq 0}$ and $\left(\frac{d\sigma_j}{dX}\right)^{\text{MG5}}_{c_i = 0}$ are the differential cross sections predicted for process j by MADGRAPH5. This reweighting approach assumes that the QCD and electroweak corrections factorize from the EFT effects [5]. The following strategy is followed in order to measure the values of the Wilson coefficients using the measured differential cross sections as an input:

- Generate samples for a certain number of benchmark configurations, corresponding to different values of the Wilson coefficients, to calculate the corresponding expected cross sections.
- Use an interpolation procedure to predict the cross sections continuously at any value of the Wilson coefficient in the search range. This procedure is detailed in Sect. 8.4.
- Define a set of measured differential cross section as inputs and compute their correlations. This procedure is described in Sect. 8.5.2.
- Compute the total experimental and theoretical covariances of the input measurements and simulated samples as detailed in Sects. 8.5.3–8.5.4.
- Using all the previous ingredients, one can extract confidence intervals for a given Wilson coefficient.

In addition to the previous procedure, there are some considerations that need to be taken into account in order to check the validity of the EFT approach and to facilitate the matching to UV-complete BSM models, as discussed below.

8.1.1 Linearized Approach

As detailed in Chap. 2, in this analysis we only consider dimension-6 terms in the expansion of the EFT Lagrangian. This truncation requires additional checks to test the self-consistency of the EFT approach within the search range. Let us consider the EFT expansion of the matrix element:

$$\mathcal{M}_{\text{EFT}} = \mathcal{M}_{\text{SM}} + \frac{c_{(6)}}{\Lambda^2}\mathcal{M}_{\text{d6}} + \frac{c_{(8)}}{\Lambda^4}\mathcal{M}_{\text{d8}} + \cdots \tag{8.3}$$

The squared matrix element will be:

$$|\mathcal{M}_{\text{EFT}}|^2 = \underbrace{|\mathcal{M}_{\text{SM}}|^2}_{\text{SM component}} + \underbrace{\frac{2c_{(6)}}{\Lambda^2}Re(\mathcal{M}_{\text{SM}}^*\mathcal{M}_{\text{d6}})}_{\text{SM-D6 Interference}} + \underbrace{\frac{c_{(6)}^2}{\Lambda^4}|\mathcal{M}_{\text{d6}}|^2}_{\text{D6 BSM-squared}} + \underbrace{\frac{2c_{(8)}}{\Lambda^4}Re(\mathcal{M}_{\text{SM}}^*\mathcal{M}_{\text{d8}})}_{\text{SM-D8 Interference}} + \cdots \tag{8.4}$$

Here $|\mathcal{M}_{\text{SM}}|^2$ is the dimension-four Standard Model squared matrix element. The term $2Re(\mathcal{M}_{\text{SM}}^*\mathcal{M}_{\text{d6}})$ represents the interference between the Standard Model and the dimension-six operators in the EFT expansion. This term is of the order $\mathcal{O}(\frac{c_{(6)}}{\Lambda^2})$, whereas the squared matrix element $|\mathcal{M}_{\text{d6}}|^2$ of the dimension-six operators is of the order $\mathcal{O}(\frac{c_{(6)}^2}{\Lambda^4})$. This term has the same order, in terms of powers of Λ, as that stemming from the interference between a dimension-8 operator and the SM part $2Re(\mathcal{M}_{\text{SM}}^*\mathcal{M}_{\text{d8}})$. The truncation of the EFT series beyond dimension-6 terms assumes that the contribution from the dimension-8 terms is sub-leading. Therefore, and following the recommendations from Ref. [6, 7], the modification $\left(\frac{d\sigma_j}{dX}\right)_{c_i}^{\text{MG5}} / \left(\frac{d\sigma_j}{dX}\right)_{c_i=0}^{\text{MG5}}$ to the state-of-the-art SM predictions are derived twice: once considering only the SM term and its interference with the dimension-6 operators, and once accounting also for the squared dimension-6 BSM matrix element. This will allow the estimation of the uncertainty due to missing terms in the EFT series depending on the relative contributions of the two terms. In addition, comparing the cross section predictions using the interference-only BSM terms to those including the contribution from the squared BSM matrix element, we can identify a *linear regime*. In this regime, the BSM-squared terms will be sub-dominant, and the interpolation procedure will be simpler, through linearization of the cross section as a function of $c_i^{(6)}$.

In cases where the cross section change resulting from the inclusion of the dimension-6 squared term is larger than the interference-only term; this might indicate that the EFT approach is not valid anymore [7]. Nevertheless, there are cases, detailed in Ref. [6], where the dimension-6 squared term has larger cross section than the interference-only terms in a valid EFT expansion. This can result from symmetries or destructive cancellation that suppress the interference term. Therefore, the validity of such results will depend on the assumptions of the UV complete coupling as they can modify the simple power counting (based on the powers of Λ) between the dimension-6 squared terms and the dimension-8 interference terms. Therefore, deriving the limits using both the interference-only and the squared dimension-6 terms allows constraining families of UV-complete BSM models.

8.1.2 CP-odd Observables

As detailed in the previous section, there are scenarios in which the interference-only cross section can be smaller than the dimension-6 squared terms. An example of such cases is that of CP-odd operators. In this case, the SM and the EFT dimension-6 CP-odd operators yield different helicity amplitudes, and hence their interference is zero [8]. Therefore, there is no contribution from the interference term to the inclusive rate, or to CP-even observables such as transverse momenta and invariant masses, and the only contribution is to CP-odd observables [9]. The Higgs boson production via VBF can provide such CP-sensitive observables via the tensor structure of the Higgs coupling to weak bosons (HVV) [10, 11]. The most general tensor structure of a HVV vertex in the limit of massless quarks can be written as [12]:

$$T^{\mu\nu}(q_1, q_2) = a_1(q_1, q_2)g^{\mu\nu} + a_2(q_1, q_2)\left[q_1 \cdot q_2 g^{\mu\nu} - q_2^\mu q_1^\nu\right] + a_3(q_1, q_2)\epsilon^{\mu\nu\rho\sigma}q_{1\rho}q_{2\sigma},$$
(8.5)

where q_1, q_2 are the four momenta of the weak bosons and $a_{1,2,3}$ are form factors. For the SM case, a_1 is constant and $a_2 = a_3 = 0$. Therefore, sizable form factors $a_{2,3}$ would represent BSM contributions. The distributions of the two jets are an important tool for the determination of the tensor structure of a HVV coupling. The $p_T^{j_i}$ or m_{jj} distributions depend strongly on the form factors, and hence can not be used to determine the tensor structure in a model-independent manner. The distribution of the azimuthal angle separation of the leading jets $|\Delta\phi_{jj}|$, on the other hand, is not sensitive to form factor effects and provides a powerful tool to determine the tensor structure as it can distinguish between the three tensor structures of Eq. (8.5). The $|\Delta\phi_{jj}|$ variable has a characteristic distribution that is distinct for CP-even and CP-odd couplings. For a pure CP-odd coupling, the cross section is suppressed at $|\Delta\phi_{jj}| = 0, \pi$, whereas for CP-even couplings it is suppressed at $|\Delta\phi_{jj}| = \pi/2$. These effects cancel each other when both CP-even and CP-odd anomalous couplings are present, and no distinction can be made. This missing information, in this case, is contained in the signed distribution $\Delta\phi_{jj,\text{signed}}$. The $\Delta\phi_{jj,\text{signed}}$ variable has a similar distribution to $|\Delta\phi_{jj}|$ with suppressed cross sections at $\Delta\phi_{jj,\text{signed}} = 0, \pm\pi$ for a pure CP-odd couplings and at $\Delta\phi_{jj,\text{signed}} = \pm\pi/2$ for a pure CP-odd coupling. However, when CP-odd and CP-even couplings are mixed with an angle α, the position of the dips due to cross section suppression are shifted with $\Delta\phi_{jj,\text{signed}} = -\alpha$ for CP-even and $\Delta\phi_{jj,\text{signed}} = -\alpha + \pi$ for CP-odd and hence a distinction can be made. An example for these different cases, from Ref. [12], is shown in Fig. 8.1 for given CP-even and CP-odd couplings values.

The distinction between the CP-even and CP-odd couplings for the $\Delta\phi_{jj,\text{signed}}$ distribution can be understood by looking at the matrix element:

$$\mathcal{M} = \mathcal{M}_{\text{SM}} + a_2\mathcal{M}_{\text{CP-even}} + a_3\mathcal{M}_{\text{CP-odd}}.$$
(8.6)

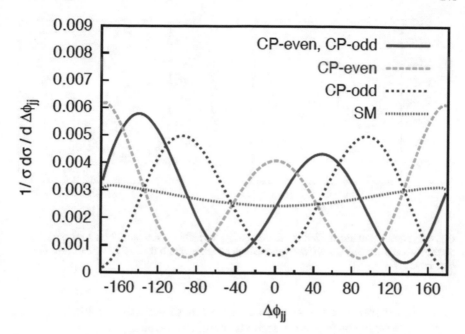

Fig. 8.1 Normalised distribution of the $\Delta\phi_{jj,\text{signed}}$ observable for $m_H = 120$ GeV. The CP-even and CP-odd couplings are characterized in terms of the parameters d and $\tilde{d}$ which are functions of the HVV Wilson coefficients. The figure shows a mixed-CP scenario ($d = \tilde{d} = 0.18$) in red solid curve, a CP-even anomalous coupling ($\tilde{d} = 0$, $d = 0.18$) in green dashed curve, CP-odd coupling ($\tilde{d} = 0.18$, $d = 0.$) in blue dotted curve, and the SM model ($\tilde{d} = 0$, $d = 0.$) case with purple dotted line [12]

This results in the squared matrix element:

$$| \mathcal{M} |^2 = | \mathcal{M}_{\text{SM}} |^2 + a_2^2 | \mathcal{M}_{\text{CP-even}} |^2 + a_3^2 | \mathcal{M}_{\text{CP-odd}} |^2 \tag{8.7}$$

$$\underbrace{+ a_2 2 Re \left(\mathcal{M}_{\text{SM}}^* \mathcal{M}_{\text{CP-even}} \right)}_{\text{SM CP-even interference}} \tag{8.8}$$

$$\underbrace{+ a_3 2 Re \left((\mathcal{M}_{\text{SM}} + a_2 \mathcal{M}_{\text{CP-even}})^* \mathcal{M}_{\text{CP-odd}} \right)}_{\text{SM CP-odd interference}} \tag{8.9}$$

$\mathcal{M}_{\text{SM}}$ and $\mathcal{M}_{\text{CP-even}}$ are even functions of $\Delta\phi_{jj,\text{signed}}$ whereas $\mathcal{M}_{\text{CP-odd}}$ is odd in $\Delta\phi_{jj,\text{signed}}$ and hence the CP-odd term interference with the SM is zero when using $|\Delta\phi_{jj}|$ as opposed to the interference between the CP-even term and the SM. An example of such interference is shown in Fig. 8.2. Therefore, the asymmetry in the distribution of $\Delta\phi_{jj,\text{signed}}$ due to its parity-odd nature can be used to probe CP-violation in the Higgs sector, as parity-odd couplings can only originate from the a_3 interference term. On the other hand, the squared CP-even or CP-odd terms will

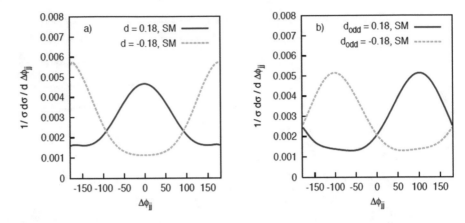

Fig. 8.2 Normalised distribution of the $\Delta\phi_{jj,\text{signed}}$ variable for $m_H = 120$ GeV for the interference between **a** the SM and CP-even anomalous coupling and **b** the SM and anomalous CP-odd couplings [12]

results in CP-even effects regardless of the nature of the operators [9]. Therefore, they will result in non-zero cross sections for all observables.

8.2 EFT Sample Generation Using MadGraph

Following Eq. (8.2), three different sets of MC samples are required in order to describe the variations of the differential cross sections as a function of the Wilson coefficients of the model we are interested in:

- State-of-the-art SM predictions that include the best known QCD and EW corrections. For this purpose, we use the Higgs boson signal samples used from the cross section measurement detailed in Sect. 7.2.2.1.
- Leading-order SM predictions generated with MADGRAPH 5 as implemented in a given EFT model by setting all the Wilson coefficients to zero.
- Leading-order variations of the cross sections generated with MADGRAPH 5 from variations of the Wilson coefficient that is being investigated.

The fiducial differential cross sections can be obtained from the generated events by interfacing the output of the MADGRAPH 5 events to a RIVET [13] routine that applies the fiducial selection detailed in Sect. 7.3.2, computes the different observables, detailed in Sect. 7.1.1, and fills corresponding histograms. In this section, we will review the event generation details of the EFT samples using MADGRAPH5 for the Higgs Effective Lagrangian (SILH basis), in Sect. 8.2.1, and the SMEFT (Warsaw basis), in Sect. 8.2.2.

8.2.1 Higgs Effective Lagrangian (SILH Basis) Event Generation

The Higgs Effective Lagrangian presented in Sect. 2.2.2.2 has been implemented in FEYNRULES [2].

Samples of ggH and VBF+VH events are produced at the parton level for specific points in EFT parameter space by interfacing the universal file output (UFO) from Feynrules to the MADGRAPH 5 event generator. The generated events are then passed through the PYTHIA8 generator [14] for modeling of parton shower and underlying event using the $A14$ tuned set of parameters [15]. For each production mode, the Higgs boson mass is set to 125 GeV and the NNPDF23LO parton distribution functions [16] are used.

Higgs boson production via gluon fusion is generated with up to two additional partons in the final state using leading-order matrix elements. The 0-parton, 1-parton, and 2-parton events are merged using the MLM matching scheme [17]. The gluon-fusion events are generated using the following commands with h denoting the Higgs boson, a a photon, and j a light parton:

```
generate p p > h NP=1  QED=1  QCD=99, h > a a  NP=1  QED=2
add process p p > h j  NP=1  QED=1  QCD=99, h > a a  NP=1  QED=2
add process p p > h j  j  NP=1  QED=1  QCD=99, h > a a  NP=1  QED=2,
```

where QED and NP specify the maximum number of electroweak and new physics couplings, respectively. The choice of NP=1 allows one new physics coupling in the production and the decay of the Higgs boson. Effectively, either the production or the decay is changed since there are no operators that affect both the production and decay at the same time since the only Wilson coefficients that affect the Higgs boson decay are $\bar{c}_\gamma$ and $\tilde{c}_\gamma$. The choice of QED=1 in the production removes the VBF and VH production modes. The first, second and third lines specify the production of a Higgs boson in association with 0-, 1-, and 2- partons, respectively.

Higgs boson production in the VBF and VH modes is generated using tree-level matrix elements. The VBF+VH events are generated using the following commands:

```
generate p p > h j  j  NP=1  QED=99  QCD=0, h > a a  NP=1  QED=2
add process p p > h b b ~ NP=1  QED=99  QCD=0, h > a a  NP=1
QED=2
add process p p > h l+ l-  NP=1  QED=99  QCD=0, h > a a  NP=1
QED=2
add process p p > h ta+ ta-  NP=1  QED=99  QCD=0, h > a a  NP=1
QED=2
add process p p > h vl vl ~ NP=1  QED=99  QCD=0, h > a a  NP=1
QED=2
add process p p > h l+ vl  NP=1  QED=99  QCD=0, h > a a  NP=1
QED=2
```

```
add process p p > h l- vl ~ NP=1 QED=99 QCD=0, h > a a NP=1
QED=2
```

The choice of QCD=0 in the production removes gluon-fusion events. The first line generates Higgs bosons in association with two gluons or light quarks, through VBF and VH, $V \to$ hadrons. The second line is the production of a Higgs boson in association with two b-quarks. The third, fourth and fifth lines are needed to generate ZH, $Z \to$ leptons events, while the sixth and seventh lines are needed to generate WH, $W \to$ leptons events.

Anomalous $t\bar{t}H$ interactions could in principle be probed by isolating a phase space region dominated by signal events from the $t\bar{t}H$ associated production mode. Anomalous $Hf\bar{f}$ interactions would also alter the Higgs boson total width and thus the $H \to \gamma\gamma$ branching ratio; however, they would manifest themselves more evidently as deviations of the $pp \to H \to f\bar{f}$ measured cross sections. Therefore, no variations to the $t\bar{t}H$ production mode are considered.

The expected impact of the coefficients $\bar{c}_g$, $\bar{c}_{HW}$ and of their CP-odd counterparts on the ggF and VBF+VH production cross sections respectively for some particular values of such coefficients is illustrated in Fig. 8.3, and the effect on the total $(ggF + VBF + VH + t\bar{t}H)$ $H \to \gamma\gamma$ differential cross sections is shown in Fig. 8.4. The chosen values correspond to the upper values expected to be excluded at the 95% confidence level. The effect of the variations of these Wilson coefficients follows directly from the SILH Lagrangian detailed in Sect. 2.2.2.2.

8.2.2   SMEFT (Warsaw Basis) Event Generation

The SMEFT Lagrangian in the Warsaw basis is implemented in FEYNRULES via the SMEFTSIM package [18]. The SMEFTsim package provides a complete implementation of the lepton- and baryon-number conserving dimension-6 Lagrangian operators in the Warsaw basis. The SM Lagrangian tree-level amplitudes are included, and extended, with the SM loop-induced Higgs couplings to gg, $\gamma\gamma$, and $Z\gamma$. The SMEFTsim package provides implementations for three different flavor symmetry assumptions and two input parameter scheme choices. The work shown in this thesis was performed using the $U(3)^5$ flavor symmetric case, with non-SM CP-violating phases, and using the α scheme which uses the input parameter set $\{\alpha_{ew}, m_Z, G_f\}$ for the electroweak sector. Additionally, the SMEFTsim package provides two alternative implementations (called "A" and "B") of the dimension-6 Lagrangian, differing in the technical implementation but producing consistent results in a set of benchmark studies [18]. For the work presented here, the implementation "A" was used.

MADGRAPH5 allows generating the Higgs boson production processes separately with the interference-only term using the option NP^2==1 or with the quadratic-only term using the option NP^2==2. On the other hand, MADGRAPH 5 cannot generate the decay process with the option NP^2==1 or 2. Therefore, for these tests in which we assess the validity of the linear regime by comparing the two cross sections,

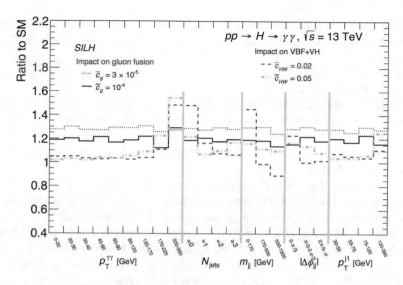

Fig. 8.3 The ratio between leading-order BSM cross section from non-zero Wilson coefficients of the SILH effective Lagrangian and the different SM production modes for the five differential distributions used in the analysis. For $\bar{c}_g$ and $\tilde{c}_g$, the effect on the gluon-fusion cross section is shown. For $\bar{c}_{HW}$ and $\tilde{c}_{HW}$, the effect on the sum of the vector-boson fusion and VH associated production cross sections is shown

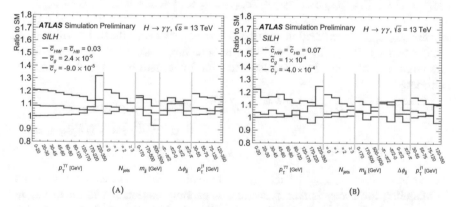

Fig. 8.4 The ratio between the BSM cross section from non-zero Wilson coefficients of the SILH effective Lagrangian and the total SM cross section for the five differential distributions used in the analysis of **A** the coefficients $\bar{c}_g$, $\bar{c}_\gamma$ and $\bar{c}_{HW}$, and **B** the coefficients $\tilde{c}_g$, $\tilde{c}_\gamma$ and $\tilde{c}_{HW}$, of the SILH effective Lagrangian for some particular values of the coefficients

the decay to $\gamma\gamma$ is performed using PYTHIA8. The following lines are added to the PYTHIA8 configuration file to turn off all the Higgs boson decays except for the one to diphotons:

```
25:onMode = off
25:onIfMatch = 22 22
```

For all SMEFTsim simulations, the decay process is not included in the MAD-GRAPH 5 generation, as the decay vertices are also affected by the same Wilson coefficients, as detailed in Sect. 2.2.2.1. The effect of these Wilson coefficients on the $H \rightarrow \gamma\gamma$ decay width and the total Higgs boson decay width is analytically modeled, as shown in Sect. 8.3.

The resulting amplitude, however, is not consistent with the truncation to the dimension-6 of the EFT expansion, as the dimension-6 Lagrangian amplitude is only defined up to terms of $\mathcal{O}(\frac{1}{\Lambda^2})$, and beyond that the amplitude is ill-defined. The inclusion of terms with more than one operator, namely one in the production and one in the decay, would require the addition of dimension-8 operators to avoid inconsistencies. Nevertheless, when the amplitudes are constructed with up to one operator it is theoretically consistent to get the dimension-6 squared terms $\mathcal{O}(\frac{1}{\Lambda^4})$. Therefore, the following formula is used to model the total change in the production cross section $\times$ branching ratio for the linear interpolation case for a generic Wilson coefficient $\bar{c}$:

$$\sigma_{\bar{c}\neq0}^{\text{linear}} \times BR_{\bar{c}\neq0}^{\text{linear}} = \sigma_{\text{SM}} \times BR_{\text{SM}} + \bar{c}\sigma_{\text{interference}} \times BR_{\text{SM}} + \bar{c}\sigma_{\text{SM}} \times \Delta BR_{\text{interference}}, \tag{8.10}$$

where $\sigma_{\text{interference}}$ is cross section from the interference between the SM and dimension-6 terms and $\Delta BR_{\text{interference}}$ is the change in branching ratio of the $H \rightarrow \gamma\gamma$ due to interference-only terms. For the case when the linear and quadratic variations are considered:

$$\sigma_{\bar{c}\neq0}^{\text{linear+quadratic}} \times BR_{\bar{c}\neq0}^{\text{linear+quadratic}} = \sigma_{\bar{c}\neq0}^{\text{linear}} \times BR_{\bar{c}\neq0}^{\text{linear}} + \bar{c}^2\sigma_{\text{quadratic}}$$
$$\times BR_{\text{SM}} + \bar{c}^2\sigma_{\text{SM}} \times \Delta BR_{\text{quadratic}}, \tag{8.11}$$

i.e. the terms $\bar{c}^2\sigma_{\text{interference}} \times BR_{\text{interference}}$ are avoided in the expansion of $\sigma \times BR$.

Modeling the decay is then performed separately as detailed in Sect. 8.3, and added analytically in the fit. For the production, the MADGRAPH 5 scripts used for gluon-fusion and $VBF + VH$ productions are the same as those used for the SILH basis in Sect. 8.2.1, excluding the decay part.

The expected impact of the different SMEFT Wilson coefficients for some particular values of such coefficients is shown in Fig. 8.5. The figure shows the effect of the different coefficients on the production cross section in addition to the change in the $H \rightarrow \gamma\gamma$ branching ratio. The chosen values correspond to the maximum value expected to be excluded at 95% CL, and their effects follow directly from the SMEFT Lagrangian (Sect. 2.2.2.1). The total effect of these variations on the total $pp \rightarrow H \rightarrow \gamma\gamma$ cross section is shown in Fig. 8.6.

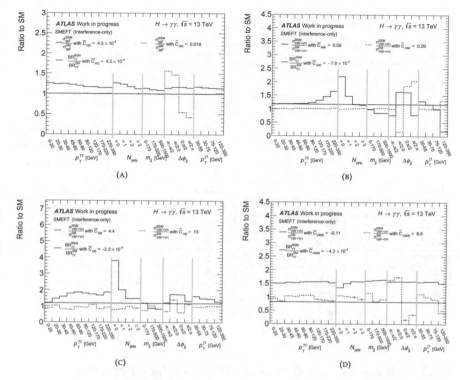

Fig. 8.5 The effect on the five differential distributions used in the analysis of the coefficients **A** $\overline{C}_{HG}$, **B** $\overline{C}_{HW}$, **C** $\overline{C}_{HB}$ and **D** $\overline{C}_{HWB}$ with their CP-odd counterparts. The figures show the effect of each Wilson coefficient on the Higgs production mode that it affects. The figures also show their effect on the change in the $H \rightarrow \gamma\gamma$ branching ratio either by changing only the total Higgs decay width ($\overline{C}_{HG}$) or by also changing the Higgs to photons decay width

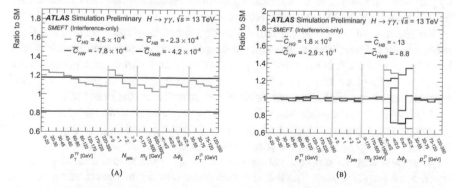

Fig. 8.6 The effect on the five differential distributions used in the analysis of the coefficients **A** $\overline{C}_{HG}$, $\overline{C}_{HW}$, $\overline{C}_{HB}$ and $\overline{C}_{HWB}$ and **B** their CP-odd counterparts. The figures show the effect of each Wilson coefficient on the total $H \rightarrow \gamma\gamma$ cross section times branching ratio. The Wilson coefficients values chosen correspond to the expected 95% CL

Table 8.1 Scanned parameter ranges for each EFT parameter

Basis	Wilson coefficient	Min value	Max value
SILH	$\bar{c}_\gamma$	−0.001	0.001
SILH	$\tilde{c}_\gamma$	−0.02	0.02
SILH	$\bar{c}_g$	−0.00100	0.00050
SILH	$\tilde{c}_g$	−0.0005	0.0005
SILH	$\bar{c}_{HW}$	−0.25	0.25
SILH	$\tilde{c}_{HW}$	−0.5	−0.5
SILH	$\bar{c}_{HB}$	−0.25	0.25
SILH	$\tilde{c}_{HB}$	−0.5	0.5
SMEFT	$\overline{C}_{HG}$	−0.005	0.005
SMEFT	$\tilde{C}_{HG}$	−0.2	0.2
SMEFT	$\overline{C}_{HW}$	−0.5	0.5
SMEFT	$\tilde{C}_{HW}$	−0.5	−0.5
SMEFT	$\overline{C}_{HB}$	−0.5	0.5
SMEFT	$\tilde{C}_{HB}$	−0.5	0.5
SMEFT	$\overline{C}_{HWB}$	−0.5	0.5
SMEFT	$\tilde{C}_{HWB}$	−0.5	0.5

8.2.3 Parameter Variations and EFT Sample Generation

To set limits on a single EFT parameter, the parameter space is scanned across a range chosen empirically, by inspecting the corresponding change induced in expected total cross section, compared to the largest deviations observed in data (or that are excluded by previous measurements). The chosen parameter ranges are shown in Table 8.1 for the SILH and SMEFT Wilson coefficients.

For the case of the linear regime, the interpolation is simplified since the cross section from the interference between the dimension-6 operators and the SM scales linearly. Therefore, it is sufficient to generate only one sample at a given benchmark value of a given Wilson coefficient. Nevertheless, additional interference-only samples were generated following Table 8.1 to cross-check the linear scaling.

To allow limits to be set on two EFT parameters at the same time, two-dimensional parameter space is scanned in a 5×5 grid for $\bar{c}_g$ versus $\tilde{c}_g$ and $\bar{c}_{HW}$ versus $\tilde{c}_{HW}$. For the scan of $\bar{c}_{HW}$ versus $\tilde{c}_{HW}$, the parameters $\bar{c}_{HB}$ and $\tilde{c}_{HB}$ are set equal to $\bar{c}_{HW}$ and $\tilde{c}_{HW}$, respectively, to suppress anomalous $H \to Z\gamma$ production, and all the other coefficients are set to zero. The two-dimensional limits for the linearized regime, on the other hand, do not require generating events in a two-dimensional grid, benefiting from the linear behavior of the interference-only cross section. Hence, deriving two-dimensional limits in this case is simplified. For each point in parameter space, 500k gluon-fusion and 500k VBF+VH events were generated for the SILH basis, and a similar procedure was performed for the SMEFT (Warsaw) basis. The generated events were passed through a RIVET routine implementing the particle-level fiducial definition of the Higgs to diphoton cross section measurement as detailed in Sect. 7.3.2.

8.3 Higgs Boson Decay Width Modelling

Non-zero values of the Wilson coefficients affecting the Higgs boson couplings can change the Higgs boson branching ratio to two photons. This change can arise from the change in the partial decay width to two photons (Higgs boson effective couplings to the photon field) or from the change in the total Higgs boson decay width (due to Higgs boson couplings to gauge bosons) as detailed in Chap. 2.

The MADGRAPH5 event generator is used to study the change in the decay width. Samples are produced for several values of the Wilson coefficients in the different bases. The partial widths are then obtained and interpolated with a second degree polynomial, to obtain a parametrization of the partial width for any arbitrary value of the Wilson coefficients. As an illustration, Fig. 8.7 shows the variation of the Higgs boson partial decay width to two photons as a function of $\overline{C}_{HW}, \overline{C}_{HB}$, and $\overline{C}_{HWB}$ and of their CP-odd counterparts, in the SMEFT Warsaw basis. The full parametrization of the variation of the diphoton decay width (in MeV) in the SMEFT Warsaw basis is:

$$\Delta\Gamma^{\gamma\gamma} = 143.1 \times 10^2\, \overline{C}^2_{HW} - 2.4 \times 10^2\, \overline{C}_{HW} + 143.1 \times 10^2\, \tilde{C}^2_{HW} \qquad (8.12)$$
$$+ 1541 \times 10^2\, \overline{C}^2_{HB} - 8 \times 10^2\, \overline{C}_{HB} + 1541 \times 10^2\, \tilde{C}^2_{HB}$$
$$+ 469.5 \times 10^2\, \overline{C}^2_{HWB} + 4.4 \times 10^2\, \overline{C}_{HWB} + 469.5 \times 10^2\, \tilde{C}^2_{HWB}$$

with the change in total width (in MeV) [19] found to be:

$$\Delta\Gamma^{\text{total}} = 50.6\, \overline{C}_{HG} - 1.21\, \overline{C}_{HW} - 1.5\, \overline{C}_{HB} + 1.21\, \overline{C}_{HWB} \qquad (8.13)$$

For the parametrization of the decay width in the SILH basis, the following analytical formulae are used [5]. The change in partial width to diphotons (in MeV) is given by:

$$\Delta\Gamma^{\gamma\gamma} = 1934.7\, \overline{c}^2_\gamma - 8.4655\, \overline{c}_\gamma + 1937.9\, \tilde{c}^2_\gamma \qquad (8.14)$$

while the change in total width (in MeV) is given by:

$$\Delta\Gamma^{\text{total}} = \Delta\Gamma^{\gamma\gamma} \qquad (8.15)$$
$$+ 6810354\, \overline{c}^2_g + 3057.12\, \overline{c}_g + 6806751\, \tilde{c}^2_g$$
$$+ 3.3611\, \overline{c}^2_{HW} + 3.368\, \overline{c}_{HW} + 0.271\, \tilde{c}^2_{HW}$$
$$+ 0.3872\, \overline{c}^2_{HB} + 0.4462\, \overline{c}_{HB}$$

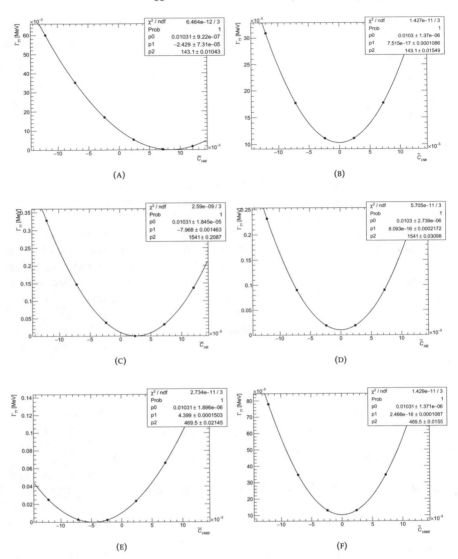

Fig. 8.7 Impact of various Wilson coefficients on the Higgs boson partial decay widths to photons, using the SMEFT parametrisation: **A** $\overline{C}_{HW}$, **B** $\tilde{C}_{HW}$, **C** $\overline{C}_{HB}$, **D** $\tilde{C}_{HB}$, **E** $\overline{C}_{HWB}$, **F** $\tilde{C}_{HWB}$

8.4   Variation of EFT Parameters

From Eq. (8.2), deriving the constraints on the Wilson coefficients requires evaluating the cross section as a function of the Wilson coefficient within the scan range $\vec{p}$, $f_{MC}(\vec{p})$. However, running a Monte-Carlo event generator and a subsequent analysis tool for each point in $\vec{p}$ is a CPU-expensive procedure. Therefore, an interpolation

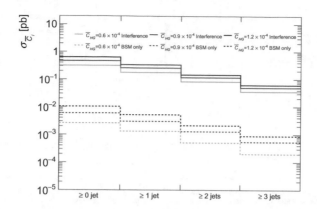

Fig. 8.8 Gluon-fusion cross section, $\left(\frac{d\sigma_j}{dX}\right)^{\text{MG5}}_{c_i}$ as a function of the number of jets with $p_T >$ 30 GeV, predicted by MADGRAPH, for different values of the Wilson coefficient $\overline{C}_{HG}$ (SMEFT Warsaw basis), when considering only the effect the interference term (solid lines) or only that of the quadratic term (dashed lines) of the matrix element expansion. All other Wilson coefficients are set to zero. The effect of the quadratic-only term is around two orders of magnitude smaller than that of the interference-only term for the chosen range of values of $\overline{C}_{HG}$

procedure is adopted in order to reduce the complexity of such computations and to ensure smooth likelihood variations as a function of the Wilson coefficient. The interpolation relies on the generated samples at given ranges in Table 8.1. The interpolation procedure is then validated against generated samples as will be detailed in this section. For samples within the linear regime, a simplified linear interpolation procedure is implemented as detailed in Sect. 8.4.1, whereas for samples with a non-negligible quadratic component a quadratic interpolation procedure is used, detailed in Sect. 8.4.2.

8.4.1 Linear Interpolation

The validity of the linear regime is tested by comparing the cross section from the interference-only terms and from the squared dimension-6 term for different values of the Wilson coefficient if the latter is much smaller than the former, then the linear regime is valid. Figures 8.8 and 8.9 show the differential cross sections, for both the interference-only terms and quadratic-only terms, generated by the variation of the Wilson coefficient $\overline{C}_{HG}$ in the SMEFT parametrisation in the range 0.001–0.002. As can be seen in Fig. 8.10, the effect of the quadratic-only term is around two orders of magnitude smaller than that of the interference-only term for the chosen range of values of $\overline{C}_{HG}$. Therefore, in that range, the linear approximation can be used. Tests of the linear behavior for some bins of the differential cross sections for different values of $\overline{C}_{HG}$ are shown in Fig. 8.11.

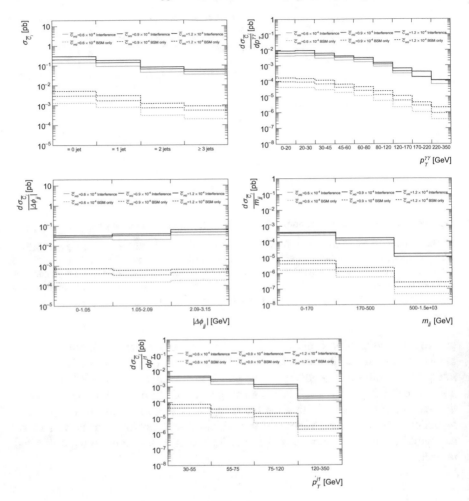

Fig. 8.9 Gluon-fusion cross section $\left(\frac{d\sigma_j}{dX}\right)^{MG5}_{c_i}$ as a function of several kinematic variables, predicted by MADGRAPH, for different values of the Wilson coefficient $\overline{C}_{HG}$ (SMEFT Warsaw basis), when considering only the effect of the interference term (solid lines) or the quadratic term (dashed lines) of the matrix element expansion. All other Wilson coefficients are set to zero. The effect of the quadratic-only term is around two orders of magnitude smaller than that of the interference-only term for the chosen range of values of $\overline{C}_{HG}$

Similarly, for the other Wilson coefficients ($\overline{C}_{HW}$, $\overline{C}_{HB}$ and $\overline{C}_{HWB}$) the linear regime can be checked by comparing the ratio of the interference-only and the quadratic dimension-6 cross sections, as shown in Fig. 8.12. It can be seen that the linear assumption is valid only for values of the coefficients up to $\mathcal{O}(0.001)$, where the interference-only term dominates; for larger variations of the Wilson coefficients

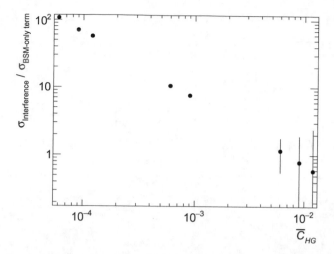

Fig. 8.10 Ratio of the non-SM contributions to the $N_{\text{jet}} = 1$ gluon-fusion cross section as predicted by MADGRAPH5, arising either from the interference term NP^2==1 or the quadratic term NP^2==2 (labelled BSM-only term), for variations of the Wilson coefficient $\overline{C}_{HG}$ between 6×10^{-5} and 0.012. The ratio decreases from more than 100 for $\overline{C}_{HG} = 6 \times 10^{-5}$ until the interference and the quadratic terms give similar contributions (ratio close to 1) for $\overline{C}_{HG} \approx \mathcal{O}(0.01)$

the quadratic term contribution to the cross section is not negligible with respect to the interference term, and a quadratic interpolation must be used instead of the simpler linear one.

8.4.2 Quadratic Interpolation with PROFESSOR

As detailed in Sect. 8.1, a quadratic interpolation is used for samples with non-negligible quadratic dimension-6 contributions. The interpolation is performed using the PROFESSOR method [20]. The Professor method is an approach that reduces the time to evaluate f_{MC} dramatically using P-dimensional polynomial parametrisations.

The key idea is to treat each bin of a histogram as an independent function of the parameter space. Once the parameterisations $f_{\text{MC}}(\vec{p})$ are known, they can be used to implement a fast pseudo-generator that yields an approximate response in milliseconds rather than hours. Furthermore, due to the usage of polynomial functions for the interpolation, the response function is steady. These properties make $f_{\text{MC}}(\vec{p})$ suitable for numeric applications. This method was used in previous ATLAS Higgs to diphoton EFT interpretation publications [5, 21].

The lowest-order polynomial function to incorporate correlations between its variables is a polynomial of second order. For a certain bin, b, at a point $\vec{p}$ in parameter space, this can be written as:

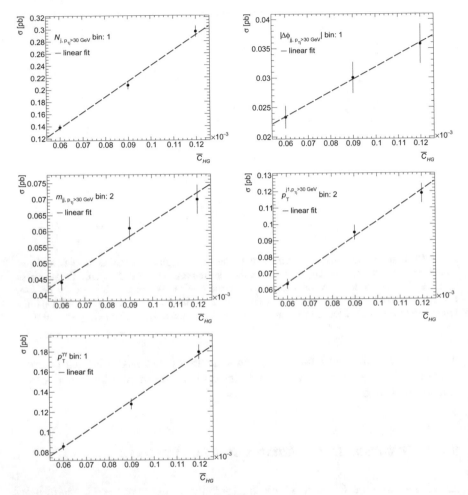

Fig. 8.11 Gluon-fusion cross section, $\left(\frac{d\sigma_j}{dX}\right)_{c_i}^{\text{MG5}}$, in some bins of the differential distributions, due to the interference-only term of the matrix element, for $\overline{C}_{HG}$ (SMEFT Warsaw basis) between 6×10^{-5} and 12×10^{-5}. All other Wilson coefficients are set to zero

$$f_{\text{MC}}^{(b)}(\vec{p}) = \alpha_0^{(b)} + \sum_i \beta_i^{(b)} p_i + \sum_{i \leq j} \gamma_{ij}^{(b)} p_i p_j, \qquad (8.16)$$

with $\vec{c}^{(b)} = \alpha_0^{(b)}, \beta_i^{(b)}, \gamma_{ij}^{(b)}$ some coefficients to be determined.

The Professor approach for determining $\vec{c}^{(b)}$ consists in constructing an over-constrained system of equations $\vec{v}_a^{(b)} = \tilde{P}_{\vec{c}}^{(b)}$ using the ensemble of bin contents $v_a^{(b)}$, $a \in [1, N]$ obtained when running the MC generator with the parameter settings $\vec{p}_a = f_{\text{MC}}^{(b)}(\vec{p}_a), a \in [1, N]$, ("anchors" of the parametrisation). Since the system

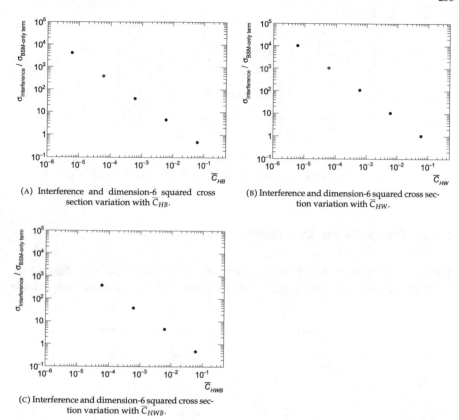

(A) Interference and dimension-6 squared cross section variation with $\overline{C}_{HB}$.

(B) Interference and dimension-6 squared cross section variation with $\overline{C}_{HW}$.

(C) Interference and dimension-6 squared cross section variation with $\overline{C}_{HWB}$.

Fig. 8.12 Ratio of the non-SM contributions to the inclusive $VBF + VH$ cross section as predicted by MADGRAPH5, arising from either the interference term NP^2==1 and the quadratic term NP^2==2 (labelled BSM-only term), for variations of the Wilson coefficients $\mathbf{A}\ \overline{C}_{HB}$, $\mathbf{B}\ \overline{C}_{HW}$, and $\mathbf{C}\ \overline{C}_{HWB}$ (between 10^{-5} and 0.1). The ratio decreases from more than 10^4 for values of the Wilson coefficients $\mathcal{O}(10^{-5})$ until the interference and the quadratic terms give similar contributions (ratio close to 1) for values of the coefficients around $\mathcal{O}(0.01)$

is overconstrained, the matrix $\tilde{P}$ can be (pseudo-)inverted using the singular value decomposition implemented in EIGEN3 [22]. Once the pseudoinverse matrix $\tilde{P}^{-1}$ is known, the coefficients $\vec{c}^{(b)}$ can be obtained calculating $\tilde{P}^{-1} \cdot \vec{v}_a$. The fast pseudogenerator is thus simply a collection of coefficients $\vec{c}^{(b)}$ for all bins b of interest. As an illustration, in the case of a bidimensional ($P = 2$) parameter space, $\vec{p} = (x, y)$, the $\vec{c}$ vector has six components to be determined ($\alpha_0, \beta_x, \beta_y, \gamma_{xx}, \gamma_{xy}, \gamma_{yy}$), requiring a system of $N \geq 6$ equations:

$$
\begin{pmatrix} v_1 \\ v_2 \\ \vdots \\ v_N \end{pmatrix} = \begin{pmatrix} 1 & x_1 & y_1 & x_1^2 & x_1 y_1 & y_1^2 \\ 1 & x_2 & y_2 & x_2^2 & x_2 y_2 & y_2^2 \\ & & & \vdots & & \\ 1 & x_N & y_N & x_N^2 & x_N y_N & y_N^2 \end{pmatrix} \begin{pmatrix} \alpha_0 \\ \beta_x \\ \beta_y \\ \gamma_{xx} \\ \gamma_{xy} \\ \gamma_{yy} \end{pmatrix} \tag{8.17}
$$

$$\underbrace{}_{v_a^{(b)}} \qquad \underbrace{}_{\tilde{P}} \qquad \underbrace{}_{\vec{c}^{(b)}}$$

Examples of the quadratic interpolation and its validation are shown in the next section for the different Wilson coefficients of the SILH and the SMEFT bases.

8.4.3   Interpolation Validation

The interpolation procedure is validated by comparing the interpolated values of the cross section with those predicted by the event generator for the same values of the

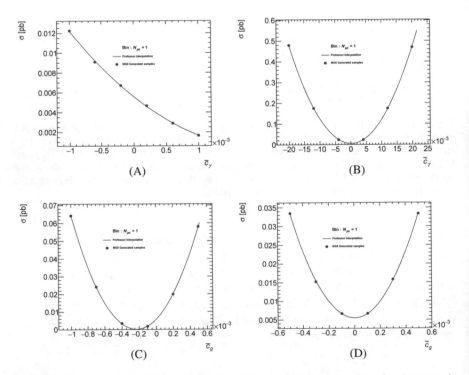

Fig. 8.13 The interpolation (red) is compared with the predicted cross section of a given sample (blue). Shown here are ggH events, and the Wilson coefficients that are varied are $\bar{c}_g$ and $\bar{c}_\gamma$, as well as their CP-odd counter-parts. All Wilson coefficients are varied independently

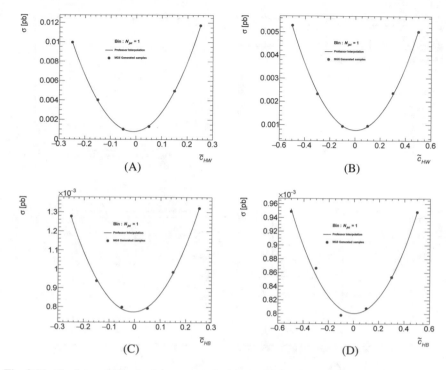

Fig. 8.14 The interpolation (red) is compared with the predicted cross section of a given sample (blue). Shown here are VBF+VH events for a representative bin of each variable, and the Wilson coefficients that are varied are $\bar{c}_{HW}$ and $\bar{c}_{HB}$, as well as their CP-odd counter-parts. All Wilson coefficients are varied independently

operator coefficients. Figures 8.13 and 8.14 show examples of this comparison for each analyzed variable and different Wilson coefficients in the SILH basis (and their CP-odd counter-parts). The interpolated values of the cross sections are in excellent agreement with those predicted by the event generator. Figure 8.15 shows the average residual, defined as the relative cross section difference between the generator prediction and the interpolation, divided by the number of generated samples, $\vec{p}_a$. The most significant deviation is of the order of 2% in the highest $p_T^{\gamma\gamma}$ bin, and the difference is due to the low statistics of events generated in this bin. The experimental uncertainty in that bin is 31%, more than an order of magnitude larger. Similar behavior is seen for the last bin of m_{jj}. Analogous comparison plots using the SMEFT basis for the different Wilson coefficients and their CP-odd counter-parts are shown in Fig. 8.16, while the average residual for the SMEFT basis is shown in Figs. 8.17 and 8.18 for the ggH and VBF+VH cross sections, respectively. Additional tests were performed, by excluding some of these points and comparing them to the interpolation without these points. The interpolation showed excellent agreement in this case as well (with differences of <1% between the average bias and the average residual obtained using all points).

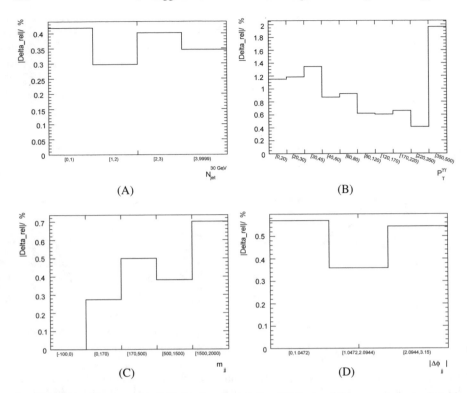

Fig. 8.15 The average residual (defined as the relative difference of the generated sample with respect to the interpolation) for each bin of the probed variables is shown

8.5 Limit Setting Procedure

Limits on Wilson coefficients are set by means of a likelihood function

$$\mathcal{L}(c_i) = \frac{1}{\sqrt{(2\pi)^k |C|}} \exp\left(-\frac{1}{2}\left(\vec{\sigma}_{\text{data}} - \vec{\sigma}_{\text{pred}}(c_i)\right)^T C^{-1}\left(\vec{\sigma}_{\text{data}} - \vec{\sigma}_{\text{pred}}(c_i)\right)\right) \tag{8.18}$$

where $\vec{\sigma}_{\text{data}}$ and $\vec{\sigma}_{\text{pred}}$ are vectors containing the measured and predicted cross sections in each bin of the five analyzed observables, and $C = C_{\text{stat}} + C_{\text{exp}} + C_{\text{pred}}$ is the total covariance matrix defined by the sum of the statistical, experimental and theoretical covariances, with $|C|$ denoting its determinant and k is total number of bins for all observables. This can also be written in a more compact form as

$$\mathcal{L}(c_i) = \mathcal{L}_{\text{max}} e^{-\frac{\chi^2(c_i)}{2}} \tag{8.19}$$

where $\chi^2(c_i) = \left(\vec{\sigma}_{\text{data}} - \vec{\sigma}_{\text{pred}}(c_i)\right)^T C^{-1}\left(\vec{\sigma}_{\text{data}} - \vec{\sigma}_{\text{pred}}(c_i)\right)$. We use the profile likelihood ratio to set confidence intervals on the parameter(s) c_i. The likelihood ratio

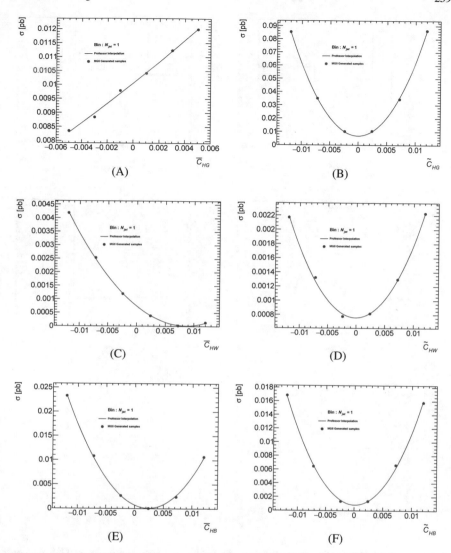

Fig. 8.16 The interpolation (red line) is compared with the predicted cross section of a given sample (blue) using the SMEFT (Warsaw) parametrization for different Wilson coefficients for the $N_{\text{jet}} = 1$ bin. All Wilson coefficients are varied independently

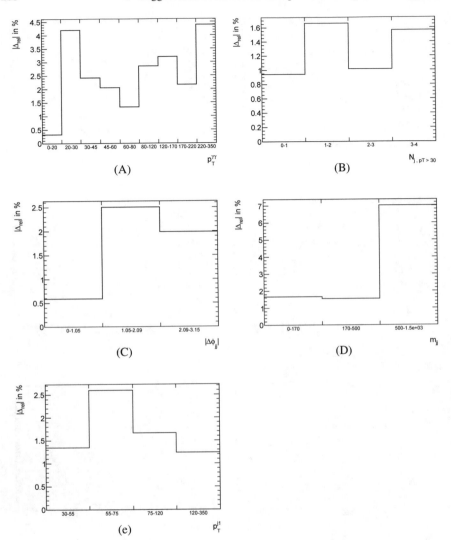

Fig. 8.17 SMEFT ggH : The average residual (defined as the relative difference of the generated sample with respect to the interpolation) for each bin of the probed variables is shown

is

$$\lambda(c_i) = \frac{\mathcal{L}(c_i)}{\mathcal{L}_{max}} = e^{-\frac{\chi^2(c_i)}{2}}$$ (8.20)

and the test statistic is

$$t_{c_i} = -2\ln\lambda(c_i) = \chi^2(c_i)$$ (8.21)

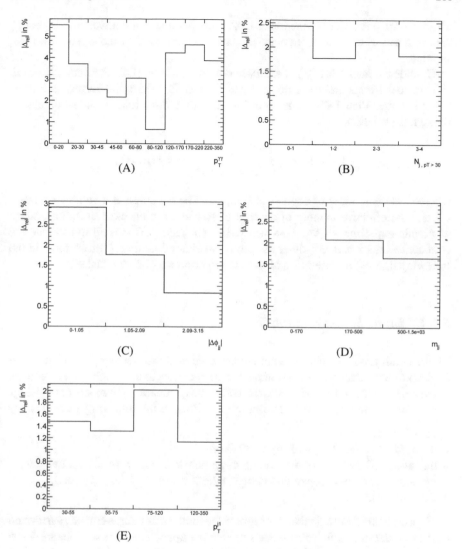

Fig. 8.18 SMEFT VBF+VH : The average residual (defined as the relative difference of the generated sample with respect to the interpolation) for each bin of the probed variables is shown

The p-value of a given hypothetical value of c_i is then

$$p_{c_i} = n \int_{-2 \ln \mathcal{L}(c_i) + 2 \ln \mathcal{L}_{\min}}^{\infty} dt_{c_i} \, f(t_{c_i}|c_i), \qquad (8.22)$$

where $n = 1$ or $\frac{1}{2}$ for two-sided or one-sided confidence interval estimation. Wilks' theorem states that in the large-sample limit $f(t_{c_i}|c_i)$ is the χ^2 probability distribution

$P(\chi^2, \nu)$ for ν degrees of freedoms. Lower $\lambda(c_i)$ values, or equivalently higher t_{c_i} values, mean worse agreement between data and the hypothesized c_i, and a smaller p-value.

Confidence intervals (CI) at a certain confidence level (CL) for one, or several, Wilson coefficients are the regions of values of c_i for which the corresponding p-value is larger than 1-CL. They are thus determined by finding their boundaries through the relation:

$$1 - CL = n \int_{-2\ln \mathcal{L}(c_i)+2\ln \mathcal{L}_{\min}}^{\infty} dx\ f(x; m), \qquad (8.23)$$

where m is the number of degrees of freedom. The coverage of a confidence interval and the effective number of degrees of freedom are checked using ensembles of pseudo-experiments. The GAMMACOMBO package [23] is used to estimate the confidence intervals using either the profile likelihood method ("Prob" label in the following figures) or ensembles of pseudo-experiments ("Plugin" label).

8.5.1   Input Measurements

The constraints on the effective Lagrangian coefficients are set using five differential Higgs boson production cross sections that are sensitive to the relative cross sections of the ggH and the VBF+VH processes, and to the CP quantum numbers of the Higgs boson. Such differential cross sections are functions of the following quantities:

- the Higgs boson p_T,
- the number of jets N_{jets} with $p_T > 30$ GeV,
- the invariant mass m_{jj} of the leading di-jet pair in events with at least two jets,
- the signed azimuthal separation between the two leading jets, $\Delta\phi_{jj}$, and
- the leading jet p_T.

Expected limits are calculated using cross section mock measurements, performed on Asimov datasets built from the SM simulated samples. The cross sections are determined by performing signal+background fits to the reconstructed diphoton invariant mass distributions to determine the signal yields and then unfolding the yields to a particle-level fiducial region close to the experimental fiducial volume. Such unfolded differential cross sections were shown in Sect. 7.7.1.

8.5.2   Statistical Uncertainties

The limits-setting procedure is based on the measurements of five differential cross sections distributions. Therefore, in order to use all of the five distributions, a measure of the shared events between the bins of the different distributions, or their *correlation*, is computed. The correlation is determined using a bootstrapping technique, in which a given data set is resampled in order to estimate the observables of interests, such as the covariance between different bins. The bootstrapping is done as follows:

i. Each event is assigned a random event weight, sampled from a Poisson distribution with variance $v = 1$.
ii. The sum of weights of events in the $m_{\gamma\gamma}$ sidebands is computed. This procedure is motivated by the fact that the $m_{\gamma\gamma}$ distribution in data, which is used to determine the signal yield, is dominated by the irreducible background, and so is the statistical uncertainty on the signal cross section.

These two steps are repeated until a large enough sample is obtained to estimate the cross-correlations. This method is applicable as long as all bins adequately sample all corners of phase space. Due to the sizeable irreducible di-photon background from $pp \rightarrow \gamma\gamma$, all bins have a population of several hundred of events. A selection of example scatter plots for the extracted yields are shown in Fig. 8.19. The correlation between two bins x and y is calculated using Pearson's product-moment correlation coefficient, i.e.

$$r = r_{xy} = \frac{1}{n-1} \sum_{i=1}^{n} \left(\frac{x_i - \bar{x}}{s_x} \right) \left(\frac{y_i - \bar{y}}{s_y} \right), \tag{8.24}$$

with $s_x = \sqrt{\frac{1}{n-1} \sum_{i=1}^{n} (x_i - \bar{x})^2}$ and $s_y = \sqrt{\frac{1}{n-1} \sum_{i=1}^{n} (y_i - \bar{y})^2}$. Here x_i and y_i are the number of background events in bins x and y in the ith bootstrap sample, and $\bar{x}$ or $\bar{y}$ are the average x and y values over the ensemble of the n bootstrap samples, i.e. $\bar{x} = \frac{1}{n} \sum_{i=1}^{n} x_i$ and $\bar{y} = \frac{1}{n} \sum_{i=1}^{n} y_i$. The corresponding 68% CI scales as a function of the number of bootstrap samples and is given by

$$r \in \left[\tanh(\operatorname{atanh}(r) - \frac{1}{\sqrt{n-3}}), \tanh(\operatorname{atanh}(r) + \frac{1}{\sqrt{n-3}}) \right] \tag{8.25}$$

This relation is derived from a Fisher-Z transformation $z = \operatorname{atanh}(r) = \frac{1}{2} \log \frac{1+r}{1-r}$, with variance $V[z] = \frac{1}{n-3}$ [24]. The full correlation matrix is obtained from 10000 bootstrap fits and is shown in Fig. 8.20.

The residual 68% statistical uncertainty from the finite number of bootstrap fits ranges from 0.02% for $r = 0.99$ to 1% for $r = 0$. In the following, extracted statistical correlations smaller than 3% are neglected, as such small contributions should have a negligible impact on the physical result and neglecting them avoids introducing small eigenvalues in the covariance matrix what can cause problems in the numerical inversion of the total covariance matrix.

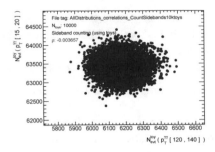
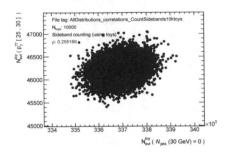

Fig. 8.19 Example scatter plots from counting the events in the sidebands from pseudo-experiments for different bins

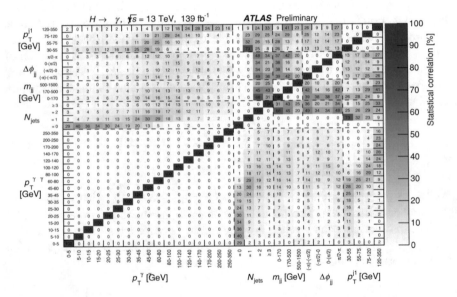

Fig. 8.20 The estimated statistical cross correlations between the bins of the 5 observables

8.5.3 Experimental Systematic Uncertainties

In addition to the statistical correlation, the systematic uncertainties from the unfolding procedure are assumed to be fully correlated between the different bins. The experimental systematical uncertainties, detailed in Sect. 7.5, are classified according to their source as follows:

a. Background modelling
b. Photon energy scale and resolution
c. Luminosity
d. Photon isolation efficiency
e. Photon identification efficiency

f. Diphoton trigger efficiency
g. Pileup modelling
h. Generator modelling
i. (Jet energy scale effective nuisance parameters (NP) 1–8)

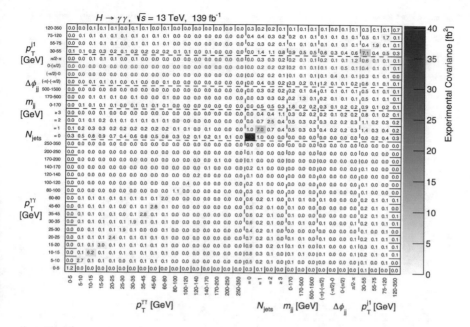

Fig. 8.21 Total experimental covariance of the measured diphoton differential cross sections using the full Run-2 dataset (139 fb^{-1})

j. (Jet energy scale η intercalibration modelling)

k. (Jet energy resolution effective nuisance parameters (NP) 1–8)

l. (Jet energy scale forward modelling)

m. (Jet energy scale high-p_T)

n. (Jet energy scale non-closure)

o. (Jet energy scale pile-up μ)

p. (Jet energy scale pile-up ρ)

q. (Jet energy scale flavour composition)

r. (Jet energy scale flavour response)

s. (JVT modelling)

The sources in parentheses are only included for the observables that involve the reconstruction of jets, such as m_{jj}, p_T^{j1}, $\Delta\phi_{jj,\mathrm{signed}}$ or N_{jets}. Identical uncertainty sources are assumed to be fully correlated across bins with the sign of the error amplitude taken into account when computing the covariance matrix. The full experimental covariance used in the analysis is given by the sum of the statistical and systematic covariances, and is shown in Fig. 8.21.

8.5.4 Theory Uncertainties

Following Eq. (8.2), variations to the SM predictions are obtained by reweighting the best-known SM predictions. Therefore, the different theoretical uncertainties

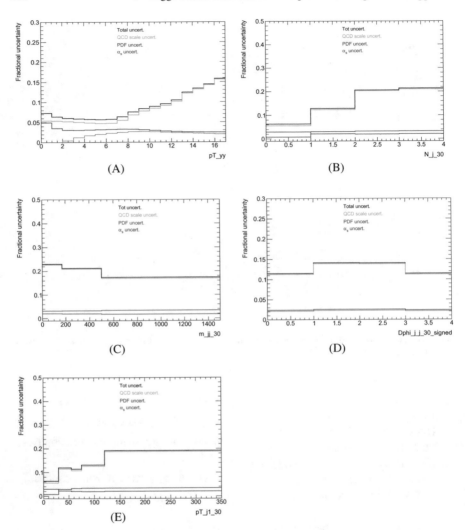

Fig. 8.22 The relative total theory uncertainties for the different differential distributions used for the limit setting

affecting the SM predictions of the different production modes are also included in the total covariance, defined in Sect. 8.5. The different uncertainties affecting the predictions are detailed in Sect. 1.3. They include uncertainties from PDF, QCD order and α_s.

For the gluon-fusion production mode, nine uncertainty sources are used to model the QCD theory uncertainties, following the recommendation of the LHC Higgs cross section working group [7]. These sources are:

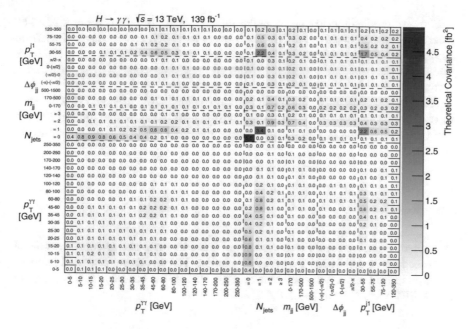

Fig. 8.23 The total theory covariance including the different PDF, QCD and α_S error sources

- two sources correspond to yield uncertainties related to the total cross section. Their magnitude is taken from the STWZ-BLPTW predictions [7, 25, 26] and their impact on the different bins is evaluated using NNLOPS.
- two sources correspond to migration uncertainties related to splitting the phase space by jet multiplicity. Their magnitude and impact are derived similarly to the yield uncertainties
- two uncertainty sources are related to the p_T^H shape and are estimated from scale variations in NNLOPS, including variations of the HNNLO input scales and the renormalization and factorization scales in Powheg [27].
- two uncertainty sources related to the enhancement of uncertainties for events with typical VBF topology (due to explicit or implicit third-jet vetos), and are estimated by scale variations in MCFM [28], and the corresponding uncertainties are estimated using the same procedure use for yield and migration uncertainties.
- one uncertainty source is related to the treatment of m_t and is most important at large p_T^H.

Following the recommendations of PDF4LHC [29], the PDF uncertainties are evaluated using the 30 eigenvectors set and treating each of them as an uncorrelated source. One additional nuisance parameter accounts for the uncertainties in α_s.

For the $VBF + VH$ production modes, QCD uncertainties are estimated as an envelope of the scale variations available in Powheg [30, 31]. Uncertainties from the choice of the PDF set and α_s are evaluated similar to the gluon-fusion case.

For $t\bar{t}H$ production, the perturbative uncertainties are taken to be $^{+5.8\%}_{-9.2\%}$. The uncertainty on the choice of the PDF set is taken to be 3%, and the uncertainty from varying α_s within its uncertainties is 2% [32–34], as detailed in Sect. 1.3.2. For $b\bar{b}H$ production, the combined uncertainty from all three sources is $^{+20\%}_{-24\%}$ [35, 36], and is treated as a separate, uncorrelated, nuisance parameter.

The relative uncertainty in the SM prediction ranges between 6% and 16% depending on the observable and bin. The breakdown of the different sources for the different distributions is shown in Fig. 8.22. The total theory covariance matrix is shown in Fig. 8.23.

8.6   Results

The observed (expected) limits are obtained from the measured (Asimov) unfolded cross sections as detailed in the previous section. The results are provided in terms of the effective Lagrangian coefficients in the HEL SILH basis (Sect. 8.6.1) and in the SMEFT Warsaw basis (Sect. 8.6.2).

8.6.1   Observed and Expected Limits in the SILH Parametrisation

Constraints are set on the Wilson coefficients of dimension-6 effective Lagrangian operators in the SILH parametrisation. The results include both 1D confidence intervals on each coefficient, as well as 2D confidence intervals on pairs of coefficients. The constraints on $\bar{c}_g$ and $\tilde{c}_g$ are derived while fixing all the remaining Wilson coefficients to zero. For $\bar{c}_{HW}$ ($\tilde{c}_{HW}$), the limits are set while setting the remaining Wilson coefficients to zero, except for $\bar{c}_{HB}$ ($\tilde{c}_{HB}$), that is set equal to $\bar{c}_{HW}(\tilde{c}_{HW})$. This ensures that the partial width for $H \rightarrow Z\gamma$ is unchanged from the SM prediction. Values of $|\bar{c}_{HW} - \bar{c}_{HB}| > 0.03$ result in a large $H \rightarrow Z\gamma$ decay rate with respect to the Standard Model ($\gg 11$), which is contradicted by the constraints from ATLAS and CMS searches for this decay, $\sigma^{Z\gamma}_{obs}/\sigma^{Z\gamma}_{SM} = 11$ at 95% CL. The constraints from $Z\gamma$ are shown in Fig. 8 of Ref. [37].

8.6.1.1   1D Confidence Intervals

Table 8.2 summarizes the 1D observed and expected limits for the different Wilson coefficients using the profile likelihood method. The limits were derived using the uncertainties from the cross section measurement using the full Run-2 data.

The $1 - CL$ curves and the coresponding $\Delta\chi^2$ ones for the CP-even Wilson coefficients are shown in Fig. 8.24, and for CP-odd in Fig. 8.25.

Table 8.2 Observed allowed ranges at 95% CL for the $\bar{c}_g, \bar{c}_{HW}, \bar{c}_\gamma$ Wilson coefficients of the SILH basis and their CP-conjugates. Limits on a coefficient are obtained by setting all others to zero. Limits on $\bar{c}_{HW}$ and $\tilde{c}_{HW}$ are derived by setting $\bar{c}_{HB} = \bar{c}_{HW}$ and $\tilde{c}_{HB} = \tilde{c}_{HW}$, respectively, with remaining coefficients set to zero

Coefficient	Observed 95% CL interval	Expected 95% CL interal
$\bar{c}_g$	$[-0.26, 0.26] \times 10^{-4}$	$[-0.25, 0.25] \cup$ $[-4.7, -4.3] \times 10^{-4}$
$\tilde{c}_g$	$[-1.3, 1.1] \times 10^{-4}$	$[-1.1, 1.1] \times 10^{-4}$
$\bar{c}_{HW}$	$[-2.5, 2.2] \times 10^{-2}$	$[-3.0, 3.0] \times 10^{-2}$
$\tilde{c}_{HW}$	$[-6.5, 6.3] \times 10^{-2}$	$[-7.0, 7.0] \times 10^{-2}$
$\bar{c}_\gamma$	$[-1.1, 1.1] \times 10^{-4}$	$[-1.0, 1.2] \times 10^{-4}$
$\tilde{c}_\gamma$	$[-2.8, 4.3] \times 10^{-4}$	$[-2.9, 3.8] \times 10^{-4}$

Overall, the measured cross sections agree with χ^2 $p_0 > 50\%$ with the SM. Therefore, no large deviations from zero for the best values of the different coefficients are observed. The observed limits agree with the expected limits. For $\bar{c}_g$, destructive interference causes the gluon-fusion cross section to be zero for $\bar{c}_g \sim -2.2 \times 10^{-4}$. This interference results in the two disjoint intervals for the expected limits in Table 8.2. The observed limits, however, do not show this behavior as the resulting $\Delta\chi^2$ for the second solution did not pass the $\Delta\chi^2 = 3.84$ threshold corresponding to the 95% CL as seen in Fig. 8.24a. This is a result of the best fit result, $\bar{c}_g = 0.2^{+1.3}_{-1.3} \times 10^{-5}$. This is a consequence of the observed inclusive cross section being slightly higher the SM expectation (details in Sect. 7.7.2). The effect of the different Wilson coefficients on the observed limits on the different observables are shown in Fig. 8.26. The figures show the ratio between data and the default SM predictions for reference. The error bars on data reflect the total covariance: statistical, experimental, and theoretical.

In order to assess the improvement in the limits with respect to previous ATLAS publications, the expected limits obtained with the full Run-2 data (139 fb^{-1}) are compared with the expected limits from cross section measurement using the 2015–2016 data set (36 fb^{-1}) [21]. The results are shown in Table 8.3. Overall, there is an improvement of approximately a factor 1.6 with the full Run-2 data set, resulting from the increased statistics with the full Run-2 data, The limits for $\bar{c}_\gamma$ and $\tilde{c}_\gamma$ are not included as the measurement was not performed using the 36 fb^{-1} dataset.

To estimate the sensitivity of the five differential cross sections to the Wilson coefficients and their constraining power, the limit setting procedure has been repeated using each differential cross section independently. The results are shown in Fig. 8.27. The results show that $\bar{c}_g$ and $\bar{c}_\gamma$, which are Wilson coefficients that result in an overall scaling of the cross section are most sensitive to differential measurements such as $p_T^{\gamma\gamma}$ and $N_{\text{jets}}^{\geq 30\text{GeV}}$ in which the full sample of diphoton candidates is analyzed, and less sensitive to measurements of variables such as p_T^{j1} or m_{jj} which require the presence of at least an additional jet with $p_T > 30$ GeV, thus reducing the number of events used in the measurement. Similar conclusions are reached for $\tilde{c}_g$ and $\tilde{c}_\gamma$, as these limits are computed using both the interference and quadratic terms with

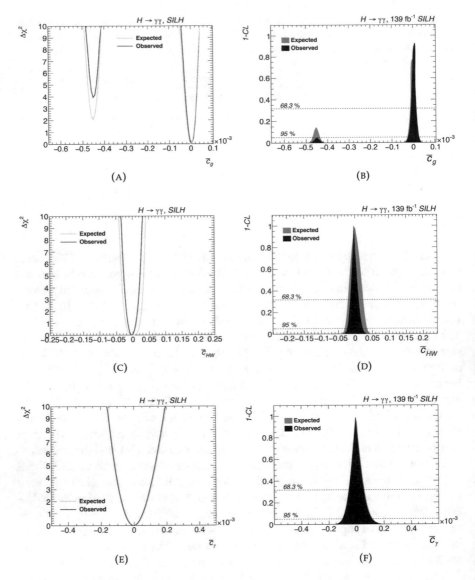

Fig. 8.24 $1 - CL$ and $\Delta\chi^2$ expected and observed curves using 139 fb^{-1} for the different SILH CP-even Wilson coefficients: **A, B** $\overline{c}_g$ **C, D** $\overline{c}_{HW}$ **E, F** $\overline{c}_\gamma$

dominating effects from the quadratic term, which has no sensitivity to CP effects. On the other hand, the coefficient $\overline{c}_{HW}$, and $\tilde{c}_{HW}$, that changes in the Higgs boson kinematics, notably in the high $p_T^{\gamma\gamma}$ and $p_T^{j_1}$ regions and $\Delta\phi_{jj,\text{signed}}$, is mostly constrained by these three differential cross section, and significantly less by the other three.

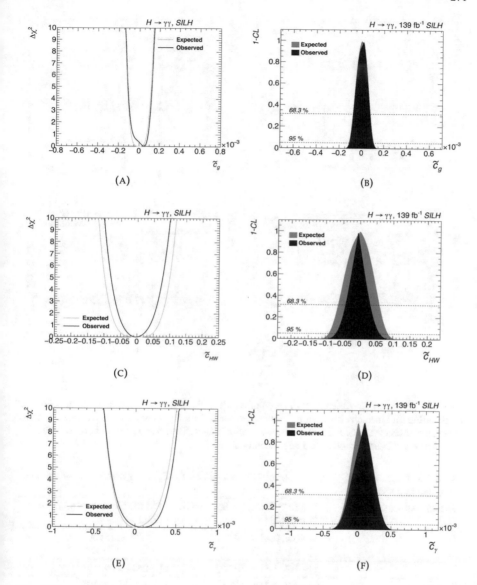

Fig. 8.25 $1 - CL$ and $\Delta\chi^2$ expected and observed curves using 139 fb^{-1} for the different SILH CP-odd Wilson coefficients: **A, B** $\tilde{c}_g$ **C, D** $\tilde{c}_{HW}$ **E, F** $\tilde{c}_\gamma$

8.6.1.2 2D Confidence Regions

The 68% and 95% confidence regions for $\overline{c}_{HW}$ versus $\tilde{c}_{HW}$ are shown in Fig. 8.28. The regions are determined setting $\overline{c}_{HB} = \overline{c}_{HW}$ and $\tilde{c}_{HB} = \tilde{c}_{HW}$ to ensure that the partial width for $H \rightarrow Z\gamma$ is unchanged from the SM prediction. The improvement

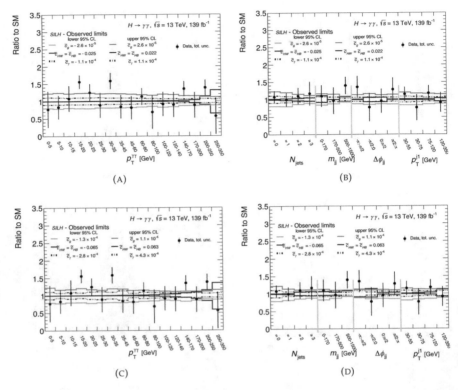

Fig. 8.26 Effect of the different Wilson coefficients at the observed upper and lower 95% CL on the different kinematic distributions relative to the default SM predictions is shown. The ratio between data and the default SM is also shown for reference. The error bars on data reflect the total covariance: statistical, experimental and theoretical

Table 8.3 Expected 95% confidence level 1D intervals on $\bar{c}_g$, $\tilde{c}_g$, $\bar{c}_{HW}$ and $\tilde{c}_{HW}$, for different integrated luminosities

Wilson coefficient	36 fb^{-1} Expected 95% 1-CL Limit [21]	140 fb^{-1} Expected 95% 1-CL Limit
$\bar{c}_g$	$[-0.4, 0.4] \times 10^{-4} \cup$ $[-4.8, -4.1] \times 10^{-4}$	$[-0.25, 0.25] \cup$ $[-4.7, -4.3] \times 10^{-4}$
$\tilde{c}_g$	$[-1.4, 1.3] \times 10^{-4}$	$[-1.1, 1.1] \times 10^{-4}$
$\bar{c}_{HW}$	$[-4.8, 4.5] \times 10^{-2}$	$[-3.0, 3.0] \times 10^{-2}$
$\tilde{c}_{HW}$	$[-13, 13] \times 10^{-2}$	$[-7.0, 7.0] \times 10^{-2}$

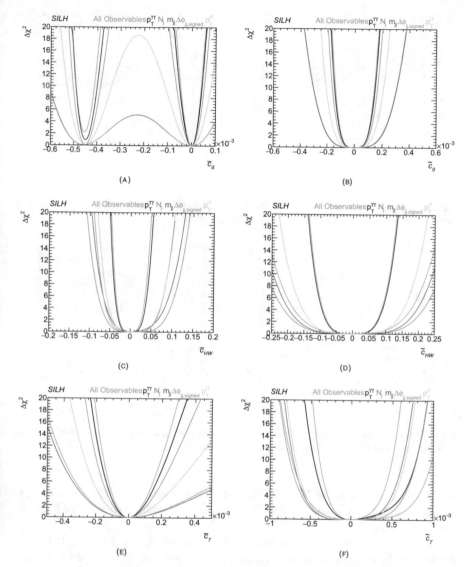

Fig. 8.27 $\Delta\chi^2$ curves for **A** $\bar{c}_g$, **B** $\tilde{c}_g$, **C** $\bar{c}_{HW}$, **D** $\tilde{c}_{HW}$, **E** $\bar{c}_\gamma$ and **F** $\tilde{c}_\gamma$ for the five differential distributions independently in the SILH basis

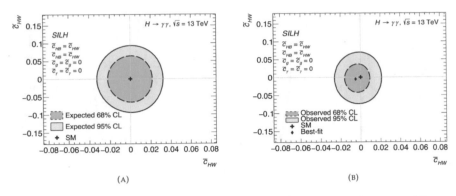

Fig. 8.28 **A** Expected and **B** observed two-dimensional confidence regions for $\bar{c}_{HW}$ and $\tilde{c}_{HW}$ at the 68% and 95% confidence level using the full Run-2 dataset

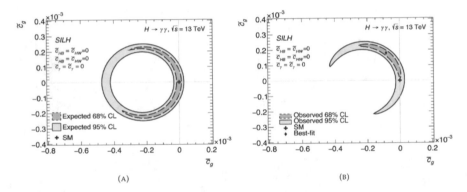

Fig. 8.29 **A** Expected and **B** observed two-dimensional confidence regions for $\bar{c}_g$ and $\tilde{c}_g$ at the 68% and 95% confidence level using the full Run-2 dataset.

of the constraints with the 139 fb^{-1} dataset, with respect to the 36 fb^{-1} dataset, is similar to what was found for the $\bar{c}_{HW}$ and $\tilde{c}_{HW}$ 1D confidence intervals in the previous section.

The 68% and 95% confidence regions for $\bar{c}_g$ versus $\tilde{c}_g$ are shown in Fig. 8.29. Destructive interference causes the gluon-fusion production cross section to be zero around $\bar{c}_g \sim -2.2 \times 10^{-4}$. The impact of this is apparent in the structure of the obtained limits in the two-dimensional parameter plane.

8.6.2 Observed and Expected Results in the SMEFT Warsaw Basis

Similar to the SILH basis, 1D and 2D limits were derived for the different Wilson coefficients in the SMEFT Warsaw basis.

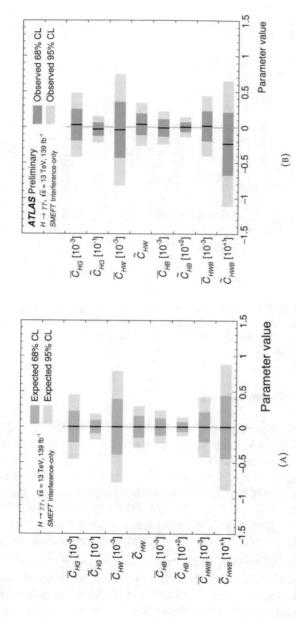

Fig. 8.30 A summary of the **A** the expected and **B** the observed 95% CL using the SMEFT model for $\overline{C}_{HG}$, $\overline{C}_{HW}$, $\overline{C}_{HB}$ and $\overline{C}_{HWB}$ along with their CP-odd counterparts using 139 fb^{-1}. The limits are the results of the individual fits, i.e. allowing only one coefficient to vary, while the remaining ones are set to zero

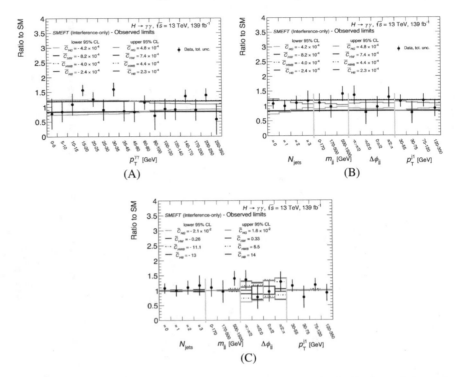

Fig. 8.31 Effect of the observed 95% CL on the different kinematic distributions relative to the default SM predictions using the 139 fb^{-1} dataset. The figures show the ratio between data and the default SM predictions for reference. The error bars on data reflect the total covariance: statistical, experimental, and theoretical

8.6.2.1   1D Confidence Intervals

A summary of the expected and observed 95% CL using the 139 fb^{-1} dataset is shown in Fig. 8.30 for 1D scans varying one Wilson coefficient at a time while setting the remaining coefficients to zero.

The results show a strong constraint for $\overline{C}_{HG}$ as it affects the gluon-fusion production cross section, scaling the overall cross section. Similarly, $\overline{C}_{HW}, \overline{C}_{HB}$, and $\overline{C}_{HWB}$ are strongly constrained as their sensitivity is driven by their effect on the $H \rightarrow \gamma\gamma$ decay width and, therefore, the branching ratio. The constraint on $\overline{C}_{HG}$ is consistent with that on $\overline{c}_g$ in the SILH basis. The $\mathcal{O}_g$ operators are similar between the two bases, and the two Wilson coefficients are related by $\overline{c}_g = \frac{m_W^2}{\Lambda^2}\overline{C}_{HG}$ [38]. Therefore, one finds a factor of 156 between the two coefficients, and hence the limit on $\overline{C}_{HG}$ translates into the interval $\overline{c}_g = [-0.27, 0.27] \times 10^{-4}$, similar to that obtained in the previous section (Table 8.2). The CP-odd Wilson coefficients, on the other hand, are loosely constrained. This is a consequence of using the interference-only cross section so that the only sensitivity for the CP-odd coefficients is from the $\Delta\phi_{jj,\text{signed}}$

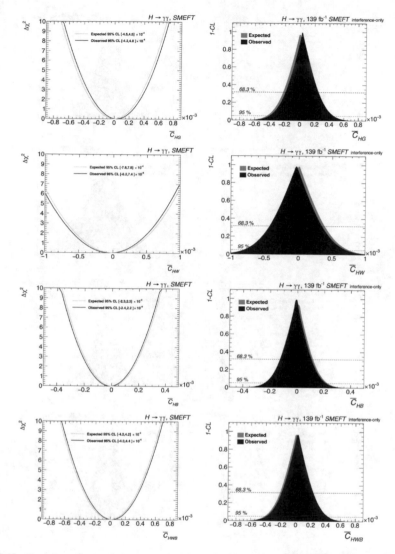

Fig. 8.32 Plots show the $1 - CL$ curves using the SMEFT model for $\overline{C}_{HG}, \overline{C}_{HW}, \overline{C}_{HB}$ and $\overline{C}_{HWB}$ along with their CP-odd counterparts using $140\ \mathrm{fb}^{-1}$

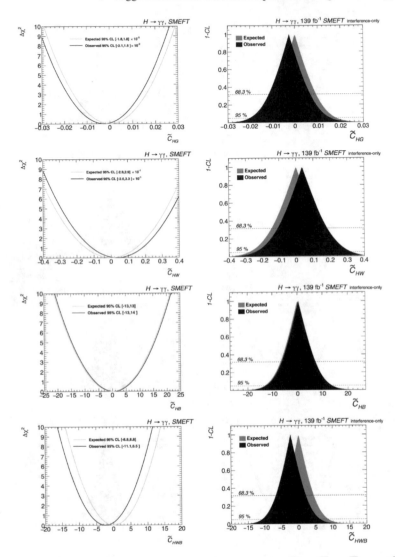

Fig. 8.33 Plots show the $1 - CL$ curves using the SMEFT model for $\overline{C}_{HG}, \overline{C}_{HW}, \overline{C}_{HB}$ and $\overline{C}_{HWB}$ along with their CP-odd counterparts using 140 fb^{-1}

variable. The effect of the different Wilson coefficients at the observed limits on the different distributions are shown in Fig. 8.31. The figures show the ratio between data and the default SM predictions for reference. The error bars on data reflect the total covariance: statistical, experimental, and theoretical.

Figures 8.32 and 8.33 shows the $1 - CL$ curves for $\overline{C}_{HG}, \overline{C}_{HW}, \overline{C}_{HB}$ and their CP-odd counterparts. The effect of these Wilson coefficients on the different production modes can be seen in Fig. 8.5.

In order to validate the linearization procedure, and following the recommendations of Ref. [6], the limits were also derived using the quadratic interpolation with Professor for samples with both the interference and the dimension-6 squared terms. This is shown in Table 8.4 for the observed limits and Table 8.5 for the expected limits. The confidence intervals obtained using both the interference plus quadratic terms were found to be in good agreement with the limits found with the linear interpolation for the CP-even operators. The constraints on the CP-odd operators, on the other hand, are significantly different between the interference-only case and the interference + quadratic terms case. This is a result of the helicity structure of these operators [8], as they do not interfere with the CP-even operators of the SM except for $\Delta\phi_{jj,\text{signed}}$ as it is a CP-sensitive variable. Therefore, $\Delta\phi_{jj,\text{signed}}$ is the only variable driving the limits, as can be seen from Fig. 8.31. This is also apparent in Fig. 8.5, where the effect of the CP-odd Wilson coefficients only results in a variation of the $\Delta\phi_{jj,\text{signed}}$ distributions.

Table 8.4 The 95% CL observed limits on the $\overline{C}_{HG}, \overline{C}_{HW}, \overline{C}_{HB}, \overline{C}_{HWB}$ Wilson coefficients of the SMEFT basis and their CP-odd counterparts using interference-only terms and using both interference and quadratic terms. Limits are derived fitting one Wilson coefficient at a time while setting the other coefficients to zero

Coefficient	95% CL, interference-only terms	95% CL, interference and quadratic terms
$\overline{C}_{HG}$	$[-4.2, 4.8] \times 10^{-4}$	$[-6.1, 4.7] \times 10^{-4}$
$\tilde{C}_{HG}$	$[-2.1, 1.6] \times 10^{-2}$	$[-1.5, 1.4] \times 10^{-3}$
$\overline{C}_{HW}$	$[-8.2, 7.4] \times 10^{-4}$	$[-8.3, 8.3] \times 10^{-4}$
$\tilde{C}_{HW}$	$[-0.26, 0.33]$	$[-3.7, 3.7] \times 10^{-3}$
$\overline{C}_{HB}$	$[-2.4, 2.3] \times 10^{-4}$	$[-2.4, 2.4] \times 10^{-4}$
$\tilde{C}_{HB}$	$[-13.0, 14.0]$	$[-1.2, 1.1] \times 10^{-3}$
$\overline{C}_{HWB}$	$[-4.0, 4.4] \times 10^{-4}$	$[-4.2, 4.2] \times 10^{-4}$
$\tilde{C}_{HWB}$	$[-11.1, 6.5]$	$[-2.0, 2.0] \times 10^{-3}$

Table 8.5 The 95% CL expected limits on the $\overline{C}_{HG}, \overline{C}_{HW}, \overline{C}_{HB}, \overline{C}_{HWB}$ Wilson coefficients of the SMEFT basis and their CP-odd counterparts using interference-only terms and using both interference and quadratic terms. Limits are derived fitting one Wilson coefficient at a time while setting the other coefficients to zero

Coefficient	95% CL, interference-only terms	95% CL, interference and quadratic terms
$\overline{C}_{HG}$	$[-4.5, 4.5] \times 10^{-4}$	$[-4.6, 4.0] \times 10^{-4}$
$\tilde{C}_{HG}$	$[-1.8, 1.8] \times 10^{-2}$	$[-1.3, 1.2] \times 10^{-3}$
$\overline{C}_{HW}$	$[-7.8, 7.8] \times 10^{-4}$	$[-8.2, 9.1] \times 10^{-4}$
$\tilde{C}_{HW}$	$[-0.29, 0.29]$	$[-3.6, 3.6] \times 10^{-3}$
$\overline{C}_{HB}$	$[-2.3, 2.3] \times 10^{-4}$	$[-2.4, 2.4] \times 10^{-4}$
$\tilde{C}_{HB}$	$[-13, 13]$	$[-1.2, 1.2] \times 10^{-3}$
$\overline{C}_{HWB}$	$[-4.2, 4.2] \times 10^{-4}$	$[-4.4, 4.0] \times 10^{-4}$
$\tilde{C}_{HWB}$	$[-8.8, 8.8]$	$[-2.0, 2.0] \times 10^{-3}$

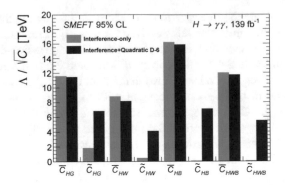

Fig. 8.34 Summary of the observed 95% CL intervals of the different CP-even and CP odd Wilson coefficients in terms of BSM scale Λ_{NP} using $C = 1$

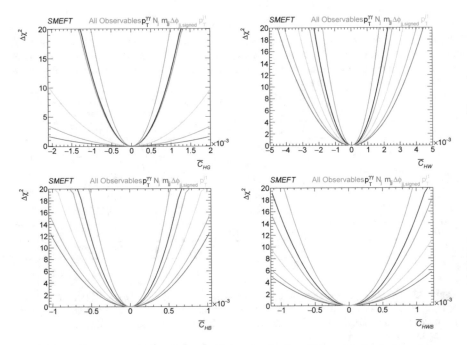

Fig. 8.35 $\Delta \chi^2$ curves for each Wilson coefficient for the five differential distributions independently in the SMEFT (Warsaw) basis

The derived limits on the different Wilson coefficients using both the interference-only and the interference+quadratic terms can be translated to BSM scales (Λ_{NP}) for the different SMEFT operators assuming $C = 1$ and using $\bar{C} = \frac{v^2}{\Lambda^2} C$. The results of such translation are shown in Fig. 8.34. The results are compatible with those found by a global EFT fit in Ref. [39].

Table 8.6 The 95% expected limits on $\overline{C}_{HG}, \overline{C}_{HW}, \overline{C}_{HB}, \overline{C}_{HWB}$ using only their effects on the Higgs boson production mode and their total effect including also the decay

Wilson coefficient	95%$1 - CL$ Limit using production cross section effects and decay effects	95%$1 - CL$ Limit using only production cross section effects
$\overline{C}_{HG}$	$[-4.5, 4.5] \times 10^{-4}$	$[-4.2, 4.2] \times 10^{-4}$
$\overline{C}_{HW}$	$[-7.8, 7.8] \times 10^{-4}$	$[-9.4, 9.4] \times 10^{-2}$
$\overline{C}_{HB}$	$[-2.3, 2.3] \times 10^{-4}$	$[-4.4, 4.4]$
$\overline{C}_{HWB}$	$[-4.2, 4.2] \times 10^{-4}$	$[-1.1, 1.1] \times 10^{-1}$

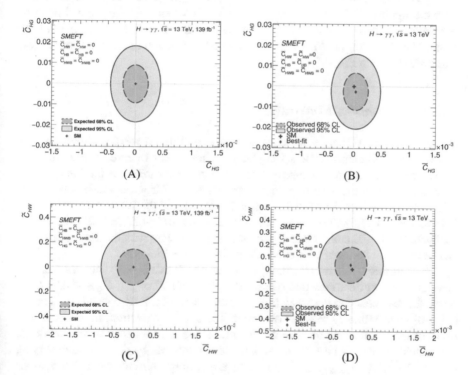

Fig. 8.36 Expected and observed 2D 68% and 95% CL obtained from scanning **A/B** $\overline{C}_{HG}$ versus $\tilde{C}_{HG}$, **C/D** $\overline{C}_{HW}$ versus $\tilde{C}_{HW}$ using the 139fb^{-1} data-set

To test the sensitivity of the different observables, the limit setting procedure is done using each of the different observables independently. The results, shown in Fig. 8.35, show that $\overline{C}_{HG}$, which is a Wilson coefficient that results in an overall scaling of the cross section, is most sensitive to $p_T^{\gamma\gamma}$ and $N_{\text{jets}}^{\geq 30\text{GeV}}$ as the total cross section is probed. Much smaller sensitivity is seen for variables with an additional selection on jet p_T. The remaining Wilson coefficients result in similar behavior as their sensitivity is due to the change in the branching ratio, which affects all the production modes and, therefore, more sensitive to inclusive variables.

1D limits using only production effects As detailed in Sect. 2.2.2.1, different Wilson coefficients affect both the Higgs boson production cross sections and the branching ratio to two photons. The confidence intervals shown so far include both effects. In order to test the sensitivity to the production, the limits were rederived including only variations of the Higgs production cross sections. The results of this derivation are summarized in Table 8.6. The results show good agreement between the limits between the two cases for $\overline{C}_{HG}$ as it mainly affects the gluon-fusion production mode. On the other hand, for the remaining Wilson coefficients, their sensitivity is mainly coming from their effect on the Higgs to diphoton decay width and hence there are significant differences between the limits when only the production cross section is considered, which are much looser, and those including the effect on the width. The effects of the Wilson coefficients on the production cross section at these limits are shown in Fig. 8.5.

8.6.2.2   2D Confidence Intervals

In addition to the 1D scans, 2D scans were also performed varying simultaneously two Wilson coefficients while setting the remaining ones to zero. These limits are derived using the interference-only cross sections as it facilitates the interpolation procedure without the need to produce samples with Wilson coefficients variations in 2D as was the case for the SILH basis. The results of the 2D scans are shown in Figs. 8.36 and 8.37 for the simultaneous variation of the CP-odd and CP-even Wilson coefficients. The 2D constraints, in this case, are consistent with the 1D ones as there is no correlation between the CP-even and CP-odd coefficients.

This is different when performing 2D scans of CP-even coefficients as there are non-negligible correlations that can change the limits. For the $\overline{C}_{HG}$ versus $\overline{C}_{HW}$ scans, the expected and observed limits are shown in Fig. 8.38. For this fit, a correlation of approximately 5% was observed. The small correlation, in this case, is a result of the small effect of $\overline{C}_{HG}$ on the total Higgs boson decay width and hence the branching ratio. On the other hand, no constraints can be derived when varying $\overline{C}_{HW}$ with $\overline{C}_{HB}$ (or $\overline{C}_{HB}$ with $\overline{C}_{HWB}$). This is resulting from substantial correlations (>98%) between $\overline{C}_{HW}$, $\overline{C}_{HB}$, and $\overline{C}_{HWB}$ as all three strongly affect the $H \rightarrow \gamma\gamma$ decay width. This is similar to degeneracies observed in Ref. [9, 39]. These operators can be disentangled in a global analysis resulting in tighter constraints [39].

8.6.3   Coverage Tests Using Toy Datasets

As detailed in *Interlude A*, one can also obtain the confidence intervals using pseudo-data. This procedure is used to check the frequentist coverage of the confidence intervals computed with the asymptotic formula. The asymptotic formula is based on the fact that in the limit of large statistics, the likelihood ratio follows a Gaussian distribution and hence $1 - CL$ is given by the χ^2 probability distribution for one

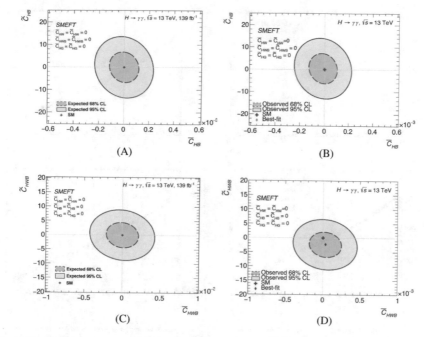

Fig. 8.37 Expected and observed 2D 68% and 95% CL obtained from scanning **A/B** $\overline{C}_{HB}$ versus $\tilde{C}_{HB}$ and **C/D** $\overline{C}_{HWB}$ versus $\tilde{C}_{HWB}$ using the 139 fb^{-1} data-set

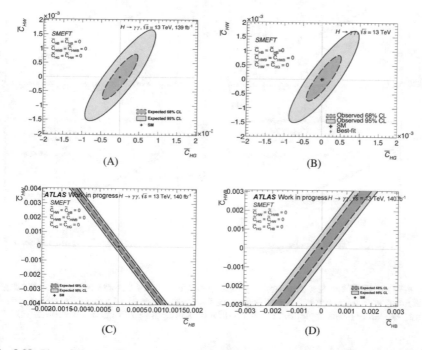

Fig. 8.38 Plots show the 2D 68% and 95% CL obtained from scanning **A, B** $\overline{C}_{HG}$ versus $\overline{C}_{HW}$, **C** $\overline{C}_{HB}$ versus $\overline{C}_{HW}$ and **D** $\overline{C}_{HB}$ versus $\overline{C}_{HWB}$ using the 139fb^{-1} datasets

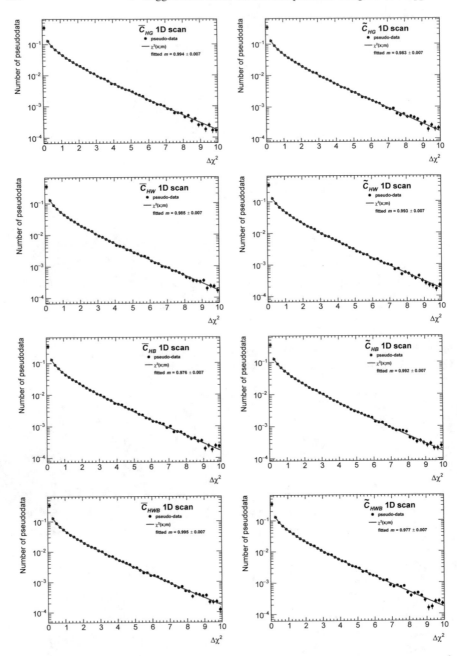

Fig. 8.39 A fit of the test statistics $\Delta\chi^2$ computed from an ensemble of pseudodata to χ^2-distribution with a floating number of degrees of freedom m. The fitted value of m are compatible with 1, validating the asymptotic formula used for obtaining the 1D limits on the coefficients

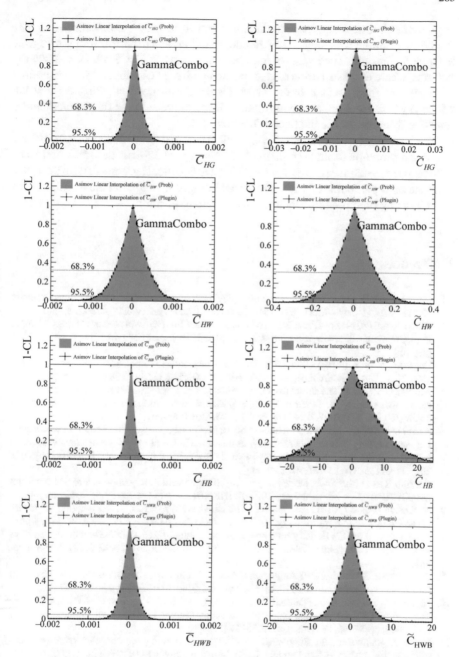

Fig. 8.40 $1 - CL$ curves obtained using the asymptotic formula (labeled "Prob") in purple and using an ensemble of pseudo-data (labeled "Plugin"). The figures show excellent agreement between the two, validating the coverage of the limits using the asymptotic Formula

degree of freedom. Using pseudo-data, one can fit the $\Delta\chi^2$ distribution, estimate the *effective* number of degrees of freedom, and cross-check the limits using the asymptotic formula. The results of this check are shown in Fig. 8.39, using 2000 toy datasets, where the distribution of $\Delta\chi^2$ is fitted with a χ^2 distribution whose number of degrees of freedom is floating. These fits are performed excluding the first bin of the $\Delta\chi^2$ distribution, as the asymptotic formula has a Dirac delta function at 0 resulting from the profile likelihood function [40].

The $1 - CL$ curves are shown in Fig. 8.40 in comparison with the curves obtained from the asymptotic formula (*Prob* method). Using pseudo-data, the 95% confidence interval were re-estimated, they are found to be in excellent agreement with the limits from the asymptotic formula as summarized in Fig. 8.40 showing the distribution of $1 - CL$ using both pseudo-data and the asymptotic formula.

References

1. Christensen ND, Duhr C (2009) FeynRules - Feynman rules made easy. Comput Phys Commun 180:1614–1641. https://doi.org/10.1016/j.cpc.2009.02.018. arXiv:0806.4194 [hep-ph]
2. Alloul A et al (2014) FeynRules 2.0 - a complete toolbox for tree-level phenomenology. Comput Phys Commun 185:2250–2300. https://doi.org/10.1016/j.cpc.2014.04.012, arXiv:1310.1921 [hep-ph]
3. Degrande C et al (2012) UFO - The Universal FeynRules Output. Comput Phys Commun 183:1201–1214. https://doi.org/10.1016/j.cpc.2012.01.022, arXiv:1108.2040 [hep-ph]
4. Alwall J et al (2014) The automated computation of tree-level and next-to-leading order differential cross sections, and their matching to parton shower simulations. JHEP 07:079. https://doi.org/10.1007/JHEP07(2014)079, arXiv:1405.0301 [hep-ph]
5. Aad G et al (2016) Constraints on non-Standard Model Higgs boson interactions in an effective Lagrangian using differential cross sections measured in the $H \rightarrow \gamma\gamma$ decay channel at $\sqrt{s} =$ 8 TeV with the ATLAS detector. Phys Lett B753:69–85. https://doi.org/10.1016/j.physletb.2015.11.071, arXiv:1508.02507 [hep-ex]
6. Contino R et al (2016) On the validity of the effective field theory approach to SM precision tests. JHEP 07:144. https://doi.org/10.1007/JHEP07(2016)144, arXiv:1604.06444 [hep-ph]
7. de Florian D et al (2016) Handbook of LHC Higgs cross sections: 4. Deciphering the nature of the Higgs sector. https://doi.org/10.23731/CYRM-2017-002, arXiv:1610.07922 [hep-ph]
8. Azatov A et al (2017) Helicity selection rules and noninterference for BSM amplitudes. Phys Rev D95(6):065014. https://doi.org/10.1103/PhysRevD.95.065014, arXiv:1607.05236 [hep-ph]
9. Bernlochner FU et al (2019) Angles on CP-violation in Higgs boson interactions. Phys Lett B790:372–379. https://doi.org/10.1016/j.physletb.2019.01.043, arXiv:1808.06577 [hep-ph]
10. Hankele V, Klamke G, Zeppenfeld D (2006) Higgs + 2 jets as a probe for CP properties. In: Meeting on CP Violation and Non-standard Higgs Physics Geneva, Switzerland, December 2–3, 2004. pp 58–62. arXiv:hep-ph/0605117 [hep-ph]
11. Plehn T, Rainwater DL, Zeppenfeld D (2002) Determining the structure of Higgs couplings at the LHC. Phys Rev Lett 88:051801. https://doi.org/10.1103/PhysRevLett.88.051801, arXiv:hep-ph/0105325 [hep-ph]
12. Hankele V et al (2006) Anomalous Higgs boson couplings in vector boson fusion at the CERN LHC. Phys Rev D74:095001. https://doi.org/10.1103/PhysRevD.74.095001, arXiv:hep-ph/0609075 [hep-ph]
13. Buckley A et al (2013) Rivet user manual. Comput Phys Commun 184:2803–2819. https://doi.org/10.1016/j.cpc.2013.05.021, arXiv:1003.0694 [hep-ph]

14. Sjstrand T, Mrenna S, Skands PZ (2008) A brief introduction to PYTHIA 8.1. Comput Phys Commun 178:852–867. https://doi.org/10.1016/j.cpc.2008.01.036, arXiv:0710.3820 [hep-ph]
15. ATLAS Run 1 Pythia8 tunes. Tech. rep. ATL-PHYS-PUB-2014-021. Geneva, CERN, 2014. https://cds.cern.ch/record/1966419
16. Ball RD et al (2013) Parton distributions with LHC data. Nucl Phys B867:244–289. https://doi.org/10.1016/j.nuclphysb.2012.10.003, arXiv:1207.1303 [hep-ph]
17. Mangano ML et al (2007) Matching matrix elements and shower evolution for top-quark production in hadronic collisions. JHEP 01:013. https://doi.org/10.1088/1126-6708/2007/01/013, arXiv:hep-ph/0611129 [hep-ph]
18. Brivio I, Jiang Y, Trott M (2017) The SMEFTsim package, theory and tools. JHEP 1712:070. https://doi.org/10.1007/JHEP12(2017)070, arXiv:1709.06492 [hep-ph]
19. Brivio I, Corbett T, Trott M (2019) The Higgs width in the SMEFT. arXiv:1906.06949 [hep-ph]
20. Buckley A et al (2010) Systematic event generator tuning for the LHC. Eur Phys J C65:331–357. https://doi.org/10.1140/epjc/s10052-009-1196-7, arXiv:0907.2973 [hep-ph]
21. Aaboud M et al (2018) Measurements of Higgs boson properties in the diphoton decay channel with 36 fb^{-1} of pp collision data at $\sqrt{s} = 13$ TeV with the ATLAS detector. Phys Rev D98:052005. https://doi.org/10.1103/PhysRevD.98.052005, arXiv:1802.04146 [hep-ex]
22. Guennebaud G, Jacob B et al (2010) Eigen v3. http://eigen.tuxfamily.org
23. GammaCombo User Manual. Tech. rep. 2017. https://gammacombo.github.io/manual.pdf
24. Cowan G (1998) Statistical data analysis. Oxford Science Publications, Clarendon Press. ISBN 9780198501558. https://books.google.ch/books?id=ff8ZyW0nlJAC
25. Stewart IW et al (2013) Jet p_T Resummation in Higgs production at NNLL' + NNLO. arXiv:1307.1808
26. Boughezal R et al (2014) Combining resummed Higgs predictions across jet bins. Phys Rev D89:074044. https://doi.org/10.1103/PhysRevD.89.074044, arXiv:1312.4535 [hep-ph]
27. Bernlochner FU et al (2016) Measurement of fiducial and differential cross sections in the $H \rightarrow \gamma\gamma$ decay channel with 36.1/fb of 13 TeV proton-proton collision data with the ATLAS detector. Tech. rep. ATL-COM-PHYS-2017-145. Geneva, CERN. https://cds.cern.ch/record/2252597
28. Campbell JM, Ellis RK (2010) MCFM for the Tevatron and the LHC. Nucl Phys Proc Suppl 205-206:10–15. https://doi.org/10.1016/j.nuclphysbps.2010.08.011, arXiv:1007.3492 [hep-ph]
29. Butterworth J et al (2016) PDF4LHC recommendations for LHC Run II. J Phys G 43:023001. https://doi.org/10.1088/0954-3899/43/2/023001, arXiv:1510.03865 [hep-ph]
30. Nason P, Oleari C (2010) NLO Higgs boson production via vector-boson fusion matched with shower in POWHEG. JHEP 02:037. https://doi.org/10.1007/JHEP02(2010)037, arXiv:0911.5299 [hep-ph]
31. Hamilton K et al (2013) Merging H/W/Z + 0 and 1 jet at NLO with no merging scale: a path to parton shower + NNLO matching. JHEP 05:082. https://doi.org/10.1007/JHEP05(2013)082, arXiv:1212.4504 [hep-ph]
32. Zhang Y et al (2014) QCD NLO and EW NLO corrections to $t\bar{t}H$ production with top quark decays at hadron collider. Phys Lett B738:1–5. https://doi.org/10.1016/j.physletb.2014.09.022, arXiv:1407.1110 [hep-ph]
33. Dawson S et al (2003) Associated Higgs production with top quarks at the large hadron collider: NLO QCD corrections. Phys Rev D68:034022. https://doi.org/10.1103/PhysRevD.68.034022, arXiv:hep-ph/0305087 [hep-ph]
34. Beenakker W et al (2003) NLO QCD corrections to t anti-t H production in hadron collisions. Nucl Phys B653:151–203. https://doi.org/10.1016/S0550-3213(03)00044-0, arXiv:hep-ph/0211352 [hep-ph]
35. Dawson S et al (2004) Exclusive Higgs boson production with bottom quarks at hadron colliders. Phys Rev D69:074027. https://doi.org/10.1103/PhysRevD.69.074027, arXiv:hep-ph/0311067 [hep-ph]
36. Dittmaier S, Krmer M, Spira M (2004) Higgs radiation off bottom quarks at the Tevatron and the CERN LHC. Phys Rev D70:074010. https://doi.org/10.1103/PhysRevD.70.074010. arXiv:hep-ph/0309204 [hep-ph]

37. ATLAS Collaboration (2015) Constraints on non-Standard Model Higgs boson interactions in an effective Lagrangian using differential cross sections measured in the $H \rightarrow \gamma\gamma$ decay channel at $\sqrt{s} = 8$ TeV with the ATLAS detector. Constraints on non-Standard Model Higgs boson interactions in an effective field theory using differential cross sections measured in the $H \rightarrow \gamma\gamma$ decay channel at $\sqrt{s} = 8$ TeV with the ATLAS detector. Phys Lett B 753.CERN-PH-EP-2015-182. CERN-PH-EP-2015-182, 69-85. 30p. https://cds.cern.ch/record/2042102
38. Falkowski A (2015) Higgs basis: proposal for an EFT basis choice for LHC HXSWG. https://cds.cern.ch/record/2001958
39. Ellis J et al (2018) Updated Global SMEFT Fit to Higgs, Diboson and Electroweak Data. JHEP 06:146. https://doi.org/10.1007/JHEP06(2018)146, arXiv:1803.03252 [hep-ph]
40. Cranmer K (2015) Practical statistics for the LHC. In: Proceedings, 2011 European School of High-Energy Physics (ESHEP 2011): Cheile Gradistei, Romania, September 7–20, 2011. [247(2015)], pp 267–308. https://doi.org/10.5170/CERN-2015-001.247, https://doi.org/10.5170/CERN-2014-003.267, arXiv:1503.07622 [physics.data-an]

Conclusions

In this thesis, measurements of Higgs boson cross sections in the two-photon decay channel are performed using pp collision data recorded by the ATLAS experiment at the LHC. The data were recorded at a center-of-mass energy of $\sqrt{s} = 13$ TeV and correspond to the full Run-2 dataset with an integrated luminosity of 139 fb^{-1}. The measurements are performed in a fiducial region defined by the detector acceptance $|\eta| < 2.37$, excluding the region of $1.37 < |\eta| < 1.52$. Selected photons satisfy identification and isolation requirements, and are required to have transverse momentum greater than 35% and 25% of the diphoton invariant mass.

The measured integrated fiducial cross section times branching ratio is:

$$\sigma_{\text{fid}} = 65.2 \pm 4.5 \,(\text{stat.}) \pm 5.6 \,(\text{exp.}) \pm 0.3 \,(\text{theory}) \,\,\text{fb}, \qquad (8.26)$$

with a total uncertainty of 11%, which is to be compared with the default Standard Model prediction for inclusive Higgs boson production of 63.5 ± 3.3 fb, resulting in the most precise $H \to \gamma\gamma$ fiducial cross section to date.

In addition, a measurement of the differential cross sections was performed as a function of different observables sensitive to the Higgs boson production kinematics. These variables probe the Higgs boson transverse momentum and rapidity distributions, the jet multiplicity, jet kinematics ($p_T^{j_1}$ and m_{jj}) and CP quantum numbers ($\Delta\phi_{jj,\text{signed}}$). The results presented in this thesis include a reduction by a factor 2 of the different photon and jet energy scale and resolution systematic uncertainties with respect to the previous ATLAS publication [1]. The results show an excellent agreement with the SM predictions. The results were also compared to several theoretical predictions thanks to the model-independent particle-level nature of the measurement, which allows a direct comparison to theoretical predictions.

The measured fiducial inclusive and differential cross sections were then used to derive full-phase space inclusive and differential cross sections for the Higgs boson transverse momentum. This resulted in the total production cross section in the $H \to \gamma\gamma$ channel of $58.6^{+6.7}_{-6.5}$ pb. In addition, a combination of the full phase-

A. Tarek Abouelfadl Mohamed, *Measurement of Higgs Boson Production Cross Sections in the Diphoton Channel*, Springer Theses, https://doi.org/10.1007/978-3-030-59516-6

space inclusive and differential Higgs boson transverse momentum cross sections was performed with the ones measured in $H \to ZZ^* \to 4l$ channel, resulting in the following inclusive cross section:

$$56.1^{+4.5}_{-4.3} \ (\ \pm 3.2(\text{stat.}) \ ^{+3.1}_{-2.8} \ (\text{sys.}) \ ) \ \text{pb}. \tag{8.27}$$

The measured total cross sections are in good agreement with the SM prediction of 55.6 ± 2.5 pb.

The measured cross sections were used to investigate the strength and tensor structure of anomalous Higgs boson interactions using the effective field theory framework (EFT). In the EFT framework, the Standard Model Lagrangian is complemented with additional CP-even and CP-odd dimension-6 operators as implemented in the SILH basis of the Higgs Effective Lagrangian [2, 3] and the Warsaw basis of the SMEFT Lagrangian [4]. Exploring these BSM effects rely on five differential observables ($p_T^{\gamma\gamma}$, N_{jets}, m_{jj}, $p_T^{j_1}$ and $\Delta\phi_{jj,\text{signed}}$) and their correlations. No significant new physics contributions are observed in the measured cross sections, and hence limits on such contributions are reported. The SILH basis limits show a significant improvement (by a factor of two) in comparison with the previous ATLAS measurements [5, 6] thanks to the larger size of the data set. The SMEFT limits presented are the first SMEFT measurement performed in the LHC. The limits are derived including contributions from the SM and the dimension-6 operators interference as well as pure dimension-6 contributions. Limits were also derived on CP-odd operators using the interference with dimension-6 operators which allows probing pure CP-violating effects using the $\Delta\phi_{jj,\text{signed}}$ distributions.

The measurements presented in this thesis rely on performance studies that were carried out and represent a key ingredient of the analyses. The precise measurement of the Higgs boson cross section in the two photons decay channel is only possible thanks to the precise calibration of the photon energies. The calibration of the electron and photon energies was performed based on a multivariate analysis (MVA). The extraction of the final energy scales from the MVA requires further corrections for discrepancies between data and simulation. Among these corrections, those of the electromagnetic calorimeter layer energies, which reduce the relative miscalibration of the longitudinal layers of the calorimeter and ensure a precise energy measurement over a large p_T range. The thesis detailed the measurement of the presampler layer energy calibration. The full calibration chain resulted in a precise energy calibration of electrons and photons with a typical uncertainty on the photon energy scales of 0.2% and up to 0.8% in the most forward regions of the detector. The resulting uncertainty on the energy resolution for the photons from the Higgs boson decays is typically between 10% and up to 20% for the high p_T photons. The calibration presented was used to perform a measurement of the Higgs boson mass in the $H \to \gamma\gamma$ and $H \to 4\ell$ channels using 36 fb^{-1} of Run-2 data at $\sqrt{s} = 13$ TeV [7] resulting in the measured mass of:

$$m_H = 124.86 \pm 0.1 \ (\text{stat.}) \pm 0.19 \ (\text{sys.}) \ \text{GeV} \tag{8.28}$$

which is in very good agreement with the Run-1 ATLAS+CMS combined mass measurement $m_H = 125.09 \pm 0.21 \ (\text{stat.}) \pm 0.11 \ (\text{sys.}) \ \text{GeV}$ [8].

Appendix A
Alternative Fiducial Definition for the $H \to \gamma\gamma$ Cross Section Measurement

As motivated in Sect. 7.1.1, the following considerations need to be taken into account when defining the fiducial volume:

- Similarity to detector-level selection to avoid extrapolating through phase space, as this assumes knowledge of the production rate in unsampled regions of phase space, and so introduces model-dependence.
- Simplicity. A simple fiducial volume definition is easier to interpret, calculate, and combine with other decay channels.
- Ability to compare to theory. Some common detector-level objects, or quantities, are difficult to include in theory calculations leading to larger uncertainties into these calculations. Since the aim of fiducial cross section measurements is to be comparable to many theory predictions, such objects should be avoided where possible.

As detailed in Sect. 7.3.3, a particle-level isolation criterion is applied in order to reduce the dependence of the bin-by-bin correction factors on the Higgs boson production mode. This criterion is chosen in order to match the detector-level photon isolation requirement that is imposed in order to separate the hard scatter photons (i.e. the Higgs photons) from photons emitted from hadronic jets (can come from electromagnetic decays of unstable hadrons).

Nevertheless, isolation is a quantity that can complicate the theoretical calculation of cross sections in a fiducial volume [9]. The main problem with isolation is sketched in Fig. A.1. In order for quantities to be infrared safe, they must not change when a parton undergoes a quasi-collinear splitting, i.e. when it emits some soft radiation which slightly perturbs its initial momentum. However, such perturbations can cause a parton to migrate across the boundary of a fixed cone such as that used for computing isolation, causing the measured isolation energy to change. This means that isolation is not infrared safe.

In this appendix, we explore an alternative method that can be used to reduce model dependence of the correction factors. This method is based on applying a

A. Tarek Abouelfadl Mohamed, *Measurement of Higgs Boson Production Cross Sections in the Diphoton Channel*, Springer Theses, https://doi.org/10.1007/978-3-030-59516-6

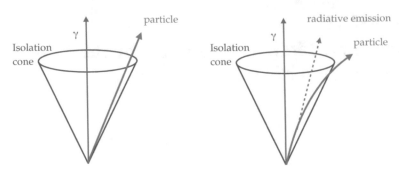

Fig. A.1 Diagram showing the reason why isolation cones are difficult to compute in theory calculations. In the case on the left: a particle is created at the same vertex as the photon and lies within its isolation cone. In the case on the right: the particle undergoes a quasi-collinear splitting. It emits some radiation and migrates across the cone boundary, and so is no longer included in the isolation cone. The measured isolation energy is therefore dependent on the modelling of real emissions in the final state, which introduces an uncertainty on the theory calculation

veto on the ΔR distance between the photons and jets defined as follows. The jets are clustered using the anti-k_t algorithm with a radius parameter of 0.2. The jets are clustered using stable truth particles (defined in Sect. 7.3.2) excluding neutrinos and the Higgs decay photons. This jet definition has similar parameters to the detector-level isolation cone. A proof-of-principle for this alternative definition is shown in Fig. A.2 with a $t\bar{t}H$ sample. The $t\bar{t}H$ is used since it is the production mode with the largest jet activity and the most sensitive production mode to particle-level isolation. The Figure shows $\Delta R_{min}^{\gamma_{LDG},jet}$ of particle-level leading-p_T photons matched to detector-level ones, divided into the different scenarios of passing and failing the detector-level isolation criteria. The plot shows that photons that fail the detector-level isolation criteria have a nearby jet, and vice versa. Therefore, an alternative way to define the fiducial region can be done in terms of a veto on $\Delta R_{min}^{\gamma_{LDG},jet}$. This veto helps to tackle some theoretical problems with the particle-level isolation, such as the migrations that can take place across the boundaries of the fixed isolation cone in the event of quasi-linear splitting of a parton.

Figure A.3 below show the effect applying jet veto on the correction factors using different thresholds for the jet p_T and $\Delta R_{min}^{\gamma_{LDG},jet}$. The plots show that some variation of jet p_T and ΔR_{min} can reduce model dependence and achieve a relatively flat behavior of the correction factor with the different Higgs production modes similar to that of the particle-level isolation.

The choice of the jet p_T and $\Delta R_{min}^{\gamma_{LDG},jet}$ is performed by mapping the different variations of these variables to the detector-level isolation. This mapping relies on the following criteria:

- A check of robustness against particle-level isolation was performed by mapping the different jet veto thresholds ($\Delta R_{min}^{\gamma_{LDG},jet}$) with different jet p_T thresholds to the p_T dependent particle-level track isolation criteria. Figure A.4 shows the mapping of the different jet vetoes with their corresponding particle-level isolation criteria.

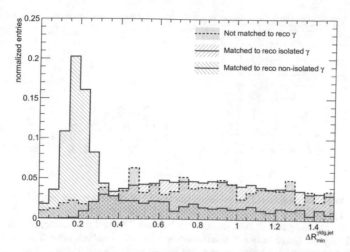

Fig. A.2 Proof-of-principle of vetoing jets with ($p_T > 15$ GeV) based on ΔR with photons from a $t\bar{t}H$ sample. The plot shows that particle-level photons matched detector-level photons that fail the isolation requirement (red) have a nearby jets (small $\Delta R_{min}^{\gamma_{LDG},jet}$). Whereas photons that pass the detector-level isolation have jets further away (blue)

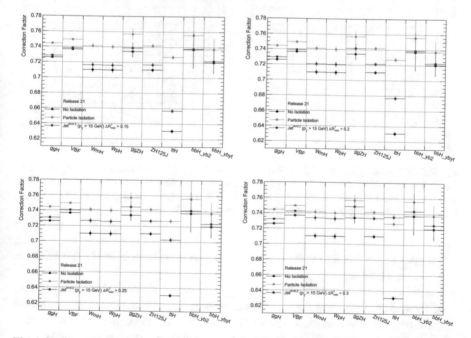

Fig. A.3 Correction factor as a function of the production mechanism for the particle-level isolation (green) and different jet vetoes (red) for jet $p_T > 15$ GeV

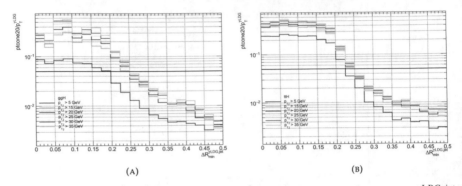

Fig. A.4 Plot shows profiling of the p_T dependent track isolation criteria to the jet veto $\Delta R_{min}^{\gamma LDG, jet}$ for different jet p_T for **A** ggH and **B** ttH. The curves define three regions; The leftmost shows a flat behavior between particle isolation and ΔR between jets and the leading photon. This is for $\Delta R < 0.2$, which comes from jets that fall entirely within the isolation cone. The second region is for jets that overlap with the isolation cone, and hence the profile of the isolation energy decreases with increasing ΔR. The third region is for jets that fall outside the isolation cone, and therefore the isolation energy is independent of ΔR between the photon and jets

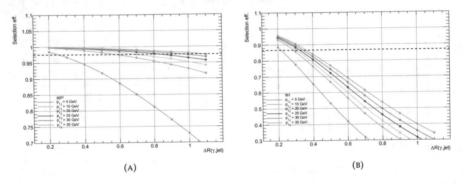

Fig. A.5 Selection efficiency of the different jet veto ΔR_{min} between jets and leading (and sub-leading) photon using different jet p_T compared to the efficiency of the mapped track isolation shown as the horizontal dotted line for ggH **A** and ttH **B**. The plot shows that the efficiency for ttH drops much faster than ggH due to the higher jet multiplicity for ttH

The plot shows that jet vetoes matched to the particle-level isolation have similar model dependence.

- The selection efficiency for using a $\Delta R_{min}^{\gamma LDG, jet}$ veto is shown in Fig. A.5 which is performed by comparing the selection efficiency for different jet vetoes compared to the particle-level isolation, and similar conclusions can be driven from the plot, where it shows similar efficiency between the isolation criteria and jet vetoes with the smallest model dependence.

The use of particle-level isolation significantly reduces the Higgs production model dependence of the correction factors. These uncertainties are quantified with the signal composition uncertainty shown in Sect. 7.5.3.1 that is reduced with the use of particle-level isolation. The use of a jet-veto was also found to produce similar results to that of the particle-level isolation. For example, for the inclusive region, the use of particle-level isolation reduces the signal composition uncertainty by approximately 88% compared to no particle-level isolation, whereas the jet-veto case with $\Delta R_{min}^{\gamma_{LDG},jet} = 0.3$ with jet $p_T > 15$ GeV results in 82% reduction. This is understood as the jet veto criteria can be mapped directly to particle-level isolation criteria (as shown in Fig. A.4). The particle-level track isolation $\texttt{ptcone20}_{particle} < 0.05 \times p_T$ was mapped to a jet veto of $\Delta R_{min}^{\gamma_{LDG},jet} = 0.3$ with jet $p_T > 15$ GeV. Therefore, both fiducial definitions will always result in compatible modeling uncertainties on the correction-factor as they can be mapped to one another, and to the detector-level isolation.

Appendix B
Higgs Boson Signal-Extraction Fits

In this appendix, the Higgs boson signal extraction fits, detailed in Sect. 7.6, are shown in Figs. B.1, B.2, B.3, B.4, B.5, B.6, B.7, B.8 and B.9. These fits are performed simultaneously to the diphoton invariant mass, $m_{\gamma\gamma}$, for the different bins of a given observable: $p_T^{\gamma\gamma}$, $|y_{\gamma\gamma}|$, N_{jets}, $p_T^{j_1}$, m_{jj}, and $\Delta\phi_{jj,\text{signed}}$. The data points are loosely binned for visualization, although the signal extraction fit is itself an unbinned maximum likelihood fit.

A. Tarek Abouelfadl Mohamed, *Measurement of Higgs Boson Production
Cross Sections in the Diphoton Channel*, Springer Theses,
https://doi.org/10.1007/978-3-030-59516-6

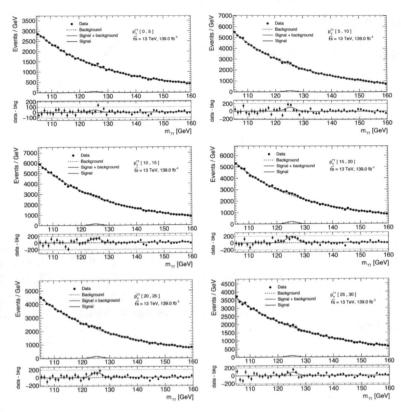

Fig. B.1 Diphoton invariant mass signal+background fits for the different bins of $p_T^{\gamma\gamma}$

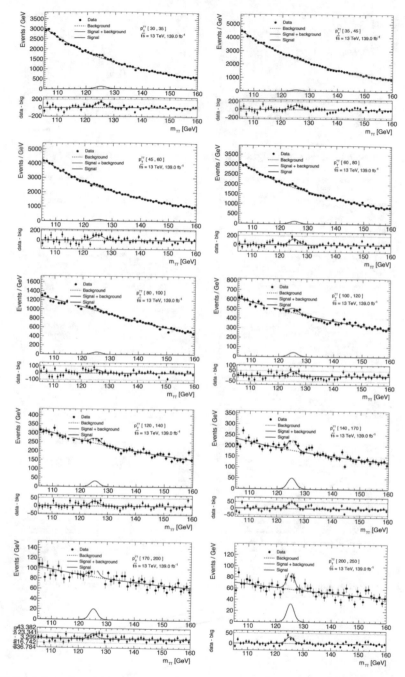

Fig. B.2 Diphoton invariant mass signal+background fits for the different bins of $p_T^{\gamma\gamma}$

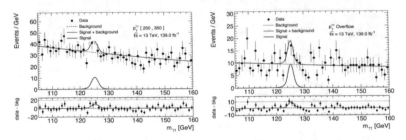

Fig. B.3 Diphoton invariant mass signal+background fits for the different bins of $p_T^{\gamma\gamma}$

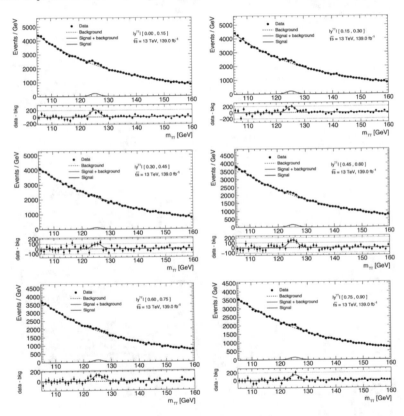

Fig. B.4 Diphoton invariant mass signal+background fits for the different bins of $|y_{\gamma\gamma}|$

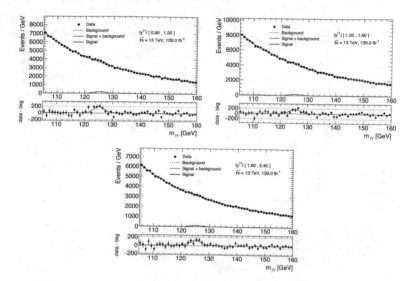

Fig. B.5 Diphoton invariant mass signal+background fits for the different bins of $|y_{\gamma\gamma}|$

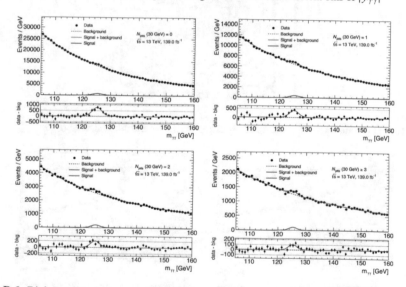

Fig. B.6 Diphoton invariant mass signal+background fits for the different bins of N_{jets}

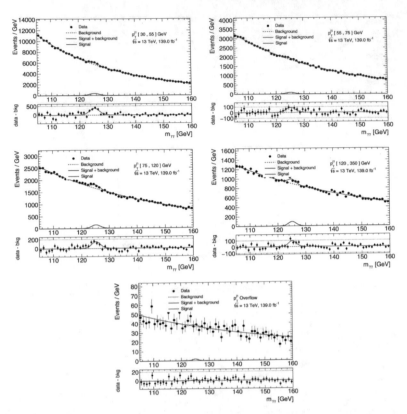

Fig. B.7 Diphoton invariant mass signal+background fits for the different bins of p_T^{j1}

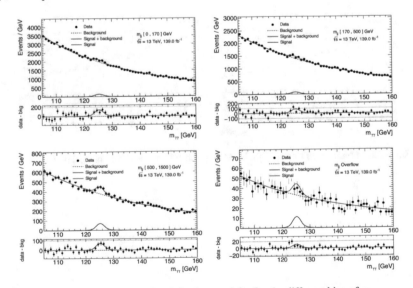

Fig. B.8 Diphoton invariant mass signal+background fits for the different bins of m_{jj}

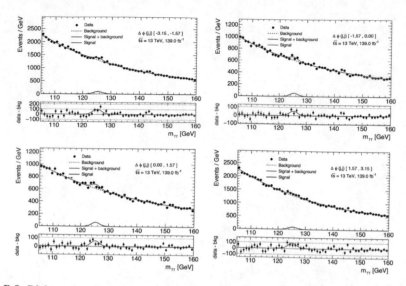

Fig. B.9 Diphoton invariant mass signal+background fits for the different bins of $\Delta\phi_{jj,\text{signed}}$

References

1. Measurements of Higgs boson properties in the diphoton decay channel using 80 fb^{-1} of pp collision data at $\sqrt{s} = 13$ TeV with the ATLAS detector. Tech. rep. ATLAS-CONF-2018-028. Geneva: CERN (2018). https://cds.cern.ch/record/2628771
2. Giudice GF et al (2007) The strongly-interacting light higgs. In: JHEP 06, p 045. https://doi.org/10.1088/1126-6708/2007/06/045. arXiv: hep-ph/0703164 [hep-ph]
3. Alloul A, Fuks B, Sanz V (2014) Phenomenology of the Higgs effective Lagrangian via FEYN-RULES. In: JHEP 04, p 110. https://doi.org/10.1007/JHEP04(2014)110. arXiv: 1310.5150 [hep-ph]
4. Brivio I, Trott M (2017) The standard model as an effective field theory. https://doi.org/10.1016/j.physrep.2018.11.002. arXiv: 1706.08945 [hep-ph]
5. Aaboud M et al (2018) Measurements of Higgs boson properties in the diphoton decay channel with 36 fb^{-1} of pp collision data at $\sqrt{s} = 13$ TeV with the ATLAS detector. Phys Rev D98:052005. https://doi.org/10.1103/PhysRevD.98.052005. arXiv: 1802.04146 [hep-ex]
6. Aad G et al (2016) Constraints on non-Standard Model Higgs boson interactions in an effective Lagrangian using differential cross sections measured in the $H \rightarrow gg$ decay channel at $\sqrt{s} = 8$ TeV with the ATLAS detector. Phys Lett B753:69–85. https://doi.org/10.1016/j.physletb.2015.11.071. arXiv: 1508.02507 [hep-ex]
7. Aaboud M et al (2018) Measurement of the Higgs boson mass in the $H \rightarrow ZZ^* \rightarrow 4\ell$ and $H \rightarrow gg$ channels with $\sqrt{s} = 13$ TeV pp collisions using the ATLAS detector. Phys Lett B784:345–366. https://doi.org/10.1016/j.physletb.2018.07.050. arXiv: 1806.00242 [hep-ex]
8. Aad G et al (2015) Combined measurement of the Higgs Boson mass in pp collisions at $\sqrt{s} = 7$ and 8 TeV with the ATLAS and CMS experiments. Phys Rev Lett 114:191803. https://doi.org/10.1103/PhysRevLett.114.191803. arXiv: 1503.07589 [hep-ex]
9. Gordon LE (1997) Isolated photons at hadron colliders at O($\alpha\alpha s2$) (I): Spin-averaged case. Nucl Phys B 501(1):175–196. https://doi.org/10.1016/S0550-3213(97)00339-8. arXiv: hep-ph/9611391 [hep-ph]

Printed in the United States
by Baker & Taylor Publisher Services